AF361820

Chemical and Technological Characterization of Dairy Products

Chemical and Technological Characterization of Dairy Products

Editor

Michele Faccia

MDPI • Basel • Beijing • Wuhan • Barcelona • Belgrade • Manchester • Tokyo • Cluj • Tianjin

Editor
Michele Faccia
University of Bari
Italy

Editorial Office
MDPI
St. Alban-Anlage 66
4052 Basel, Switzerland

This is a reprint of articles from the Special Issue published online in the open access journal *Foods* (ISSN 2304-8158) (available at: https://www.mdpi.com/journal/foods/special_issues/Chemical_Technological_Characterization_Dairy).

For citation purposes, cite each article independently as indicated on the article page online and as indicated below:

LastName, A.A.; LastName, B.B.; LastName, C.C. Article Title. *Journal Name* **Year**, *Volume Number*, Page Range.

ISBN 978-3-0365-0218-2 (Hbk)
ISBN 978-3-0365-0219-9 (PDF)

Cover image courtesy of Michele Faccia.

Contents

About the Editor . . . . **vii**

Michele Faccia
Chemical and Technological Characterization of Dairy Products
Reprinted from: *Foods* **2020**, *9*, 1475, doi:10.3390/foods9101475 . . . . **1**

Piero Franceschi, Massimo Malacarne, Paolo Formaggioni, Claudio Cipolat-Gotet,
Giorgia Stocco and Andrea Summer
Effect of Season and Factory on Cheese-Making Efficiency in Parmigiano Reggiano Manufacture
Reprinted from: *Foods* **2019**, *8*, 315, doi:10.3390/foods8080315 . . . . **5**

Camillo Martino, Andrea Ianni, Lisa Grotta, Francesco Pomilio and Giuseppe Martino
Influence of Zinc Feeding on Nutritional Quality, Oxidative Stability and Volatile Profile of
Fresh and Ripened Ewes' Milk Cheese
Reprinted from: *Foods* **2019**, *8*, 656, doi:10.3390/foods8120656 . . . . **15**

Michele Faccia, Giuseppe Gambacorta, Giovanni Martemucci, Graziana Difonzo and
Angela Gabriella D'Alessandro
Chemical-Sensory Traits of Fresh Cheese Made by Enzymatic Coagulation of Donkey Milk
Reprinted from: *Foods* **2020**, *9*, 16, doi:10.3390/foods9010016 . . . . **27**

Ali Saleh, Abdellatif A. Mohamed, Mohammed S. Alamri, Shahzad Hussain,
Akram A. Qasem and Mohamed A. Ibraheem
Effect of Different Starches on the Rheological, Sensory and Storage Attributes of Non-fat
Set Yogurt
Reprinted from: *Foods* **2020**, *9*, 61, doi:10.3390/foods9010061 . . . . **41**

Sofie Buhler, Ylenia Riciputi, Giuseppe Perretti, Maria Fiorenza Caboni, Arnaldo Dossena,
Stefano Sforza and Tullia Tedeschi
Characterization of Defatted Products Obtained from the Parmigiano–Reggiano Manufacturing
Chain: Determination of Peptides and Amino Acids Content and Study of the Digestibility and
Bioactive Properties
Reprinted from: *Foods* **2020**, *9*, 310, doi:10.3390/foods9030310 . . . . **55**

Paolo Formaggioni, Massimo Malacarne, Piero Franceschi, Valentina Zucchelli,
Michele Faccia, Giovanna Battelli, Milena Brasca and Andrea Summer
Characterisation of Formaggella della Valle di Scalve Cheese Produced from Cows Reared in
Valley Floor Stall or in Mountain Pasture: Fatty Acids Profile and Sensory Properties
Reprinted from: *Foods* **2020**, *9*, 383, doi:10.3390/foods9040383 . . . . **67**

Andrea Serra, Giuseppe Conte, Leonor Corrales-Retana, Laura Casarosa, Francesca Ciucci
and Marcello Mele
Nutraceutical and Technological Properties of Buffalo and Sheep Cheese Produced by the
Addition of Kiwi Juice as a Coagulant
Reprinted from: *Foods* **2020**, *9*, 637, doi:10.3390/foods9050637 . . . . **83**

Marcello Alinovi, Germano Mucchetti, Ulf Andersen, Tijs A. M. Rovers, Betina Mikkelsen,
Lars Wiking and Milena Corredig
Applicability of Confocal Raman Microscopy to Observe Microstructural Modifications of
Cream Cheeses as Influenced by Freezing
Reprinted from: *Foods* **2020**, *9*, 679, doi:10.3390/foods9050679 . . . . **99**

Han Chen, Haotian Zheng, Margaret Anne Brennan, Wenpin Chen, Xinbo Guo and Charles Stephen Brennan
Effect of Black Tea Infusion on Physicochemical Properties, Antioxidant Capacity and Microstructure of Acidified Dairy Gel during Cold Storage
Reprinted from: *Foods* 2020, 9, 831, doi:10.3390/foods9060831 113

Faten Dhawi, Hossam S. El-Beltagi, Esmat Aly and Ahmed M. Hamed
Antioxidant, Antibacterial Activities and Mineral Content of Buffalo Yoghurt Fortified with Fenugreek and *Moringa oleifera* Seed Flours
Reprinted from: *Foods* 2020, 9, 1157, doi:10.3390/foods9091157 133

Mahmoud Abdel-Hamid, Zizhen Huang, Takuya Suzuki, Toshiki Enomoto, Ahmed M. Hamed, Ling Li and Ehab Romeih
Development of a Multifunction Set Yogurt Using *Rubus suavissimus* S. Lee (Chinese Sweet Tea) Extract
Reprinted from: *Foods* 2020, 9, 1163, doi:0.3390/foods9091163 149

Mu Tian, Cuina Wang, Jianjun Cheng, Hao Wang, Shilong Jiang and Mingruo Guo
Preparation and Characterization of Soy Isoflavones Nanoparticles Using Polymerized Goat Milk Whey Protein as Wall Material
Reprinted from: *Foods* 2020, 9, 1198, doi:10.3390/foods9091198 161

Michele Manoni, Chiara Di Lorenzo, Matteo Ottoboni, Marco Tretola and Luciano Pinotti
Comparative Proteomics of Milk Fat Globule Membrane (MFGM) Proteome across Species and Lactation Stages and the Potentials of MFGM Fractions in Infant Formula Preparation
Reprinted from: *Foods* 2020, 9, 1251, doi:10.3390/foods9091251 175

Karina A. Parra-Ocampo, Sandra T. Martín-del-Campo, José G. Montejano-Gaitán, Rubén Zárraga-Alcántar and Anaberta Cardador-Martínez
Evaluation of Biological, Textural, and Physicochemical Parameters of Panela Cheese Added with Probiotics
Reprinted from: *Foods* 2020, 9, 1507, doi:10.3390/foods9101507 193

About the Editor

Michele Faccia is an Associate Professor of Food Science and Technology in the Department of Soil, Plant, and Food Sciences at the University of Bari, Italy. Since 2007 he has been led the Department's Dairy Science and Technology research group. In 2020, he obtained the National Scientific Qualification as Full Professor. Since 1988, he has carried out research regarding Dairy Science and Technology, dealing with product/process innovation, chemical characterization of traditional products, and problem solving in the dairy industry. He has authored/coauthored more than 160 scientific papers in journals, international books, and congress proceedings, 90 of which are indexed in Scopus and/or the WOS database. He has obtained research funds both from public institutions and private dairy companies, and has been scientifically responsible for several peer reviewed projects. He belongs to the Italian Society of Food Science and Technology (SISTAL), is an academic member of the Italian National Association of Cheese Tasters (ONAF), and is on the Editorial Board for the following journals indexed in Scopus/WOS: *Foods* (MDPI, Switzerland), *Journal of Dairy Research* (Cambridge University Press, UK), and *Journal of Ethnic Foods* (Springer Nature, UK).

 *foods*

Editorial

Chemical and Technological Characterization of Dairy Products

Michele Faccia

Department of Soil, Plant and Food Sciences, University of Bari, Via Amendola 165/A, 70126 Bari, Italy; Michele.faccia@uniba.it

Received: 27 September 2020; Accepted: 13 October 2020; Published: 16 October 2020

The dairy sector is facing a decisive challenge in developed countries, which could deeply influence its future and its historical status of being a pillar for human nutrition. The most challenging issue is to give suitable answers to the demand for nutritionally balanced and environmentally sustainable products, the two main aspects of the new "food paradigm" that increasingly sees foods as drugs (specifically renamed as "nutraceuticals") and imposes a stringent eco-friendly approach in their production ("green foods") [1,2]. In this context, all animal products are often met with hostility that is not always justified under the scientific point of view, particularly when it results in imposing their complete elimination [3,4]. As expected, this challenge has soon been met by researchers in dairy and food sciences, who are giving suitable scientific answers and are driving dairy farms and industries to develop new products and processes that can better satisfy the new requirements. In this context, the Special Issue "Chemical and Technological Characterization of Dairy Products" has collected 13 articles (12 original researches and one review) that give an interesting contribution to the field. The articles can be grouped into three categories: the first one concerns product innovation and includes eight papers reporting technological and compositional details on innovative dairy products developed in line with the nutritional and/or environmental requirements mentioned above; the second one has an interdisciplinary approach (animal husbandry-food technology) and is made of three studies aimed to deepen the influence of the cattle rearing conditions on cheese quality; the third one contains two papers dealing with different aspects of dairy science.

Product Innovation. Abdel-Hamid et al. [5] and Dhawi et al. [6] developed different types of functional buffalo milk yogurt and investigated their chemical, microbiological, organoleptic and bioactivity characteristics. In the first case, a functional yogurt was obtained by fortifying milk with Chinese sweet tea extract (*Rubus suavissimus* S. Lee leaves). The phenolic compounds included as a consequence of extract addition improved biological activity in terms of antioxidant and antihypertensive activity and inhibition of the Caco-2 carcinoma cell line; on the other hand, the viability of the yogurt starter cultures during refrigerated storage was not significantly affected. Finally, the sensory analysis demonstrated a high acceptability of the product and allowed for establishing the most suitable level of fortification. In the second paper, two different plant-based ingredients were used as yogurt-fortifying agents: fenugreek (*Trigonella foenum-graecum*) seed flour and *Moringa oleifera* seed flour. *Moringa oleifera* samples had higher values of phenolic compounds and antioxidant activity as compared to fenugreek, and exerted higher antibacterial activity against several undesired species. On the contrary, the viability of *Streptococcus thermophilus* and *Lactobacillus bulgaricus* was improved. Incorporation of the flours caused a modification of the concentration of mineral compounds, with connected increase of some valuable elements, and of the sensory characteristics. Saleh et al. [7] investigated the use of different types of starch (corn, sweet potato, potato, Turkish beans, and chickpea) as fat replacer in set yogurt. The results show that all starches used reduced syneresis and improved firmness of the product, but at different level due to their origin and amylose content. Yogurts with corn and tuber starches had the highest viscosity, chickpea starch exhibited the highest storage moduli. Different starches behave differently during yogurt storage, and some of them performed better at the beginning of the storage period. Sensory evaluation showed a preference for starch-containing samples as compared to the control, regardless of the starch type. The work of Chen et al. [8] focused on a stirred acidified dairy gel used as a model system for studying ingredient functionalities in yogurt. The study regarded the effect of including black tea infusion on the physicochemical properties, antioxidant capacity and microstructure of the system during a 28-day cold storage period. The results suggest that tea improved antioxidant capacity but significantly altered the texture of gel. The papers published by Serra et al. [9] and Faccia et al. [10] dealt with two types of innovative fresh cheeses. The first one regarded a cheese obtained both by sheep and buffalo milk by using kiwi juice as a coagulant, and it was addresses to assess the influence on flavor and on the presence of nutraceutical substances in comparison with calf rennet. Although the kiwifruit extract caused a longer coagulation and syneresis time than calf rennet, a positive effect on the nutraceutical properties of the cheese was found due to a higher presence of polyphenols and phytosterols. Contrastingly, the profile of volatile organic compounds was not deeply affected, since the typical odorants of the kiwi aroma were poorly transferred to the cheese; the authors suggested the need for further study to evaluate the impact on the sensory characteristics. Faccia et al. described the production technology and the compositional/sensory characteristics of a cheese obtained by the enzymatic coagulation (microbial rennet) of donkey milk. Donkey milk coagulated rapidly, but the curd remained soft, and was only suitable for making fresh cheese; contrastingly, cow milk (used as control under the same conditions) coagulated almost instantaneously and gave rise to a semi-hard curd. Higher yields than those reported in the literature were obtained, probably due to the high protein content of the milk used. The main compositional and sensory characteristics of the cheese were assessed and discussed. Solid Phase Micro Extraction-Gas Chromatography-Mass Spectrometry (SPME-GC-MS) analysis allowed for identifying 11 volatile compounds in milk and 18 in cheese. The other two papers included in this category regard two very innovative products. Buhler et al. [11] investigated the effect of defatting cheese (instead of milk) as an alternative way for obtaining a low-fat Parmigiano Reggiano cheese type. Two defatting procedures were tested, and the composition of the nitrogen fraction of the obtained products were compared. Moreover, the nitrogen compounds were extracted and subjected to simulated gastrointestinal digestion in order to test the antioxidant and angiotensin converting enzyme (ACE) inhibition capacities of the digests. The results obtained show that the defatted products kept the same nutritional properties of the whole cheese. Tian et al. [12] made use of the emulsifying property of polymerized goat whey proteins (PGWP) to prepare soy isoflavones (SIF) nanoparticles. High encapsulation efficiencies were ascertained, and the inclusion of isoflavones increased the particle size and lower zeta potential compared with PGWP alone. The secondary structure of the proteins changed after interacting with SIF, with the transformation of α-helix and β-sheet to disordered structures. The authors concluded that PGWP might be a good carrier material for the delivery of SIF in functional foods.

The influence of rearing conditions on cheese quality. Formaggioni et al. [13] compared the fatty acid profile and the sensory properties of a traditional cheese manufactured from milks of cows reared and fed under different conditions. The experimentation demonstrated that the fat of the cheese obtained from cows fed indoors was richer in medium-chain fatty acids, whereas grazing positively influenced the concentrations of long-chain and unsaturated fatty acids, such as oleic, Conjugated Linoleic Acids (CLAs) and omega 3 fatty acids. Nevertheless, the sensory analysis showed that the tasters were not always able to find significant differences among the cheese samples. Martino et al. [14] dealt with the relationships between feeding dairy sheep with zinc (a key mineral that is not stored in the animal body) supplementation and cheese characteristics. The feeding strategy induced a significant increase in zinc concentration in milk, but it also seemed to be connected to an increase in vaccenic, rumenic and total polyunsaturated fatty acids, in both milk and cheese. The study suggests possible positive effects of dietary zinc supplementation in improving the nutritional characteristics of fresh and ripened dairy products. Franceschi et al. [15] deepened the effect of season and factory on cheese-making efficiency in Parmigiano Reggiano cheese manufacture. They focused particularly on the relationship between cheese-making losses and protein and fat content by considering the production process of 288 Parmigiano Reggiano cheese moulds manufactured in three different cheese factories. The authors reported that estimated cheese losses strictly depend on milk characteristics, in particular, milk fat, casein contents, and rennet coagulation properties; they concluded that estimated cheese-making losses of protein and fat can be used as an instrument for controlling the manufacturing process.

Two further papers complete the Special Issue: an Original Article and a Review Article. The first one [16] is a very interesting study in which confocal Raman microscopy was applied to investigate the microstructure of high moisture cream cheese after freezing and thawing in comparison with confocal laser scanning microscopy. Raman spectroscopy is very interesting since it allows for observing different classes of molecules in situ, in complex food matrices, without modifying them. The results show that it was possible to identify and map the large water domains formed during freezing and thawing and that this technique could be complementary to confocal laser scanning microscopy. In addition, the microscopy data complemented the information derived from low-resolution Nuclear Magnetic Resonance (NMR), suggesting that NMR and Raman microscopy can be complementary to distinguish between different commercial formulations, and different destabilization levels. Finally, the Review by Manoni et al. [17] supplies exhaustive information about the formation of bovine milk fat globules and the milk fat globule membrane (MFGM), highlighting the main similarities and differences across the MFGM proteomes of the most-studied mammal species. Moreover, the potential supplementation of MFGM fractions in infant formula in order to underline the beneficial effects exerted by MFGM bioactive components was investigated.

In summary, the Special Issue "Chemical and Technological Characterization of Dairy Products" offers readers a series of innovative information that can be useful both for developing new research ideas and for developing new types of dairy products.

Funding: This research received no external funding.

Conflicts of Interest: The author declare no conflict of interest.

References

1. Augustin, M.A.; Udabage, P.; Juliano, P.; Clark, P.T. Towards a More Sustainable Dairy Industry: Integration Across the Farm–Factory Interface and the Dairy Factory of the Future. *Int. Dairy J.* **2013**, *31*, 2–11. [CrossRef]

2. Malcata, X.F. Critical Issues Affecting the Future of Dairy Industry: Individual Contributions in the Scope of a Global Approach. *J. Dairy Sci.* **1999**, *82*, 1595–1611. [CrossRef]

3. Rice, B.H.; Cifelli, C.J.; Pikosky, M.A.; Miller, G.D. Dairy Components and Risk Factors for Cardiometabolic Syndrome: Recent Evidence and Opportunities for Future Research. *Adv. Nutr.* **2011**, *2*, 396–407. [CrossRef]

4. Michaelidou, M. Factors Influencing Nutritional and Health Profile of Milk and Milk Products. *Small Rum. Res.* **2008**, *79*, 42–50. [CrossRef]

5. Abdel-Hamid, M.; Huang, Z.; Suzuki, T.; Enomoto, T.; Hamed, A.M.; Li, L.; Romeih, E. Development of a Multifunction Set Yogurt Using Rubus suavissimus S. Lee (Chinese Sweet Tea) Extract. *Foods* **2020**, *9*, 1163. [CrossRef] [PubMed]

6. Dhawi, F.; El-Beltagi, H.S.; Aly, E.; Hamed, A.M. Antioxidant, Antibacterial Activities and Mineral Content of Buffalo Yoghurt Fortified with Fenugreek and Moringa oleifera Seed Flours. *Foods* **2020**, *9*, 1157. [CrossRef] [PubMed]

7. Saleh, A.; Mohamed, A.A.; Alamri, M.S.; Hussain, S.; Qasem, A.A.; Ibraheem, M.A. Effect of Different Starches on the Rheological, Sensory and Storage Attributes of Non-fat Set Yogurt. *Foods* **2020**, *9*, 61. [CrossRef] [PubMed]

8. Chen, H.; Zheng, H.; Anne Brennan, M.; Chen, W.; Guo, X.; Brennan, C.S. Effect of Black Tea Infusion on Physicochemical Properties, Antioxidant Capacity and Microstructure of Acidified Dairy Gel during Cold Storage. *Foods* **2020**, *9*, 831. [CrossRef] [PubMed]

9. Serra, A.; Conte, G.; Corrales-Retana, L.; Casarosa, L.; Ciucci, F.; Mele, M. Nutraceutical and Technological Properties of Buffalo and Sheep Cheese Produced by the Addition of Kiwi Juice as a Coagulant. *Foods* **2020**, *9*, 637. [CrossRef] [PubMed]

10. Faccia, M.; Gambacorta, G.; Martemucci, G.; Difonzo, G.; D'Alessandro, A.G. Chemical-Sensory Traits of Fresh Cheese Made by Enzymatic Coagulation of Donkey Milk. *Foods* **2020**, *9*, 16. [CrossRef]

11. Buhler, S.; Riciputi, Y.; Perretti, G.; Caboni, M.F.; Dossena, A.; Sforza, S.; Tedeschi, T. Characterization of Defatted Products Obtained from the Parmigiano–Reggiano Manufacturing Chain: Determination of Peptides and Amino Acids Content and Study of the Digestibility and Bioactive Properties. *Foods* **2020**, *9*, 310. [CrossRef]

12. Tian, M.; Wang, C.; Cheng, J.; Wang, H.; Jiang, S.; Guo, M. Preparation and Characterization of Soy Isoflavones Nanoparticles Using Polymerized Goat Milk Whey Protein as Wall Material. *Foods* **2020**, *9*, 1198. [CrossRef]

13. Formaggioni, P.; Malacarne, M.; Franceschi, P.; Zucchelli, V.; Faccia, M.; Battelli, G.; Brasca, M.; Summer, A. Characterisation of Formaggella della Valle di Scalve Cheese Produced from Cows Reared in Valley Floor Stall or in Mountain Pasture: Fatty Acids Profile and Sensory Properties. *Foods* **2020**, *9*, 383. [CrossRef] [PubMed]

14. Martino, C.; Ianni, A.; Grotta, L.; Pomilio, F.; Martino, G. Influence of Zinc Feeding on Nutritional Quality, Oxidative Stability and Volatile Profile of Fresh and Ripened Ewes' Milk Cheese. *Foods* **2020**, *8*, 656. [CrossRef] [PubMed]

15. Franceschi, P.; Malacarne, M.; Formaggioni, P.; Cipolat-Gotet, C.; Stocco, G.; Summer, A. Effect of Season and Factory on Cheese-Making Efficiency in Parmigiano Reggiano Manufacture. *Foods* **2020**, *8*, 315. [CrossRef] [PubMed]

16. Alinovi, M.; Mucchetti, G.; Andersen, U.; Rovers, T.A.M.; Mikkelsen, B.; Wiking, L.; Corredig, M. Applicability of Confocal Raman Microscopy to Observe Microstructural Modifications of Cream Cheeses as Influenced by Freezing. *Foods* **2020**, *9*, 679. [CrossRef] [PubMed]

17. Manoni, M.; Di Lorenzo, C.; Ottoboni, M.; Tretola, M.; Pinotti, L. Comparative Proteomics of Milk Fat Globule Membrane (MFGM) Proteome across Species and Lactation Stages and the Potentials of MFGM Fractions in Infant Formula Preparation. *Foods* **2020**, *9*, 1251. [CrossRef] [PubMed]

Publisher's Note: MDPI stays neutral with regard to jurisdictional claims in published maps and institutional affiliations.

Article

Effect of Season and Factory on Cheese-Making Efficiency in Parmigiano Reggiano Manufacture

Piero Franceschi, Massimo Malacarne *, Paolo Formaggioni *, Claudio Cipolat-Gotet, Giorgia Stocco and Andrea Summer

Department of Veterinary Science, University of Parma, Via del Taglio 10, I-43126 Parma, Italy
* Correspondence: massimo.malacarne@unipr.it (M.M.); paolo.formaggioni@unipr.it (P.F.);
  Tel.: +39-0521032614 (P.F.)

Received: 28 June 2019; Accepted: 1 August 2019; Published: 3 August 2019

Abstract: The assessment of the efficiency of the cheese-making process (ECMP) is crucial for the profitability of cheese-factories. A simple way to estimate the ECMP is the measure of the estimated cheese-making losses (ECL), expressed by the ratio between the concentration of each constituent in the residual whey and in the processed milk. The aim of this research was to evaluate the influence of the season and cheese factory on the efficiency of the cheese-making process in Parmigiano Reggiano cheese manufacture. The study followed the production of 288 Parmigiano Reggiano cheese on 12 batches in three commercial cheese factories. For each batch, samples of the processed milk and whey were collected. Protein, casein, and fat ECL resulted in an average of 27.01%, 0.72%, and 16.93%, respectively. Both milk crude protein and casein contents were negatively correlated with protein ECL, r = −0.141 ($p \leq 0.05$), and r = −0.223 ($p \leq 0.001$), respectively. The same parameters resulted in a negative correlation with casein ECL ($p \leq 0.001$) (r = −0.227 and −0.212, respectively). Moreover, fat ECL was correlated with worse milk coagulation properties and negatively correlated with casein content (r = −0.120; $p \leq 0.05$). In conclusion, ECLs depend on both milk characteristics and season.

Keywords: milk composition; Parmigiano Reggiano cheese; cheese-making efficiency; curd fines; cheese-making losses

---

1. Introduction

The cheese-making process of rennet coagulated cheeses consists in the formation of a three-dimensional network of paracasein, in which fat globules and part of the milk whey are entrapped. The quantity of milk constituents recovered into cheese is strictly dependent on the quality of the milk (casein content, casein micelle structure, and integrity) and conditions of the cheese-making process (for example, pre-acidification of processed milk, type and quantity of rennet, cooking temperature, acidification of the cheese mass), and thus varies depending on the cheese type.

The assessment of the efficiency of the cheese-making process (ECMP) is crucial for the profitability of cheese-factories. The best way to quantify the ECMP is to measure the recovery of milk constituents into cheese through a mass balance determination. This can be obtained by measuring the quantity of a constituent in the processed milk and the resulting cheese. However, to perform this kind of analysis, it is necessary to also measure the weight of both the processed milk and cheeses. This is not always possible, especially in artisanal cheese factories, as those involved in Parmigiano Reggiano production, where the weight of the processed milk is estimated using a wooden measuring stick with a sensitivity, are not acceptable for research purposes.

Although less reliable than a mass balance, a rough and simple alternative way to estimate the ECMP is to measure the estimated cheese-making losses (ECL). In this method, the estimated loss of a milk constituent is expressed by the ratio between its concentration in the cheese whey (C-whey) (that remains in the vat after the extraction of the cheese mass) and that in the processed vat milk (V-milk). Consequently, the determination of ECL is easier and faster to perform, since for a single milk constituent it is necessary to only measure its concentration in V-milk and C-whey. No weight of milk, cheese, or whey is needed, and it is not necessary to sample and analyze the cheese. Moreover, to date, the concentration of most milk constituents can be rapidly assessed by applying the mid infrared technology (MIR) [1–3]. The MIR technology is widely used by a lot of laboratories which provide analysis and technological consulting services to dairy farms producing Parmigiano Reggiano cheese. In these laboratories, this kind of analysis is routine, cheap, and certified ISO/IDF. The ECLs have been used in several studies to estimate the effect of breed [4], storage conditions [5], and somatic cells [6,7] on efficiency in Parmigiano Reggiano cheese-making.

Parmigiano Reggiano is a hard, cooked, and long-ripened protected designation of origin (PDO) cheese made from raw milk, following a strict manufacture procedure [8]. In case of raw milk cheese, the quality of milk-in terms of chemical composition and microbial characteristics is one of the main factors influencing the efficiency of the cheese-making process. Seasonal variations of milk characteristics at the herd level were reported in several studies [9,10]. Significant variations of the chemical and microbial quality of milk employed for Parmigiano Reggiano cheese throughout the year have also been reported [11,12]. These variations could have repercussions on ECLs and curd fines, as reported by Formaggioni et al. [13] and by Franceschi et al. [5]. However, these papers considered a limited number of cheese-making trials and milk traits, since minerals were not taken into account and only two periods of the year were considered. Moreover, in the majority of PDO cheese manufacture, such as Parmigiano Reggiano, where milk standardization and automation of the processes are not implemented, a strong variability in the ECMP is expected among dairy farms [14]. However, the quantification of this variability has never been carried out.

The aim of this research was to evaluate the influence of the season and cheese-factory on ECLs on the quantity of curd particles lost in the whey (curd fines) in Parmigiano Reggiano cheese manufacture carried out in field conditions.

2. Materials and Methods

2.1. Cheese-Making Process

Cheeses were produced by the approved method of the Consortium [8,15]. A natural whey starter culture (about 2.5–3 L for every 100 kg V-milk), obtained by the spontaneous acidification of previous day milk whey (C-whey), was added to the V-milk before coagulation. It was then heated to 33 °C and clotted in 10–12 min with 2.5 g for every 100 kg V-milk of calf rennet (1:125,000 units). The curd was broken up into small granules (approximately the size of a rice grain) and cooked. During this operation, the temperature was increased in two steps to 55 °C within 10–15 min; during this phase the curd was stirred continuously. After cooking the broken-up curd particles, thy were deposited by simple decantation at the bottom of the vat, where they aggregated and blended together spontaneously. In this step, the temperature was 55–53 °C and the process lasted 45–60 min. The cheese mass was then removed from the vat, divided into two parts, and placed in special molds called "fascere" for two days. During this period, the cheese wheels were naturally cooled and periodically turned over to allow an homogeneous drying. Furthermore, pH decreased from about 6.0 (at the extraction from the vat) to about 5.1 at the end of the two days. This is related to the activity of thermophilic lactic bacteria added with the natural whey starter, which converts lactose to lactic acid. The cheese wheels were then placed into a saturated brine for a period of 20–25 days. Finally, the cheese entered the ripening phase, a process that lasts about 24 months, and, at the end of the ripening, the cheese wheels resulted in a cylindrical in shape, with a slightly convex side, 22–24 cm high, 40–45 cm diameter, weight 35–36 kg.

2.2. Experimental Design and Sampling Procedure

The research involved 288 Parmigiano Reggiano cheese-making trials, carried out in 3 cheese factories (CF1, CF2, and CF3), throughout two years.

Briefly, every cheese factory vat (that was filled with the milk of the same herd throughout the two years of the experimental period) was selected (6 vats in CF1, 3 in CF2, and 3 CF3). Each selected vat was sampled once a month, and all the vats of the same cheese factory were sampled on the same day. Each vat always contained the milk from the same farm. From each cheesemaking, samples of V-milk and C-whey were collected, following the International Dairy Federation standard [16]. V-milk samples were collected at the beginning of the cheese making process, before the addition of the natural whey starter culture. C-whey samples were collected after the extraction of the cheese mass and stirring of the whey.

2.3. Analytical Methods

The following traits were determined or calculated on both V-milk and C-whey. Total N (TN) and non casein N (NCN) were measured on milk and acid whey at pH4.6, respectively, by Kjeldahl, from which the values of crude protein (TN × 6.38/1000) and casein ((TN-NCN) × 6.38/1000) were calculated [7]. Fat content was assessed by the mid infrared method using a FT 6000 (Foss Electric, DK-3400 Hillerød Denmark); dry matter was measured after oven drying at 102 °C and ash was measured after muffle calcination at 530 °C [17]; total Ca and Mg were determined on a chloridric ash solution by atomic absorption spectrometry (AAS) with a wavelength reading at 422.7 and 285.2 nm, respectively; and total P was assessed on a chloridric ash solution following the colorimetric method [18] with a wavelength reading at 750.0 nm. Titratable acidity was measured only on V-milk by titration of 50 mL of milk with 0.25 N sodium hydroxide with the Soxhlet–Henkel method [7]. Rennet coagulation properties were also measured on V-milk, using Formagraph (Foss Electric, DK-3400 Hillerød Denmark) [4]. The analysis was performed adding 0.2 mL (1:100) of rennet solution (1:19,000; Chr. Hansen, I-20094 Corsico MI, Italy) to milk samples (10 mL). The rennet coagulation properties, milk clotting time, curd firming time, and curd firmness, were measured at 35 °C. Milk clotting time is the time from the addition of rennet to the onset of gelation. Curd firming time is the time from the onset of gelation till the signal attains a width of 20 mm. Curd firmness is the width of the signal 30 min after the addition of rennet. To record curd firming time values in milk samples that do not reach a width of 20 mm within 30 min, the analysis was prolonged to 60 min. The curd fines were determined in C-whey by the gravimetric method proposed by van den Berg et al. [19]. In this method, 250 of C-whey were centrifuged at 2000 g for 30 min. The pellet was resuspended in distilled water and filtered on a Whatman 40 filter paper. The filter was dried at 102 °C for 2 h and weighed.

ECLs of dry matter, protein, casein, fat, calcium, phosphorus, and magnesium were calculated as follows:

$$ECL = [C\text{-whey}] \times 100/[V\text{-milk}]$$

where ECL is expressed as percentage; C-whey = concentration in whey, expressed as g/100 g (mg/100 g for Ca, P, Mg); V-milk = concentration in milk, expressed as g/100 g (mg/100 g for Ca, P, Mg).

2.4. Statistical Analysis

The significance of the differences between seasons and cheese-factories were tested by analysis of variance, using the software for statistical analysis SPSS (IBM SPSS Statics 23, Armonk, New York 10504-1722, NY, USA), according to the following univariate model:

$$Y_{ijk} = \mu + S_i + C_j + \varepsilon_{ijk}$$

where Y_{ijk} = dependent variable; μ = overall mean; S_i = effect of season (i = 1, … , 4; winter, from January to March; spring, from April to June; summer, from July to September, Autumn, from October to

December); C_j = effect of cheese-factory (j = 1, ... , 3; CF1, CF2, CF3); ε_{ijk} = residual error. The Bonferroni post-hoc test was employed to evaluate the significance of the differences between means.

Data were also processed by the Pearson product moment correlation coefficient to measure the degree of linear relationship between milk constituents and ECLs.

3. Results

3.1. Overall Average Values and Descriptive Statistics

The descriptive statistics of V-milk characteristics, ECLs values, and curd fines content in whey are reported in Table 1. The Pearson product moment coefficient of correlations between the milk characteristics, ECLs values, and curd fines content are reported in Table 2.

Table 1. Descriptive statistics of vat milk characteristics and estimated cheese-making loss (ECL) values from 288 Parmigiano Reggiano cheese-making trials.

		Mean		SD [1]	Minimum	Maximum	CV [2] (%)
Vat milk characteristics							
Dry matter	g/100 g	11.73	±	0.32	10.67	12.45	2.70
Crude protein	g/100 g	3.18	±	0.12	2.76	3.47	3.77
Casein	g/100 g	2.46	±	0.10	2.11	2.66	3.94
Fat	g/100 g	2.68	±	0.21	2.02	3.13	7.86
Fat to casein ratio	Value	1.09	±	0.08	0.85	1.27	7.04
Ash	g/100 g	0.73	±	0.01	0.69	0.78	2.02
Calcium	mg/100 g	119.59	±	5.31	109.22	138.46	4.44
Phosphorus	mg/100 g	88.62	±	3.29	77.90	97.90	3.71
Magnesium	mg/100 g	10.67	±	0.76	9.23	15.14	7.17
Titratable acidity	°SH/50 mL	3.29	±	0,11	2.95	3.60	3.37
Clotting time	min	18.52	±	2.20	11.50	24.00	11.89
Curd firming time	min	7.01	±	2.80	2.75	11.25	39.94
Curd firmness	mm	26.25	±	6.10	9.44	43.48	23.23
ECLs [3]							
Dry matter	%	66.91	±	3.12	58.77	75.18	4.66
Protein	%	27.01	±	0.93	22.44	31.59	3.45
Casein	%	0.72	±	0.05	0.10	3.50	6.94
Fat	%	16.93	±	3.59	10.31	27.78	21.21
Ash	%	75.42	±	1.57	70.07	84.57	2.08
Calcium	%	36.51	±	2.73	28.07	44.08	7.48
Phosphorus	%	50.87	±	2.25	44.43	58.76	4.42
Magnesium	%	76.54	±	4.59	54.83	88.57	6.00
Curd fines	mg/kg	122.01	±	66.63	9.30	428.00	54.61

[1] Standard deviation; [2] Coefficient of variation; [3] Estimated cheese-making losses, expressed as the % of ratio between the concentrations in the residual cheese whey and vat milk.

The average contents of crude protein, casein, and fat in V-milk results were consistent with those reported by Formaggioni et al. [14] in a research carried out on 89 vat milk samples. Both contents results of the crude protein and casein in V-milk were negatively correlated with protein ECL and casein ECL. Moreover, casein content negatively correlated with fat ECL. This is in agreement with Malacarne et al. [4], who observed how milk with high casein content gives rise to a rennet curd with an improved capacity to entrap fat globules in the cheese matrix during coagulation. The casein ECL is lower if compared to those reported by Franceschi et al. [5], who found a casein ECL value of 1.25% for V-milk that was stored at 20 °C before processing. However, it is worth noting that Franceschi et al. [5] analysed only three samples collected in the winter season and three samples collected in the summer season. Protein ECL results were higher with respect to casein ECL, but showed a lower variability. The protein ECL average value was consistent with those reported by Franceschi et al. [5] (27.81%)

and Summer et al. [7] (27.33%). The difference between the average values of protein ECL and casein ECL is due to milk whey proteins, which remain in the C-whey. Fat ECL showed a higher variability with respect to protein ECL and casein ECL. In this case, the average value found was consistent with the data reported by Franceschi et al. [5] (14.75%) and Summer, et al. [7] (14.95%). Fat ECL results correlated with the rennet coagulation parameters of V-milk. In particular, positive correlations were found with clotting time and curd firming time, while a negative correlation was evidenced with curd firmness. In fact, faster coagulating milk and firming curd give rise to higher curd firmness and, consequently, have an improved capacity to entrap fat globules into the paracasein matrix.

Table 2. Pearson product moment correlation coefficient (r) between the milk characteristics and the estimated cheese-making loss (ECL) values and curd fines. Only significant correlations ($p < 0.05$) are reported.

	ECLs [1] (%)							
	Protein		Casein		Fat		Curd Fines	
	r	p [2]	r	p [2]	r	p [2]	r	p [2]
Dry matter			−0.112	*			−0.114	*
Crude protein	−0.141	*	−0.227	***			−0.185	**
Casein	−0.223	***	−0.212	***	−0.120	*	−0.195	***
Fat								
Clotting time					0.141	*		
Curd firming time					0.169	**	0.109	*
Curd firmness					−0.176	**	−0.152	*

[1] Cheese-making loss for a milk constituent is expressed as the % ratio between the concentrations in the residual cheese whey and vat milk; [2] p-value: * $p \leq 0.05$; ** $p \leq 0.01$; *** $p \leq 0.001$.

The curd fines are cheese particles that are too small to precipitate on the vat bottom, and therefore remain in suspension in the C-whey [20]. Consequently, they are not included in the cheese wheels. The curd fines content results were approximately twice than that reported by Franceschi et al. [5] (66.40 mg/kg), but this difference could be expected because, in their investigation, these Authors considered only six samples and the quantity of curd fines is generally small and its variability very high. The curd fines quantity in the C-whey results negatively correlated with the contents of crude protein and casein in V-milk and with the curd firmness, and positively correlated with the curd firming time. This is due to the fact that the higher the casein content is in the milk, the higher the crude protein content is [21] and the lower the results for the curd firming time [22], with a consequent higher curd firmness [21,22].

3.2. Seasonal Variations of ECLs

Seasonal variations of ECLs are shown in Table 3. The ECLs of dry matter, casein, fat, and calcium and the content of curd fines in the C-whey showed statistically significant differences among the seasons.

The estimated loss of dry matter result were lower in summer and higher in winter. It is worth noting that, although statistically significant, the differences were very small, amounting to approximately 1.5 percentage units. Additionally, casein ECL showed a very small variation, and result were higher in spring and lower in autumn. This is mainly due to the lactation stage of cows, which affects milk casein content with repercussions on casein ECL values. In fact, as reported by Summer et al. [11], during the spring season, most of cattle are in the early stage of lactation, which is characterized by a progressive increase of milk production and a decrease in milk protein content. In this season, milk samples showed the lowest average value of casein (2.40 g/100 g, data not shown in table). On the contrary, during the autumn season, most of cattle are in late lactation, which is characterised by a progressive decrease of milk production and an increase of milk protein and casein contents [11]. In this season, milk samples showed the highest casein content (2.52 g/100 g, data not shown in table).

Table 3. Seasonal variation of estimated cheese-making loss values and curd fines (least square means values).

		Winter $n^1 = 72$		Spring $n^1 = 72$		Summer $n^1 = 72$		Autumn $n^1 = 72$		SE[2]	p[3]
Dry matter	%	67.61	b	66.94	a,b	66.08	a	67.01	a,b	0.35	*
Crude protein	%	27.05		26.82		26.95		27.21		0.11	NS
Casein	%	0.79	b,c	0.87	c	0.63	a,b	0.58	a	0.06	**
Fat	%	15.94	a	16.51	a,b	17.52	b,c	17.74	c	0.35	**
Ash	%	75.43		75.76		75.42		75.06		0.19	NS
Phosphorus	%	50.84		51.18		50.91		50.54		0.28	NS
Calcium	%	36.96	b	37.13	b	36.83	b	35.00	a	0.42	***
Magnesium	%	76.70		76.69		76.43		76.41		0.83	NS
Curd fines	mg/kg	104.17	a	137.27	b	124.89	a,b	121.68	a,b	5.81	*

[1] Number of samples; [2] Standard error of the mean; [3] p-value: a, b, and c are different for $p \leq 0.05$; NS, $p > 0.05$; * $p \leq 0.05$; ** $p \leq 0.01$; *** $p \leq 0.001$.

Fat ECL results were lower during winter and spring and higher in summer and autumn. The high value of fat ECL during the summer season is due to the general worsening of the milk characteristics in this season, as reported by Summer et al. [11] and Bertocchi et al. [12]. The production area of Parmigiano Reggiano cheese during the summer period is characterised by a high temperature-humidity index [12]. This could induce heat stress conditions for the cow with an increase of the milk somatic cell content [11,12]. The increase of somatic cell content leads to a decrease of milk casein content [7], titratable acidity value [7,11], and alteration of milk mineral content and salts distribution [7,23], with a worsening of rennet coagulation properties [24,25] and an increase of fat losses [7,13].

Curd fines results were lowest in winter and highest in spring. This is in contrast with reports from Summer et al. [20], who did not find significant differences among seasons. However, Summer et al. [20] collected the samples from May to January and considered only two seasonal categories, namely Spring–Summer and Autumn–Winter. The seasonal trend of curd fines, although significant, is difficult to explain because the quantity of curd fines that remain in the C-whey is affected by many factors, such as milk casein content, curd firming time, and curd firmness [5,13,20]. For example, in a research carried out on 102 milk samples, Formaggioni et al. [13] showed a significant correlation between curd fines content of C-whey, curd firming time, and curd firmness of milk. In the present research, milk casein content and rennet coagulation properties (data not shown) showed different seasonal trends and, it is likely that, since all these parameters influenced the curd fines content in different measures, curd fines did not show a clear seasonal trend.

3.3. Difference of ECLs among Cheese Factories

Difference of ECL values among cheese factories are shown in Table 4. The cheese-making losses of dry matter, fat, protein, casein, phosphorus, ash, and curd fines showed significant differences between the cheese factories. Compared to the other two cheese factories, CF3 showed higher values of protein, casein, fat, and ash ECLs. This in turn affected the dry matter ECL that was significantly higher in this cheese-factory. It is not easy to explain the difference of the ECLs based on CF3 milk characteristics, as they were not always the worst among cheese factories. Even when considering the level of cheese production and the location of the three cheese factories, differences in ECLs are hard to explain. In fact, CF1 is characterized by a higher production and is located in the plain zone; CF2 is characterised by a small production and is located in the hill zone; and the CF3 is characterised by a small production, similar to the CF2, and is located in the plain zone, like the CF1. The difference in ECLs among cheese factories can be explained with differences in the technological process. Parmigiano Reggiano cheese is still a craftsmanship product, not a standardized product, and only the crucial technological steps are reported in the Parmigiano Reggiano cheese disciplinary. A lot of variability can be observed between different cheese factories on the duration of technological steps, the dimensions of curd granules during the curd broken step, and the quantity of whey starter cultures that are added to the V-milk in the pre-acidification milk step.

Table 4. Differences among cheese factories in estimated cheese-making loss values and curd fines.

		Cheese Factory 1 n^1 = 144			Cheese Factory 2 n^1 = 72			Cheese Factory 3 n^1 = 72			p^4
		LSMean[2]	SE[3]		LSMean[2]	SE[3]		LSMean[2]	SE[3]		
Dry matter	%	67.68	0.25	b	65.84	0.36	a	66.40	0.36	b	***
Crude protein	%	26.97	0.07	a	26.89	0.10	a	27.25	0.10	b	*
Casein	%	0.63	0.04	a	0.75	0.06	a,b	0.85	0.06	b	*
Fat	%	15.53	0.25	a	16.71	0.36	b	19.99	0.36	c	***
Ash	%	75.75	0.13	b	74.83	0.18	a	75.33	0.18	a,b	***
Phosphorus	%	51.24	0.19	b	50.53	0.27	a	50.45	0.27	a	*
Calcium	%	36.90	0.26		35.94	0.36		36.19	0.36		NS
Magnesium	%	77.21	0.46		75.74	0.64		76.09	0.64		NS
Curd fines	mg/kg	111.06	5.42	a	127.21	7.78	a,b	139.07	7.78	b	*

[1] Number of samples; [2] Least square means values; [3] Standard error of the mean; [4] p-value: a, b, and c are different for $p \leq 0.05$; NS, $p > 0.05$; * $p \leq 0.05$; *** $p \leq 0.001$.

4. Conclusions

In conclusion, the two most relevant parameters of the cheese-making losses are protein and fat ECLs; they strictly depend on milk characteristics, in particular, milk fat, casein contents, and rennet coagulation properties. Since the season affects the milk composition and the rennet coagulation properties, it also influences the ECL. Finally, the differences in the technology of milk transformation in cheese that exists among the cheese factories strongly affected the ECLs. The estimated cheese-making losses of protein and fat can be used as an instrument for the control of the technological process.

Author Contributions: Conceptualization, A.S., P.F. (Piero Franceschi) and M.M.; methodology, A.S., P.F. (Piero Franceschi) and M.M.; software, P.F. (Piero Franceschi) and P.F. (Paolo Formaggioni); validation, A.S., M.M., P.F. (Paolo Formaggioni), C.C.–G., and P.F. (Piero Franceschi); formal analysis, P.F. (Piero Franceschi), P.F. (Paolo Formaggioni), G.S., and C.C.–G.; investigation, M.M. and A.S.; resources, A.S. and M.M.; data curation, P.F. (Piero Franceschi), P.F. (Paolo Formaggioni), G.S., and C.C.–G.; writing—original draft preparation, P.F. (Piero Franceschi), M.M., and P.F. (Paolo Formaggioni); writing—review and editing, P.F. (Piero Franceschi), M.M. and P.F. (Paolo Formaggioni); visualization, P.F. (Piero Franceschi), A.S. and M.M.; supervision, A.S. and P.F. (Piero Franceschi); project administration, A.S. and M.M.

Funding: This research received no external funding.

Conflicts of Interest: The authors declare that there are no conflict of interest in this research article.

References

1. Lugibühl, W. Evaluation of designed calibration samples for casein calibration in Fourier transform infrared analysis of milk. *Lebensm. Wiss. Technol.* **2002**, *35*, 554–558. [CrossRef]
2. De Marchi, M.; Fagan, C.C.; O'Donnell, C.P.; Cecchinato, A.; Dal Zotto, R.; Cassandro, M.; Penasa, M.; Bittante, G. Prediction of coagulation properties, titratable acidity, and pH of bovine milk using mid-infrared spectroscopy. *J. Dairy Sci.* **2009**, *92*, 423–432. [CrossRef] [PubMed]
3. De Marchi, M.; Toffanin, V.; Cassandro, M.; Penasa, M. Invited review: Mid-infrared spectroscopy as phenotyping tool for milk traits. *J. Dairy Sci.* **2014**, *97*, 1171–1186. [CrossRef] [PubMed]
4. Malacarne, M.; Summer, A.; Fossa, E.; Formaggioni, P.; Franceschi, P.; Pecorari, M.; Mariani, P. Composition, coagulation properties and Parmigiano-Reggiano cheese yield of Italian Brown and Italian Friesian herd milks. *J. Dairy Res.* **2006**, *73*, 171–177. [CrossRef] [PubMed]

5.   Franceschi, P.; Sandri, S.; Pecorari, M.; Vecchia, P.; Sinisi, F.; Mariani, P. Effects of milk storage temperature at the herd on cheesemaking losses in the manufacture of Parmigiano-Reggiano cheese. *Vet. Res. Comm.* **2008**, *32* (Suppl. 1), 339–341. [CrossRef]

6.   Franceschi, P.; Summer, A.; Sandri, S.; Formaggioni, P.; Malacarne, M.; Mariani, P. Effects of the full cream milk somatic cell content on the characteristics of vat milk in the manufacture of Parmigiano-Reggiano cheese. *Vet. Res. Comm.* **2009**, *33* (Suppl. 1), 281–283. [CrossRef]

7.   Summer, A.; Franceschi, P.; Formaggioni, P.; Malacarne, M. Influence of milk somatic cell content on Parmigiano-Reggiano cheese yield. *J. Dairy Res.* **2015**, *82*, 222–227. [CrossRef]

8.   Council Regulation 1992 (EEC) No 2081/92 of 14 July 1992 on the protection of geographical indications and designations of origin for agricultural products and foodstuffs. *Off. J. Eur. Union* **1992**, *208*, 1–8.

9.   Chen, B.; Lewis, M.J.; Grandison, A.S. Effect of seasonal variation on the composition and properties of raw milk destined for processing in the UK. *Food Chem.* **2014**, *158*, 216–223. [CrossRef]

10.   O'Connell, A.; Mc Parland, S.; Ruegg, P.L.; O'Brien, B.; Gleeson, D. Seasonal trends in milk quality in Ireland between 2007 and 2011. *J. Dairy Sci.* **2014**, *98*, 3778–3790. [CrossRef]

11.   Summer, A.; Franceschi, P.; Formaggioni, P.; Malacarne, M. Characteristics of raw milk produced by free-stall or tie-stall cattle herds in the Parmigiano-Reggiano cheese production area. *Dairy Sci. Tech.* **2014**, *94*, 581–590. [CrossRef]

12.   Bertocchi, L.; Vitali, A.; Lacetera, N.; Varisco, G.; Bernabucci, U. Seasonal variations in the composition of Holstein cow's milk and temperature-humidity index relationship. *Animal* **2014**, *8*, 667–674. [CrossRef]

13.   Formaggioni, P.; Sandri, S.; Franceschi, P.; Malacarne, M.; Mariani, P. Milk acidity, curd firming time, curd firmness and protein and fat losses in the Parmigiano Reggiano cheesemaking. *Ital. J. Anim. Sci.* **2005**, *4* (Suppl. 2), 239–241. [CrossRef]

14.   Formaggioni, P.; Summer, A.; Malacarne, M.; Franceschi, P.; Mucchetti, G. Italian and Italian-style hard cooked cheeses: Predictive formulas for Parmigiano-Reggiano 24 h cheese yield. *Int. Dairy J.* **2015**, *51*, 52–58. [CrossRef]

15.   Malacarne, M.; Summer, A.; Formaggioni, P.; Franceschi, P.; Sandri, S.; Pecorari, M.; Vecchia, P.; Mariani, P. Dairy maturation of milk used in the manufacture of Parmigiano-Reggiano cheese: Effects on chemical characteristics, rennet coagulation aptitude and rheological properties. *J. Dairy Res.* **2008**, *75*, 218–224. [CrossRef]

16.   IDF Standard. *Milk and Milk Products, Guidance on Sampling*; International Dairy Federation Standard 50/ISO707: Brussels, Belgium, 2008.

17.   Malacarne, M.; Franceschi, P.; Formaggioni, P.; Pisani, G.M.; Petrera, F.; Abeni, F.; Soffiantini, C.S.; Summer, A. Mineral content and distribution in milk from red deer (*Cervus elaphus*) fallow deer (*Dama dama*) and roe deer (*Capreolus capreolus*). *Small Rum. Res.* **2015**, *13*, 208–215. [CrossRef]

18.   Malacarne, M.; Criscione, A.; Franceschi, P.; Tumino, S.; Bordonaro, S.; Di Frangia, F.; Donata, M.; Summer, A. Distribution of Ca, P and Mg and casein micelle mineralisation in donkey milk from the second to ninth month of lactation. *Int. Dairy J.* **2017**, *66*, 1–5. [CrossRef]

19.   Van den Berg, G.; de Vries, E.; Arentzen, A.G.J. Which sampling method should be used for the exact determination of the curd-fines content of first whey? *Nizo Nieuws* **1973**, *7*, 825–828.

20.   Summer, A.; Franceschi, P.; Bollini, A.; Formaggioni, P.; Tosi, F.; Mariani, P. Seasonal variation of milk characteristics and cheesemaking losses in the manufacture of Parmigiano-Reggiano cheese. *Vet. Res. Comm.* **2003**, *27* (Suppl. 1), 663–666. [CrossRef]

21.   Comin, A.; Cassandro, M.; Chessa, S.; Ojala, M.; Dal Zotto, R.; De Marchi, M.; Carnier, P.; Gallo, L.; Pagnacco, G.; Bittante, G. Effects of composite β-and k-casein genotypes on milk coagulation, quality, and yield traits in Italian Holstein cows. *J. Dairy Sci.* **2008**, *91*, 4022–4027. [CrossRef]

22.   Cassandro, M.; Comin, A.; Ojala, M.; Dal Zotto, R.; De Marchi, M.; Gallo, L.; Carnier, P.; Bittante, G. Genetic parameters of milk coagulation properties and their relationships with milk yield and quality traits in Italian Holstein cows. *J. Dairy Sci.* **2008**, *91*, 371–376. [CrossRef]

23.   Summer, A.; Franceschi, P.; Malacarne, M.; Formaggioni, P.; Tosi, F.; Tedeschi, G.; Mariani, P. Influence of somatic cell count on mineral content and salt equilibria of milk. *Ital. J. Anim. Sci.* **2009**, *8* (Suppl. 2), 435–437. [CrossRef]

24. Bijl, E.; van Valenberg, H.J.F.; Huppertz, T.; van Hooijdonk, A.C.M. Protein, casein, and micellar salts in milk: Current content and historical perspectives. *J. Dairy Sci.* **2013**, *96*, 5455–5464. [CrossRef]
25. Malacarne, M.; Franceschi, P.; Formaggioni, P.; Sandri, S.; Mariani, P.; Summer, S. Influence of micellar calcium and phosphorus on rennet coagulation properties of cows milk. *J. Dairy Res.* **2014**, *81*, 129–136. [CrossRef]

Article

Influence of Zinc Feeding on Nutritional Quality, Oxidative Stability and Volatile Profile of Fresh and Ripened Ewes' Milk Cheese

Camillo Martino [1], Andrea Ianni [2], Lisa Grotta [3], Francesco Pomilio [4] and Giuseppe Martino [3,*

[1]   Specialist Diagnostic Department, Istituto Zooprofilattico Sperimentale dell'Abruzzo e del Molise "G. Caporale" Via Campo Boario, 64100 Teramo, Italy; c.martino@izs.it
[2]   Department of Medical, Oral and Biotechnological Sciences, "G. d'Annunzio" University of Chieti-Pescara, Via dei Vestini 31, 66100 Chieti, Italy; andreaianni@hotmail.it
[3]   Faculty of Bioscience and Technology for Food, Agriculture and Environment, University of Teramo, Via Renato Balzarini 1, 64100 Teramo, Italy; lgrotta@unite.it
[4]   Food Hygiene Unit, NRL for L. monocytogenes, Istituto Zooprofilattico Sperimentale dell'Abruzzo e del Molise "G. Caporale" Via Campo Boario, 64100 Teramo, Italy; f.pomilio@izs.it
*   Correspondence: gmartino@unite.it

Received: 23 October 2019; Accepted: 5 December 2019; Published: 7 December 2019

Abstract: Zinc represents a ubiquitous element in cells with relevant roles in the metabolism of essential nutrients in animals. The aim of this study was to investigate the effect of dietary zinc supplementation on nutritional and aromatic properties of milk and Pecorino cheeses obtained from lactating ewes. Fifty-two commercial ewes were randomly assigned to two groups. The control group was fed with a conventional complete diet, while the experimental group received a daily supplementation of 375 mg/head of zinc oxide. At the end of the trial, which lasted 30 days, samples of milk and related cheese were collected in order to obtain information about the chemical composition and volatile profile. The experimental feeding strategy induced a significant increase in zinc concentration in milk. Furthermore, both in milk and cheese, was observed an increase in vaccenic, rumenic and total polyunsaturated fatty acids, with the consequent significant reduction of atherogenic and thrombogenic indices. The volatile profile of dairy products was also positively affected by dietary zinc intake, with an increase in concentration of hexanoic acid and ethyl esters. The present study suggests interesting possible effects of dietary zinc supplementation of ewes in improving the nutritional characteristics of fresh and ripened dairy products, although more specific and in-depth assessments should be performed on these new products, in order to characterize potential variations on consumers acceptability.

Keywords: zinc; ewes' milk cheese; rumenic acid; zinc-dependent enzyme; volatile compound

---

1. Introduction

Zinc (Zn) belongs to the family of transition metals and the considerable importance of this microelement lies in its fundamental role in the correct execution of several biochemical mechanisms which mostly provide for the activity of zinc-dependent enzymes [1]. Zn is not stored in animal body, for that reason a constant dietary supply is necessary in order to avoid the onset of a wide range of pathological conditions, such as skin parakeratosis, reduced or cessation of growth, general debility, lethargy and increased susceptibility to infection [2].

It has been reported that almost the half of the soils in the world may be zinc deficient, causing decreased Zn content in plant. In many of these areas, where grazing livestock is widespread, zinc deficiency is prevented by zinc fertilization of pastures. For livestock under more defined conditions, such as poultry, swine, and dairy cattle, feeds are enriched with zinc salts to prevent deficiency [3].

The essentiality of Zn in livestock nutrition is well established, for this reason several feeding strategies have been tested over time to ensure the adequate dietary intake of all the necessary trace elements. In addition to this, the concentration of these elements in milk and dairy products has been reported to be heavily influenced by the feeding strategy. Regarding Zn, the chemical form mostly used for the industrial preparation of animal feeds is represented by zinc oxide (ZnO), although it has recently been introduced the nano zinc oxide (nZnO) in an attempt to improve solubility and Zn availability, without inducing toxicity [4].

Nutritional requirements of ruminants are different from those of monogastric animals. Several studies showed that the bioavailability of specific trace elements is of primary relevance in supporting an adequate ruminal fermentation and digestion. Sonawane and Arora [5] conducted an in vitro study in which observed an increased synthesis of microbial proteins as a consequence of ruminal fluid incubation with additional Zn as $ZnCl_2$ or $ZnSO_4$; more recently, has been demonstrated the Zn ability to inhibit the ruminal hydrolysis of urea when fed to steers consuming low quality hay, therefore avoiding the excessive increase in NH_3 concentration which could negatively interfere with protein synthesis by ruminal microbes [6]. In ewes, the extra dietary Zn supplementation was also reported to induce the transcriptional modulation of protein mediators of cellular signaling, cardiac contractility and immune response [7].

Over time different studies have been performed with the aim to evaluate productive and qualitative parameters of milk obtained from ruminants fed a dietary supplementation of organic and inorganic Zn. Salama et al. [8] reported that milk yield was not significantly affected by Zn-methionine intake in dairy goats [8], furthermore no variations were observed concerning the percentages of protein, lactose, fat, solid non-fat, total solid, and density of milk. Recently, Ianni et al. [9] confirmed this finding in lactating dairy cows supplemented with ZnO, also highlighting an improvement in the nutraceutical properties of milk, due to the increased concentration of conjugated linoleic acids (CLA). In similar studies has been also found an increase in Zn concentration both in milk and in bovine cheese, as evidence of the fact that changes in animal feeding represent promising approaches to modify Zn amount in milk and related dairy products [10,11].

The objective of this study was to assess the influence of a dietary zinc oxide supplementation in lactating ewes on nutritional characteristics, fatty acids composition, lipid peroxidation and volatile profile of fresh and 90-days ripened ewes' milk cheese. There are adequate evidences that would support a positive role for Zn in influencing the biochemical mechanisms directly involved in defining the qualitative parameters of the animal production.

2. Materials and Methods

2.1. Experimental Design, Cheese Manufacturing Protocol and Sampling

Fifty-two half-bred ewes have been randomly divided into two groups: a control group (CG) and an experimental group (EG) whose diet was supplemented with Zn. Individual milk samples were collected before the trials to obtain information about milk yield, chemical composition and fatty acid profile. This approach was useful to verify the eventual presence of variations among the selected groups.

For 30 days, the CG received a complete diet that was prepared in accordance with the sheep nutritional needs, and guaranteeing each animal the daily Zn requirement of about 79 mg. The EG received the same complete food, formulated according to the same requirements and prepared in the same way, however enriching the daily ration of each sheep with additional 296 mg Zn in order to obtain a total intake of about 375 mg. The management of Zn doses was executed according to the Regulation (EC) No. 1831/2003 of the European Parliament and of the Council of 22 September 2003 on additives for use in animal nutrition [12].

With regard to the Pecorino cheese, the production was performed by the same company in which the trial was conducted, located in the province of Teramo (Abruzzo, Italy). The manufacturing protocol provided the bulk ewes' milk pasteurization at 70 ± 1 °C for 15 s, followed by cooling at 40 °C and inoculation with a freeze-dried starter culture (*Streptococcus thermophilus*, *Lactobacillus casei* and *Lactobacillus delbrueckii* subsp. *bulgaricus*) produced by FERM IN (ChemiFerm s.r.l., Livraga, Italy). Then, milk was coagulated by adding kid rennet paste (75% chymosin and 25% pepsin; 1:18,000 strength; Clerici, Cadorago, Italy) and the curd was subsequently cut into small pieces by stirring with a spatula, heated to 42 ± 1 °C and manually pressed. The resulting cheeses, of about 1 kg each, were held at 10 °C until the next day, when they were salted in aqueous solution containing 18% of sodium chloride for 12 h. Ripening was conducted at 12 ± 1 °C.

With the purpose of evaluating variations in chemical composition and quality attributes due to ripening, the sampling of ewes' milk cheese was carried out after 1 (T_1) and 90 (T_{90}) days from the cheese-making. Samples, collected in triplicate from three different cheese-makings, were partly immediately analyzed and partly packed under vacuum and frozen at −20 °C until analysis.

2.2. Chemical Analysis of Milk and Cheese

MilkoScan FT 6000 (Foss Integrator IMT; Foss, Hillerød, Denmark) was used to determine the chemical composition of milk (fat, protein, casein, lactose, and urea), while somatic cells count (SCC) and total bacterial count (TBC) were performed using respectively the Fossomatic TM FC and the BactoScan FC (Foss, Hillerød, Denmark). In cheese, the evaluation of pH, moisture, total proteins, lipids and ash were performed according to AOAC methods (1990) [13]; water-soluble nitrogen (WSN) and trichloroacetic acid-soluble nitrogen (TCA-SN) were determined according to the International Standard ISO 27871 IDF 224 (2011) [14], and results have been reported as percentage of total nitrogen, following appropriate calibration.

The total amount of Zn in milk and cheese was determined by inductively coupled plasma mass spectrometry (ICP-MS) by using an Agilent 7500ce (Agilent Technologies, Palo Alto, CA, USA) and following the procedure reported by Gerber et al. [15] with slight modifications. Samples, 5 mL of milk or 5 g of cheese, were accurately inserted into quartz digestion vessels. At this point, 3 mL of 30 % hydrogen peroxide (Sigma Aldrich, Milan, Italy) and 10 mL 65 % nitric acid (Sigma Aldrich, Milan, Italy) were added to each tube, which was then closed for sample digestion at 95 °C for 2 h. After the vessels had cooled down, the digests were transferred into 50 mL volumetric flasks and filled to the mark using ultrapure water. One milliliter of the solution was added with 9 mL of distilled nitric acid (1%) and analyzed. The Zn determination was performed by referring to a calibration and results were expressed in mg/kg.

2.3. Evaluation of Fatty Acid Profile in Milk and Cheese

Extraction of the milk lipid fraction was made according to the AOAC official method [16], while in cheese was used a mix of chloroform and methanol (2:1, *v/v*; Sigma Aldrich, Milan, Italy). Trans-methylation of lipid extracts and separation of fatty acyl methyl esters (FAMEs) was performed according to the procedure reported by Ianni et al. [17]. Individual FAMEs were identified by comparing the retention time of a standard mixture (FIM-FAME7-Mix; Matreya LLC, Pleasant Gap, PA, USA), and individual C18:1 *trans*-11 and C18:2 *cis*-9, *trans*-11 (Matreya LLC). The ChromeCard software was used for the quantification of peak areas, and each FAME was expressed as a percentage of the total FA. These values were used to obtain the sum of saturated (SFA), monounsaturated (MUFA) and polyunsaturated fatty acids (PUFA). Furthermore, atherogenic and thrombogenic indices (AI and TI, respectively) were calculated in milk and ewes' cheese using the formulas proposed by Ulbricht and Southgate [18], whereas the desaturation index (DI) was defined as proposed by Mele et al. [19].

2.4. Evaluation of Lipid Peroxidation by TBARS-Test

Lipid peroxidation in Pecorino cheese was determined by evaluating the amount of thiobarbituric acid reactive substances (TBARS). The analysis was performed in accordance with the procedure described by Bennato et al. [20] with slight modifications. Five grams of frozen cheese were mixed, within 2 min of sample withdrawal from the freezer, with 500 µL of 0.1% of butylated hydroxytoluene (BHT; Sigma Aldrich, Milan, Italy) in methanol to block the oxidation process. The mixture was homogenized with Ultra Turrax T-25 high speed homogenizer (IKA, Staufen, Germany) in 50 mL of an aqueous solution containing 7% trichloroacetic acid (TCA; Sigma Aldrich, Milan, Italy), and then distilled (ASTORI Tecnica s.n.c., Poncarale, BS, Italy). For each distillate, 2 mL were mixed with an equal volume of a 0.02 M thiobarbituric acid (TBA; Sigma Aldrich, Milan, Italy) solution in 90% acetic acid and then heated up to 80 °C in a thermostated bath, keeping the temperature constant for 1 hour. The absorbance at 534 nm was evaluated with a spectrophotometer (Jenway, Essex, UK) after cooling. The malondialdehyde (MDA) amount in each sample was calculated by referring to a calibration curve ranging from 0 to 100 ppm ($R^2 = 0.989$), and results were expressed in µg of MDA per g of cheese.

2.5. Analysis of Volatile Compounds

Volatile compounds (VOC) were extracted from Pecorino cheese samples through solid-phase microextraction (SPME), and the analysis was performed with a gas chromatograph (Clarus 580; Perkin Elmer, Waltham, MA, USA) coupled with a mass spectrometer (SQ8S; Perkin Elmer, Waltham, MA, USA). The gas chromatograph was equipped with an Elite-5MS column (length × internal diameter: 30 × 0.25 mm; film thickness: 0.25 µm; Perkin Elmer, Waltham, MA, USA). The samples preparation and the settings relating to the thermal program used for the analysis were performed as previously reported by Ianni et al. [21]. Five grams of cheese previously grated were mixed with 10 mL of saturated NaCl solution (360 g/L). After the addition of 10 µL of internal standard solution (4-methyl-2-heptanone; 10 mg/kg in ethanol), the vials were sealed and stirred at 50 °C; VOCs were extracted from the headspace with a divinylbenzene-carboxen-polydimethylsiloxane SPME fiber (length: 1 cm; film thickness: 50/30 µm; Supelco, Bellefonte, PA, USA) with an exposition time of 60 min. VOCs were identified by comparison with mass spectra of a library database (NIST Mass Spectral library, Search Program version 2.0, National Institute of Standards and Technology, U.S. Department of Commerce, Gaithersburg, MD, USA) and by comparing the eluting order with Kovats indices.

2.6. Statistical Analysis

All analyses were performed at least in triplicate and results were reported as mean ± standard deviation. The SigmaPlot 12.0 software (Systat software, Inc., San Jose, CA, USA) for Windows operating system was used to analyze the statistical significance of the differences between the averages for each group (ANOVA, Student's *t*-test); *p* values lower than 0.05 were considered statistically significant.

3. Results

3.1. Chemical Composition of Milk and Cheese

Taking into account the milk production over the entire duration of the dietary zinc enrichment, no significant differences were evidenced between the two groups, reflecting the fact that such parameter was not affected by diet. Regarding the chemical quality of milk (Table 1), all the analyzed parameters did not undergo variations during the experimental period. Similarly, no significant differences were observed as regards the ureic content and pH, whereas the EG samples showed a lower SCC with respect to the CG ($p < 0.01$). In regards to the amount of Zn, in the experimental group were found higher average values (4.82 ± 0.23 vs 5.42 ± 0.34 mg/kg, in CG and EG respectively; $p < 0.05$).

Table 1. Milk yield and chemical composition of milk obtained from the control group (CG) and the experimental group (EG).

	Diet		p
	CG	**EG**	
Animal Parameters			
Milk yield (mL/day)	778 ± 38	810 ± 45	n.s.
Item			
Fat (%)	7.53 ± 0.58	7.49 ± 0.41	n.s.
Protein (%)	6.03 ± 0.48	6.17 ± 0.45	n.s.
Casein (%)	4.78 ± 0.37	4.89 ± 0.29	n.s.
Lactose (%)	4.64 ± 31	4.63 ± 0.26	n.s.
Urea (mg/100 mL)	57.68 ± 4.21	57.72 ± 3.74	n.s.
SCC (LS) [1]	5.51 ± 0.12	5.07 ± 0.09	**
Total bacterial count (TBC, UFC/mL 103)	686 ± 72	633 ± 54	n.s.
pH	6.54 ± 0.13	6.55 ± 0.09	n.s.
Zinc (mg/kg)	4.82 ± 0.23	5.42 ± 0.34	*

Data are expressed as mean ± S.D. [1] Somatic cell count (SCC) is reported in linear score (LS): LS = $\log_2[(\text{cells}/\mu L)/100]$ + 3. *$p < 0.05$; ** $p < 0.01$; n.s. = not significant.

As evidenced for milk, the dietary supplementation did not influence cheese yield ($p > 0.05$). Regardless of the feeding strategy, no significant differences in composition of cheeses were evidenced (Table 2). Regarding the ripening time, as expected a significant reduction in moisture was found in T_{90} samples ($p < 0.05$); protein and lipids were not influenced by ripening, as well as the zinc amount which maintained similar values between the two groups. Furthermore, in T_{90} samples obtained from EG, the nitrogen fractions were significantly higher (0.94% vs. 0.63%, $p < 0.05$, for WSN; 0.56% vs. 0.35%, $p < 0.05$, for TCA-SN).

Table 2. Chemical composition of Pecorino cheese obtained from the control group (CG) and the experimental group (EG), analyzed after 1 (T_1) and 90 (T_{90}) days after the cheese-making.

	T_1		T_{90}	
Item	**CG**	**EG**	**CG**	**EG**
Moisture (%)	46.35 [A] ± 2.12	49.47 [A] ± 2.23	37.66 [B] ± 1.54	39.31 [B] ± 1.68
Fat [1] (%)	28.44 ± 2.03	27.86 ± 1.87	25.08 ± 1.93	23.98 ± 1.79
Protein[1] (%)	22.03 ± 2.14	19.89 ± 1.54	21.18 ± 1.64	19.04 ± 1.65
WSN (%N)	0.58 ± 0.06	0.63[A] ± 0.07	0.64 [a] ± 0.06	0.94 [b,B] ± 0.08
12% TCA-SN (%N)	0.27 ± 0.03	0.35[A] ± 0.04	0.33 [a] ± 0.04	0.56 [b,B] ± 0.05
pH	6.41 ± 0.15	6.64 ± 0.17	5.86 [a] ± 0.15	5.46 [b] ± 0.15
Zinc (mg/kg)	2.21 ± 0.18	2.16 ± 0.19	1.96 ± 0.21	1.88 ± 0.20

Data are expressed as mean percentage ± S.D. [1] Data are reported on a dry matter (DM) basis. [a,b] Means with different superscripts are significantly different by diet ($p < 0.05$). [A,B] Means with different superscripts are significantly different by ripening time ($p < 0.01$).

3.2. Fatty acid Profile of Milk and Cheese

The fatty acid composition of individual milk samples collected at the beginning and at the end of the trial by both the experimental groups is reported in Table 3. At T_0 not significant variations between CG and EG were evidenced testifying to the homogeneity of the animals selected for the study. At the end of the experimental period, samples of milk obtained from EG evidenced an increase in the content of vaccenic acid (C18:1 *trans*11; $p < 0.05$), rumenic acid (RA; $p < 0.01$) and total polyunsaturated fatty acids (PUFA, $p < 0.05$).

Table 3. Fatty acid profile of milk and fresh cheese obtained from the control group (CG) and the experimental group (EG).

	Milk						Fresh Cheese		
	T_0			T_{30}					
	CG	EG	*p*	CG	EG	*p*	CG	EG	*p*
C4:0	2.21 ± 0.19	2.03 ± 0.18	n.s.	2.02 ± 0.17	1.98 ± 0.16	n.s.	2.16 ± 0.18	2.08 ± 0.16	n.s.
C6:0	2.14 ± 0.18	2.18 ± 0.16	n.s.	2.06 ± 0.16	1.94 ± 0.16	n.s.	1.98 ± 0.16	1.89 ± 0.17	n.s.
C8:0	2.23 ± 0.20	1.91 ± 0.16	n.s.	2.28 ± 0.17	2.16 ± 0.18	n.s.	2.33 ± 0.19	2.25 ± 0.20	n.s.
C10:0	7.45 ± 0.61	7.33 ± 0.56	n.s.	7.89 ± 0.54	7.50 ± 0.61	n.s.	7.76 ± 0.64	7.41 ± 0.59	n.s.
C11:0	0.27 ± 0.03	0.33 ± 0.03	n.s.	0.31 ± 0.03	0.29 ± 0.03	n.s.	0.35 ± 0.03	0.32 ± 0.03	n.s.
C12:0	5.16 ± 0.49	4.61 ± 0.42	n.s.	5.00 ± 0.36	4.49 ± 0.34	n.s.	5.21 ± 0.44	4.71 ± 0.41	n.s.
C14:0	13.08 ± 1.12	12.11 ± 0.97	n.s.	12.99 ± 1.03	11.76 ± 0.98	n.s.	13.26 ± 0.99	12.02 ± 1.02	n.s.
C15:0	1.29 ± 0.11	1.22 ± 0.12	n.s.	1.20 ± 0.15	1.18 ± 0.09	n.s.	1.16 ± 0.12	1.11 ± 0.11	n.s.
C16:0	25.12 ± 2.14	26.53 ± 2.09	n.s.	26.28 ± 1.83	27.00 ± 1.65	n.s.	26.58 ± 2.07	27.09 ± 2.24	n.s.
C17:0	0.61 ± 0.06	0.52 ± 0.05	n.s.	0.53 ± 0.04	0.49 ± 0.04	n.s.	0.55 ± 0.05	0.43 ± 0.04	n.s.
C18:0	6.78 ± 0.62	7.04 ± 0.68	n.s.	7.97 ± 0.61	7.29 ± 0.63	n.s.	7.59 ± 0.65	6.88 ± 0.58	n.s.
SFA	66.34 ± 3.67	65.81 ± 3.11	n.s.	68.53 ± 2.98	66.08 ± 3.02	n.s.	68.93 ± 3.05	66.19 ± 2.27	n.s.
C14:1	0.85 ± 0.08	0.88 ± 0.08	n.s.	0.82 ± 0.07	0.86 ± 0.09	n.s.	0.83 ± 0.06	0.89 ± 0.08	n.s.
C16:1	1.02 ± 0.09	0.95 ± 0.09	n.s.	0.96 ± 0.07	1.12 ± 0.09	n.s.	1.15 ± 0.09	1.21 ± 0.10	n.s.
C18:1 *trans*11	0.72 ± 0.07	0.75 [a] ± 0.05	n.s.	0.67 ± 0.06	0.88 [b] ± 0.07	*	0.57 ± 0.06	0.79 ± 0.07	**
C18:1 *cis*9	18.13 ± 1.24	17.89 ± 1.31	n.s.	17.40 ± 1.12	17.94 ± 1.19	n.s.	17.73 ± 1.24	19.85 ± 1.41	n.s.
C18:1 *cis*11	0.55 ± 0.06	0.71 ± 0.07	n.s.	0.51 ± 0.06	0.62 ± 0.06	n.s.	0.44 ± 0.04	0.57 ± 0.06	n.s.
MUFA	21.27 ± 1.78	21.18 ± 1.23	n.s.	20.36 ± 1.17	21.42 ± 1.24	n.s.	20.72 ± 1.21	23.31 ± 2.13	n.s.
C18:2	1.78 ± 0.14	1.85 ± 0.16	n.s.	1.84 ± 0.13	2.09 ± 0.16	n.s.	1.66 ± 0.14	1.71 ± 0.16	n.s.
C18:3	1.01 ± 0.09	0.92 ± 0.09	n.s.	0.86 ± 0.07	0.96 ± 0.08	n.s.	0.75 ± 0.07	0.83 ± 0.08	n.s.
RA	1.71 ± 0.11	1.67 [a] ± 0.13	n.s.	1.73 ± 0.14	2.13 [b] ± 0.16	**	1.56 ± 0.15	1.89 ± 0.17	*
C20:4	0.05 ± 0.00	0.06 ± 0.00	n.s.	0.07 ± 0.01	0.06 ± 0.01	n.s.	0.04 ± 0.00	0.04 ± 0.01	n.s.
PUFA	4.55 ± 0.38	4.50 [a] ± 0.31	n.s.	4.50 ± 0.32	5.24 [b] ± 0.39	*	4.01 ± 0.19	4.47 ± 0.26	*
Others	7.84 ± 0.67	8.34 ± 0.71	n.s.	7.71 ± 0.44	7.36 ± 0.53	n.s.	6.44 ± 0.62	5.03 ± 0.48	*
DI	0.06 ± 0.00	0.07 ± 0.01	n.s.	0.06 ± 0.00	0.07 ± 0.01	n.s.	0.06 ± 0.00	0.07 ± 0.01	n.s.
AI	3.20 ± 0.27	3.10 ± 0.16	n.s.	3.35 ± 0.21	2.94 ± 0.16	*	3.43 ± 0.31	2.77 ± 0.25	*
TI	2.97 ± 0.21	3.09 ± 0.24	n.s.	3.33 ± 0.27	3.05 ± 0.25	n.s.	3.41 ± 0.31	2.88 ± 0.26	*

Analysis on milk have been performed at the beginning (T_0) and at the end of the trial (T_{30}). SFA = saturated fatty acid; MUFA = monounsaturated fatty acid; PUFA = polyunsaturated fatty acid; RA = rumenic acid; AI = atherogenic index; TI = thrombogenic index; DI = desaturation index. Data are expressed as mean (%) ± S.D. * $p < 0.05$; ** $p < 0.01$; n.s. = not significant. [a,b] Means with different superscripts (in milk) are significantly different by time ($p < 0.05$).

Similarly, the evaluation of the total fatty acids profile in cheese evidenced modifications already evident in milk (Table 3), with an increase in concentration, in EG, of vaccenic acid (C18:1 trans11; $p < 0.01$), rumenic acid (RA; $p < 0.05$) and total PUFA ($p < 0.05$). Based on the obtained FA profile, calculations of desaturation, thrombogenic and atherogenic indices were performed. Dietary supplementation with zinc did not induce significant modifications of desaturation index ($p > 0.05$) both in milk and cheese; thrombogenic index was significantly lower only in cheese samples obtained from EG ($p < 0.05$), whereas atherogenic index decreased both in milk and cheese obtained from the EG ($p < 0.05$).

3.3. Analysis of the Oxidative Stability in Pecorino Cheese

Diet enrichment with zinc did not induce alterations of the oxidative stability in samples of fresh cheese (Figure 1). Very interesting is instead the result obtained after 90 days from the cheese making, as expected lipid peroxidation increased in all the analyzed samples, but in ewes' milk cheese obtained from EG the value of malondialdehyde was stood at significantly lower values if compared to samples of the control group (0.092 vs 0.066 μg MDA/g of cheese, in CG and EG respectively; $p < 0.05$).

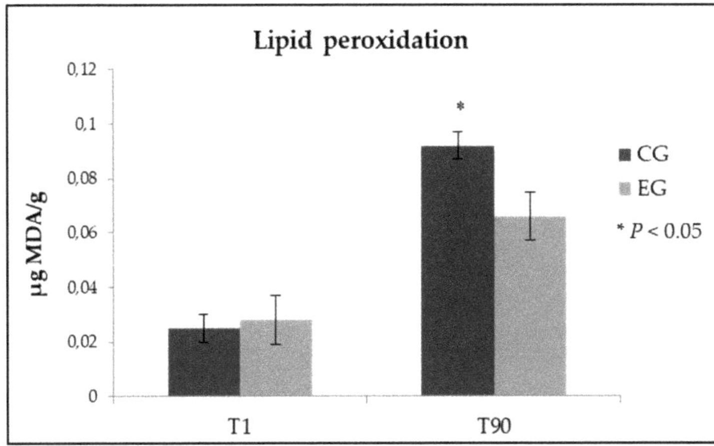

Figure 1. Lipid peroxidation in fresh and ripened Pecorino cheese samples obtained from control group (CG) and experimental group (EG).

3.4. Volatile Profile of Cheese

The analysis of the volatile profile allowed to identify 24 volatile compounds (VOC) in samples of T_1 and T_{90} cheese obtained from CG and EG: 6 carboxylic acids, 5 ethyl esters, 3 aldehydes, 2 alcohols, 2 lactones, 2 ketones and 4 classified as aromatic hydrocarbons (Table 4).

Table 4. Volatile compounds (VOCs) detected in cheese samples obtained from control group (CG) and experimental group (EG).

	VOC	T_1			T_{90}		
		CG	EG	*p*	CG	EG	*p*
Acids	acetic acid	1.98 ± 0.14	1.76 ± 0.11	n.s.	2.29 ± 0.17	2.48 ± 0.19	n.s.
	hexanoic acid	2.12 ± 0.17	1.99 ± 0.16	n.s.	2.06 ± 0.18	6.75 ± 0.45	**
	octanoic acid	3.83 ± 0.28	3.56 ± 0.29	n.s.	7.40 ± 0.45	7.74 ± 0.56	n.s.
	nonanoic acid	1.76 ± 0.14	1.34 ± 0.10	*	1.09 ± 0.08	1.24 ± 0.10	n.s.
	decanoic acid	9.63 ± 0.73	10.23 ± 0.88	n.s.	9.67 ± 0.62	5.09 ± 0.39	**
	dodecanoic acid	13.63 ± 1.04	15.23 ± 1.29	n.s.	9.57 ± 0.63	6.67 ± 0.44	**
Aldehydes	hexanal	4.11 ± 0.22	4.73 ± 0.37	n.s.	7.46 ± 0.53	9.86 ± 0.24	*
	heptanal	6.88 ± 0.43	8.07 ± 0.67	n.s.	n.d.	n.d.	n.s.
	nonanal	13.29 ± 1.15	10.21 ± 0.77	*	1.68 ± 0.14	1.12 ± 0.09	*
Alcohols	3-methylbutanol	4.87 ± 0.29	5.31 ± 0.34	n.s.	5.98 ± 0.36	3.27 ± 0.22	*
	2,3-butanediol	n.d.	n.d.	n.s.	2.72 ± 0.23	2.15 ± 0.18	n.s.
Esters	butanoic acid, ethyl ester	n.d.	n.d.	n.s.	4.21 ± 0.31	9.29 ± 0.76	**
	hexanoic acid, ethyl ester	6.57 ± 0.41	6.79 ± 0.42	n.s.	8.96 ± 0.64	11.66 ± 0.98	*
	octanoic acid, ethyl ester	8.32 ± 0.52	7.67 ± 0.39	n.s.	11.85 ± 0.78	7.31 ± 0.56	**
	decanoic acid, ethyl ester	n.d.	n.d.	n.s.	4.10 ± 0.29	4.78 ± 0.38	n.s.
	dodecanoic acid, ethyl ester	n.d.	n.d.	n.s.	0.40 ± 0.05	0.31 ± 0.04	n.s.
Lactones	δ-decalactone	3.89 ± 0.26	3.83 ± 0.29	n.s.	1.22 ± 0.10	1.09 ± 0.07	n.s.
	δ-dodecalactone	4.44 ± 0.32	3.86 ± 0.34	n.s.	0.98 ± 0.08	0.85 ± 0.07	n.s.
Ketones	2-heptanone	1.91 ± 0.09	1.39 ± 0.08	**	3.18 ± 0.25	2.95 ± 0.22	n.s.
	2-nonanone	n.d.	n.d.	n.s.	0.82 ± 0.06	1.03 ± 0.09	n.s.
Aromatic hydrocarbons	ethylbenzene	n.d.	n.d.	n.s.	2.22 ± 0.19	1.94 ± 0.15	n.s.
	1,3-dimethylbenzene	n.d.	n.d.	n.s.	0.27 ± 0.03	0.32 ± 0.03	n.s.
	1,2,3,4-tetramethylbenzene	n.d.	n.d.	n.s.	0.84 ± 0.06	0.95 ± 0.08	n.s.
	p-cymene	n.d.	n.d.	n.s.	0.62 ± 0.06	0.77 ± 0.07	n.s.

Data are expressed as mean (%) ± S.D. * $p < 0.05$; ** $p < 0.01$; n.s.: not significant; n.d.: not detectable.

Regarding the carboxylic acids, is interesting the increase in concentration of the hexanoic acid in the EG samples after 90 days of ripening (6.75% vs. 2.06% in EG and CG samples respectively; $p < 0.01$); in T_{90} samples should also be underlined the significant reduction of the longer chain acids:

decanoic acid (9.67% vs. 5.09% in CG and EG respectively; $p < 0.01$) and dodecanoic acid (9.57% vs. 6.67% in CG and EG respectively; $p < 0.01$).

In the case of esters, the dietary zinc supplementation induced the increase of butanoic acid ethyl ester ($p < 0.01$) and of hexanoic acid ethyl ester ($p < 0.05$) at the end of the ripening period. Different resulted the behavior of octanoic acid ethyl ester, which is higher in the T_{90} samples obtained from CG (11.85% vs. 7.31%, $p < 0.01$).

In the most ripened samples was also found the hexanal increase in EG samples (9.86% vs. 7.46%, $p < 0.05$), and variations of both nonanal (1.68% vs. 1.12% in CG and EG respectively; $p < 0.05$) and 3-methylbunanol (5.98% vs. 3.27% in CG and EG respectively; $p < 0.05$).

Additionally, in T_1 samples have been evidenced some significant differences, specifically in EG cheeses have been observed lower concentrations of nonanoic acid ($p < 0.05$), nonanal ($p < 0.05$) and 2-heptanone ($p < 0.01$).

4. Discussion

In the present study, Zn enrichment of ewes' diet did not induce significant changes on milk composition, and this finding is consistent with what has been previously reported in dairy cows [22,23] and dairy goats [8]. According to what was observed in milk, no variations were evidenced in the chemical composition of Pecorino cheese samples, both in relation to dietary treatment and ripening time. The analysis on milk showed instead the capacity of the experimental diet to markedly reduce the number of somatic cells, a datum already observed in dairy cows by Pechová et al. [24] who tested the effect of dietary Zn at a daily dose of 440 mg/animal, therefore by administering a Zn dose similar to that used in the present work but on animals with a much greater body weight. Such phenomenon was justified by assuming an increased Zn supply into the mammary gland with a consequent improvement of the immune response. A precise and exhaustive evaluation of the influence of this parameter on the quality of dairy products is not feasible, however, it is known that somatic cells contribute to proteolysis in milk and cheese because of their tendency to release proteolytic enzymes in the extracellular environment [25].

A finding deserving special attention concerns the Zn content, which is higher in EG milk samples, whereas no variations were highlighted in cheese. The Zn concentration in cheese could be directly related to the presence of caseins, with which the microelement interacts through a complex kinetic that develops in two phases: an initial rapid stage in which about the 70% of Zn interacts with polar amino acids of casein, and a significantly slower second stage in which the equilibrium is reached [26]. Since after the cheese-making both CG cheese and EG cheese showed similar protein concentrations, it could be supposed that in EG samples the Zn excess, not associated with caseins, may have been lost with serum after rennet breakage. In addition to this, it should be also noted that in this study Zn seems to show a moderate capacity for association with caseins, as can be proved by the reduced Zn concentration in cheeses as compared with milk (considering the casein concentration by cheesemaking).

The Zn enrichment of ewes' diet showed effective in inducing variations in the FA profile, both in milk and its derived cheese; in all EG samples, the amount of vaccenic acid (C18:1 *trans*11), rumenic acid (RA) and total PUFA significantly increased. With regard to vaccenic acid, the result is explainable, at least in part, by taking into account the study of Szczechowiak et al. [27], in which the increase in concentration of this compound was justified by the action of bioactive compounds taken through the diet which tend to slow down or totally inhibit the terminal steps of ruminal biohydrogenation, thus avoiding the formation of stearic acid (C18:0). The relevance of vaccenic acid is its role as substrate of the mammary gland stearoyl coenzyme A desaturase (SCD), an endoplasmic reticulum-bound enzyme which is responsible for the catalytic mechanism that gives origin to CLA [28–30]. In this study the RA concentration effectively increased in both milk and cheese obtained from the experimental group. A similar finding was recently reported by Ianni et al. [9] who tested a dietary zinc supplementation in lactating dairy cows. In that case, the phenomenon was partly justified by advancing the hypothesis of a role of the Zn supplemented diet in promoting the catalytic function of site-2 protease (S2P), a Zn-dependent metalloprotease which contributes to the activation of the sterol response element binding protein (SREBP), a transcription factor responsible for the regulation of several genes encoding for SCD and other lipogenic factors.

RA has been indicated as a factor with a strong antioxidant function which is able to improve the mammary gland functionality by protecting the mammary epithelial cells from lipoperoxidation through the reduction of the reactive oxygen species [31]. For humans, the ruminant products represent the primary dietary source of CLA, which are credited of several health benefits; the most relevant examples concern the modulation of the immune system response [32], and their potential activity in slowing down the progression of different pathological conditions, as in the case of atherosclerosis [33].

Both in milk and its derived cheese, the feeding strategy based on Zn enrichment also induced an overall increase of PUFA concentration, at the expense of SFA. As a direct consequence of this evidence, the atherogenic and thrombogenic indices decreased, testifying a noteworthy improvement of the health functionality of animal productions.

The oxidative damage in cheese was determined by evaluating the amount of thiobarbituric acid reactive substances. Zn has been reported to act as a free radical scavenger in biological systems, with the consequent inhibition of free radical lipid peroxidation [34]. Additionally, in this study, Zn appeared to provide antioxidant protection since at the end of the 90 days of ripening (T_{90}) the TBA values, although rising in both experimental groups, were found to be significantly lower in cheese samples deriving from the dietary Zn supplementation.

The analysis performed with the purpose to characterize the volatile profile of fresh (T_1) and ripened (T_{90}) ewes' milk cheese, led to the identification of different families of compounds, the most represented of which are those of free fatty acid (FFA) and ethyl esters. The biochemical mechanism behind the increased concentration of FFA could relate to the degradation of triglycerides by enzymes of both endogenous and microbial origin. Among FFA the hexanoic acid was present at higher concentrations in EG samples after 90 days of ripening. This compound can only originate from lipolysis [35], and the datum concerning its increase in concentration, should have a significant influence in the determination of cheese flavor, due to its association with strong odors, described as cheesy, rancid and sweaty. In addition to this, the increase of hexanoic acid could also derive from the lipolytic action on longer chain acids, decanoic and dodecanoic, which in fact undergo a marked reduction in the EG samples [36]. What has just been described would suggest a condition characterized by an increase in lipolysis during the ripening period. Some authors explain this phenomenon with an increase in the autolysis of bacterial starters, with the consequent release in the extracellular environment of a wide range of enzymes capable to attack and degrade both the protein component and fatty acids, contributing to the development of the organoleptic properties in ripened cheeses [37]. The autolysis process is, in turn, mediated by peptidoglycan hydrolases, named autolysins, that in the presence of certain stimuli degrade the cellular envelope which separates the cytoplasmic compartment from the extracellular matrix [38,39].

The FFAs, in addition, to directly contribute to the cheese flavor, represent the substrate for the biosynthesis of other classes of compounds: ethyl esters, methyl ketones, secondary alcohols, aldehydes, and lactones [36]. With regard to esters, the EG samples analyzed at the end of the ripening period were characterized by higher concentrations of butanoic acid ethyl ester and hexanoic acid ethyl ester. Generally, this class of compounds is responsible for the supply of sweet, fruity, and floral notes in surface ripened cheese [40], and is considered to be decisive in defining the cheese flavor because of the low odor threshold. Among aldehydes, was instead evidenced a marked increase of hexanal after ripening in EG samples; this compound gives what is called "green grass-like" aroma, with lemon, herbal and slightly fruity notes. A strange behavior is associated to alcohols which mainly derive from FFAs catabolism; after the 90 days of ripening, these compounds are not well represented, and their concentration tends to decrease in the EG samples. Specifically, we found a significant reduction of 3-methylbutanol, which is generally found on surface-ripened cheese and is responsible for alcoholic, fruity, grainy, and solvent-like notes [34]. No variations were evidenced for lactones and ketones, therefore further analysis could be necessary in order to better characterize the biochemical mechanisms that contribute to their biosynthesis.

5. Conclusions

The experimented feeding strategy showed able to modify the nutritional properties of ewes' milk and its derived Pecorino cheese. The main finding concerns the increase in concentration of vaccenic acid, rumenic acid and total PUFA both in milk and cheese, as evidence of a presumable improvement in the health functionality of the products. In addition to this, a clear improvement of the oxidative stability was also shown in the 90-days aged cheese, an aspect that deserves interesting implications for the preservation of the cheese quality during ripening. The volatile profile of ewes' milk cheese was also positively affected by the dietary Zn intake. Of particular interest, with a view to improving the cheese aroma, is the increase in concentration in hexanal and 3-methylbutanol at the end of ripening. However, it would be necessary to plan a specific investigation aimed at assessing the approval rate of these new products by consumers.

Author Contributions: Conceptualization, G.M.; Methodology, C.M., and L.G.; Validation, A.I., and L.G.; Formal analysis, A.I. and C.M.; Investigation, G.M., A.I. and C.M.; Resources, G.M. and F.P.; Data curation, A.I., L.G. and C.M.; Writing—original draft preparation, A.I.; Writing—review and editing, G.M., C.M. and F.P.; Supervision, G.M.; Project administration, G.M.; Funding acquisition, G.M.

Funding: This research is part of the project "Caratterizzazione e miglioramento degli indici salutistici e sicurezza alimentare delle produzioni ovine tipiche abruzzesi a marchio di origine" supported by a grant from Rural Development Plan 2007–2013–MISURA 1.2.4, Regione Abruzzo (Italy).

Conflicts of Interest: The authors declare no conflicts of interest. The funders had no role in the design of the study; in the collection, analyses, or interpretation of data; in the writing of the manuscript, or in the decision to publish the results.

References

1. Coleman, J.E. Zinc enzymes. *Curr. Opin. Chem. Biol.* **1998**, *2*, 222–234. [CrossRef]
2. Miller, W.J. Zinc nutrition of cattle: A review. *J. Dairy Sci.* **1970**, *53*, 1123–1135. [CrossRef]
3. Nielsen, F.H. History of zinc in agriculture. *Adv. Nutr.* **2012**, *3*, 783–789. [CrossRef] [PubMed]
4. Singh, K.K.; Maity, S.B.; Maity, A. Effect of nano zinc oxide on zinc bioavailability and blood biochemical changes in pre-ruminant lambs. *Indian J. Anim. Sci.* **2018**, *88*, 805–807.
5. Sonawane, S.N.; Arora, S.P. Influence of zinc supplementation on rumen microbial protein synthesis in in vitro studies. *Indian J. Anim. Sci.* **1976**, *46*, 13–18.
6. Arelovich, H.M.; Owens, F.N.; Horn, G.W.; Vizcarra, J.A. Effects of supplemental zinc and manganese on ruminal fermentation, forage intake, and digestion by cattle fed prairie hay and urea. *J. Anim. Sci.* **2000**, *78*, 2972. [CrossRef]
7. Elgendy, R.; Palazzo, F.; Castellani, F.; Giantin, M.; Grotta, L.; Cerretani, L.; Dacasto, M.; Martino, G. Transcriptome profiling and functional analysis of sheep fed with high zinc-supplemented diet: A nutrigenomic approach. *Anim. Feed Sci. Technol.* **2017**, *234*, 195–204. [CrossRef]
8. Salama, A.A.K.; Caja, G.; Albanell, E.; Such, X.; Casals, R.; Plaixats, J. Effects of dietary supplements of zinc-methionine on milk production, udder health and zinc metabolism in dairy goats. *J. Dairy Res.* **2003**, *70*, 9–17. [CrossRef]
9. Ianni, A.; Innosa, D.; Martino, C.; Grotta, L.; Bennato, F.; Martino, G. Zinc supplementation of Friesian cows: Effect on chemical-nutritional composition and aromatic profile of dairy products. *J. Dairy Sci.* **2019**, *102*, 2918–2927. [CrossRef]
10. Ianni, A.; Iannaccone, M.; Martino, C.; Innosa, D.; Grotta, L.; Bennato, F.; Martino, G. Zinc supplementation of dairy cows: Effects on chemical composition, nutritional quality and volatile profile of Giuncata cheese. *Int. Dairy J.* **2019**, *94*, 65–71. [CrossRef]
11. Ianni, A.; Martino, C.; Innosa, D.; Bennato, F.; Grotta, L.; Martino, G. Zinc supplementation of lactating dairy cows: Effects on chemical-nutritional quality and volatile profile of Caciocavallo cheese. *Asian Australas J. Anim. Sci.* **2019**, in press. [CrossRef] [PubMed]

12. European Commission Regulation (EC) No. 1831/2003 of the European Parliament and of the Council of 22 September 2003 on Additives for Use in Animal Nutrition. *Off. J. Eur. Union* **2003**, *50*, 29–43.
13. AOAC International. *Official Methods of Analysis*, 15th ed.; Association of Official Analytical Chemists (AOAC): Arlington, VA, USA, 1990; Volume 1.
14. International Standard ISO 27871 IDF 224. *Cheese and Processed Cheese–Determination of the Nitrogenous Fractions*; International Dairy Federation: Brussels, Belgium, 2011.
15. Gerber, N.; Brogioli, R.; Hattendorf, B.; Scheeder, M.R.L.; Wenk, C.; Günther, D. Variability of selected trace elements of different meat cuts determined by ICP-MS and DRC-ICPMS. *Animal* **2009**, *3*, 166–172. [CrossRef] [PubMed]
16. AOAC International. *Official Methods of Analysis*, 17th ed.; Association of Official Analytical Chemists (AOAC): Washington, DC, USA, 2000.
17. Ianni, A.; Di Maio, G.; Pittia, P.; Grotta, L.; Perpetuini, G.; Tofalo, R.; Cichelli, A.; Martino, G. Chemical–nutritional quality and oxidative stability of milk and dairy products obtained from Friesian cows fed with a dietary supplementation of dried grape pomace. *J. Sci. Food Agric.* **2019**, *99*, 3635–3643. [CrossRef] [PubMed]
18. Ulbricht, T.L.V.; Southgate, D.A.T. Coronary heart disease: Seven dietary factors. *Lancet* **1991**, *338*, 985–992. [CrossRef]
19. Mele, M.; Conte, G.; Castiglioni, B.; Chessa, S.; Macciotta, N.P.P.; Serra, A.; Buccioni, A.; Pagnacco, G.; Secchiari, P. Stearoylcoenzyme A desaturase gene polymorphism and milk fatty acid composition in Italian Holsteins. *J. Dairy Sci.* **2007**, *90*, 4458–4465. [CrossRef]
20. Bennato, F.; Ianni, A.; Martino, C.; Di Luca, A.; Innosa, D.; Fusco, A.M.; Pomilio, F.; Martino, G. Dietary supplementation of Saanen goats with dried licorice root modifies chemical and textural properties of dairy products. *J. Dairy Sci.* **2019**, in press. [CrossRef]
21. Ianni, A.; Innosa, D.; Martino, C.; Bennato, F.; Martino, G. Compositional characteristics and aromatic profile of caciotta cheese obtained from Friesian cows fed with a dietary supplementation of dried grape pomace. *J. Dairy Sci.* **2019**, *102*, 1025–1032. [CrossRef]
22. Cope, C.M.; Mackenzie, A.M.; Wilde, D.; Sinclair, L.A. Effects of level and form of dietary zinc on dairy cow performance and health. *J. Dairy Sci.* **2009**, *92*, 2128–2135. [CrossRef]
23. Wang, R.L.; Liang, J.G.; Lu, L.; Zhang, L.Y.; Li, S.F.; Luo, X.G. Effect of zinc source on performance, zinc status, immune response, and rumen fermentation of lactating cows. *Biol. Trace Elem. Res.* **2013**, *152*, 16–24. [CrossRef]
24. Pechová, A.; Pavlata, L.; Lokajová, E. Zinc supplementation and Somatic Cell Count in milk of dairy cows. *Acta Vet. Brno* **2006**, *75*, 355–361. [CrossRef]
25. Marino, R.; Considine, T.; Sevi, A.; McSweeney, P.L.H.; Kelly, A.L. Contribution of proteolytic activity associated with somatic cells in milk to cheese ripening. *Int. Dairy J.* **2005**, *15*, 1026–1033. [CrossRef]
26. Pomastowski, P.; Sprynskyy, M.; Buszewski, B. The study of zinc ions binding to casein. *Colloids Surf. B Biointerfaces* **2014**, *120*, 21–27. [CrossRef] [PubMed]
27. Szczechowiak, J.; Szumacher-Strabel, M.; El-Sherbinya, M.; Pers-Kamczyc, E.; Pawlakd, P.; Cieslaka, A. Rumen fermentation, methane concentration and fatty acid proportion in the rumen and milk of dairy cows fed condensed tannin and/or fish-soybean oils blend. *Anim. Feed Sci. Technol.* **2016**, *216*, 93–107. [CrossRef]
28. Lock, A.L.; Bauman, D.E. Modifying milk fat composition of dairy cows to enhance fatty acids beneficial to human health. *Lipids* **2004**, *39*, 1197–1206. [CrossRef]
29. Miyazaki, M.; Ntambi, J.M. Role of stearoyl-coenzyme A desaturase in lipid metabolism. *Prostaglandins Leukotrienes and Essent. Fat. Acids* **2003**, *68*, 113–121. [CrossRef]
30. Smith, S.B.; Lunt, D.K.; Chung, K.J.; Choi, C.B.; Tume, R.K.; Zembayashi, M. Adiposity, fatty acid composition, and delta-9 desaturase activity during growth in beef cattle. *Anim. Sci. J.* **2006**, *77*, 478–486. [CrossRef]
31. Basiricò, L.; Morera, P.; Dipasquale, D.; Tröscher, A.; Bernabucci, U. Comparison between conjugated linoleic acid and essential fatty acids in preventing oxidative stress in bovine mammary epithelial cells. *J. Dairy Sci.* **2017**, *100*, 2299–2309. [CrossRef]
32. Song, H.J.; Grant, I.; Rotondo, D.; Mohede, I.; Sattar, N.; Heys, S.D.; Wahle, K.W.J. Effect of CLA supplementation on immune function in young healthy volunteers. *Eur. J. Clin. Nutr.* **2005**, *59*, 508–517. [CrossRef]

33. Lock, A.L.; Corl, B.A.; Barbano, D.M.; Bauman, D.E.; Ip, C. The anticarcinogenic effect of trans-11 18:1 is dependent on its conversion to cis-9, trans-11 CLA by Δ9-desaturase in rats. *J. Nutr.* **2004**, *134*, 2698–2704. [CrossRef]

34. Fang, Y.Z.; Yang, S.; Wu, G. Free radicals, antioxidants, and nutrition. *Nutrition* **2002**, *18*, 872–879. [CrossRef]

35. Faccia, M.; Trani, A.; Natrella, G.; Gambacorta, G. Chemical-sensory and volatile compound characterization of ricotta forte, a traditional fermented whey cheese. *J. Dairy Sci.* **2018**, *101*, 5751–5757. [CrossRef] [PubMed]

36. Bertuzzi, A.S.; McSweeney, P.L.H.; Rea, M.C.; Kilcawley, K.N. Detection of volatile compounds of cheese and their contribution to the flavor profile of surface-ripened cheese. *Compr. Rev. Food Sci. Food Saf.* **2018**, *17*, 371–390. [CrossRef]

37. Collins, Y.F.; McSweeney, P.L.H.; Wilkinson, M.G. Lipolysis and free fatty acid catabolism in cheese: A review of current knowledge. *Int. Dairy J.* **2003**, *13*, 841–866. [CrossRef]

38. Buist, G.; Kok, J.; Leenhouts, K.J.; Dabrowska, M.; Venema, G.; Haandrikman, A.J. Molecular cloning and nucleotide sequence of the gene encoding the major peptidoglycan hydrolase of *Lactococcus lactis*, a muramidase needed for cell separation. *J. Bacteriol.* **1995**, *177*, 1554–1563. [CrossRef]

39. Huard, C.; Miranda, G.; Wessner, F.; Bolotin, A.; Hansen, J.; Foster, S.J.; Chapot-Chartier, M.P. Characterization of AcmB, an N-acetylglucosaminidase autolysin from *Lactococcus lact.* *Microbiology* **2003**, *149*, 695–705. [CrossRef]

40. Niimi, J.; Eddy, A.I.; Overington, A.R.; Silcock, P.; Bremer, P.J.; Delahunty, C.M. Sensory interactions between cheese aroma and taste. *J. Sens. Stud.* **2015**, *30*, 247–257. [CrossRef]

Article

Chemical-Sensory Traits of Fresh Cheese Made by Enzymatic Coagulation of Donkey Milk

Michele Faccia [1], Giuseppe Gambacorta [1,*], Giovanni Martemucci [2], Graziana Difonzo [1] and Angela Gabriella D'Alessandro [2]

[1] Department of Soil, Plant and Food Sciences, University of Bari, Via Amendola 165/A, 70126 Bari, Italy; michele.faccia@uniba.it (M.F.); graziana.difonzo@uniba.it (G.D.)

[2] Department of Agro-Environmental and Territorial Sciences, University of Bari, Via Amendola 165/A, 70126 Bari, Italy; giovanni.martemucci@uniba.it (G.M.); angelagabriella.dalessandro@uniba.it (A.G.D.)

* Correspondence: giuseppe.gambacorta@uniba.it

Received: 11 December 2019; Accepted: 20 December 2019; Published: 23 December 2019

Abstract: Making cheese from donkey milk is considered unfeasible, due to difficulties in coagulation and curd forming. Two recent studies have reported the protocols for making fresh cheese by using camel chymosin or calf rennet, but the chemical and sensory characteristics of the products were not thoroughly investigated. The present paper aims to give a further contribution to the field, by investigating cheesemaking with microbial rennet and evaluating the chemical composition, total fatty acid, volatile organic compounds (VOCs) and sensory profile of the resultant product. Six trials were undertaken at laboratory scale on donkey milk from a Martina Franca ass, by applying the technological scheme as reported for calf rennet, with some modifications. Bulk cow milk was used as a control. Donkey milk coagulated rapidly, but the curd remained soft, and was only suitable for making fresh cheese; differently, cow milk coagulated almost instantaneously under these strong technological conditions, giving rise to a semi-hard curd in very short time. The moisture level of donkey cheese was almost the same as reported in the literature, whereas the yield was higher, probably due to the high protein content of the milk used. The total fatty acid composition of cheese presented some differences with respect to milk, mostly consisting in a higher presence of saturated compounds. A connection with a better retention of the large sized fat globules into the curd was hypothesised and discussed. The VOC analyses, performed by solid-phase micro extraction gas chromatography-mass spectrometry, allowed the identification of 11 compounds in milk and 18 in cheese. The sensory characteristics of donkey cheese were strongly different with respect to the control, and revealed unique and pleasant flavours.

Keywords: cheesemaking; donkey milk; fatty acids; sensory analysis; VOC

---

1. Introduction

The use of equid milk for human nutrition and wellness has an ancient history. The use of mare milk for producing koumiss, a lactic–alcoholic beverage, belongs to the food tradition of several ethnic groups from Central Asia [1]. Ancient Romans and Greeks were already aware of the cosmetic and therapeutic properties of donkey milk, as evidenced by the legendary milk bath of Cleopatra and Poppea, and Hippocrates' indications as a remedy for several diseases [2,3]. After a long and slow decline, donkey rearing is now undergoing a revival, due to an interest in innovative and nutraceutical foods. Donkey milk is not in competition with cow's milk, and represents an added value functional food with high market potential. Besides, European policy encourages the diversification of agricultural production and the valorisation of local genetic resources. Donkeys are able to use marginal areas unsuitable for specialised dairy cows, and they are better adapted to warm climates.

For these reasons, rearing donkeys for milk is gaining increasing attention in the Mediterranean area, in particular Italy [4], where milk is sold as a raw product directly at farms. Other forms of marketing at niche level also exist, such as selling online or through pharmacies as pasteurised or, rarely, as UHT-treated or freeze-dried product [5]. The interest of the food researchers towards donkey milk is also increasing; several protocols for manufacturing innovative fermented milks and functional beverages at laboratory level are already available [6–8]. Unfortunately, studies on cheesemaking are very rare, since rennet coagulation is commonly considered as unfeasible. Apart from economic considerations connected to the high cost and low availability of the raw matter, the main technical obstacles seem to be the low casein concentration (0.64 to 1.03 g/100 g milk) and different micelle structure with respect to ruminants [9]. Recent studies have reported that donkey casein micelles are much larger than bovine micelles (298.5 ± 18.9 nm at pH 6.8 versus 186.5 ± 1.2 nm), have a lower absolute zeta potential (−15.4 ± 0.5 mV), poor colloidal stability and different abundance of the casein fractions [10,11]. This latter aspect has long been debated, until the work of Chianese et al. [12] definitively clarified the matter using electrophoresis and immunostaining of the protein bands with polyclonal antibodies against the four 'classic' caseins. The authors ascertained the presence of κ-casein, and demonstrated that β-casein was the most abundant fraction, followed by αs1 and αs2.

Despite the difficulty in milk coagulation, some successful efforts in making fresh donkey cheese have been recently reported. In particular, Iannella [13] and Faccia et al. [14] developed two different cheesemaking protocols, based on the use of camel chymosin or calf rennet, respectively. Even though the results obtained were encouraging, the two studies had some conflicting points, and the products were only poorly investigated. The aim of the present study was to provide a more in-depth insight on the chemical and sensory traits of a cheese made from donkey milk by using microbial rennet.

2. Materials and Methods

2.1. Milk and Cheesemaking Trials

Milk was derived from 10 donkeys (Martina Franca ass) at the mid-period of lactation routinely milked by a mechanical milker at a farm located in the Apulia region (Southern Italy). The gross composition (fat, protein and lactose) was assessed by Infrared Spectroscopy by MilkoScan™ FT1 (FOSS, Hillerød, Denmark). Cheesemaking with donkey milk, using bulk cow milk as the control, was carried out at laboratory level (six replicates) by following the protocol reported by Faccia et al. [14] with some modifications (Figure 1). The experimentation was performed in small pots using 5 L milk, which was heated and maintained at the desired temperatures in a thermostatic water bath. The starter used was *Streptococcus termophilus* (Choozit Star, Danisco/Dupont, Cernusco sul Naviglio, Italy), whereas microbial rennet was Fromase from *Rhizomucor miehei* (normal strength, DSM, Delft, The Netherlands). The curd cutting time was established by using a vibro-viscometer (SV-10; A & D Company, Limited, Tokyo, Japan) applied to a 40 mL aliquot acidified and renneted milk sample, taken from the pot and kept at 42 °C. Cutting started when the viscosity value did not change within 3 min; the control milk coagulum was also cut at the same time. After moulding, samples of whey were taken for determination of the fat content (Gerber method) and of the total fatty acids profile. The cheeses obtained, weighing about 300 g, were kept at 25 °C for 2 h, and then were dry salted and stored overnight at 4 °C. The following morning (about 24 h after production), they were weighted for yield calculation and subjected to chemical and sensory analyses.

Milk

Heating to 42 °C (temperature kept constant from now on)

Acidification to pH 6.3 by starter fermentation

Addition of 0.3 g/L^{-1} calcium chloride

Addition of 1.0 mL/ L^{-1} microbial rennet

Coagulation

Curd cutting (1.0 cm cubes) while heating to 46 °C

Settling for 10 min

Gentle whey drainage (30 min)

Moulding

Storage at 4 °C

Figure 1. Technological scheme used for donkey milk cheesemaking.

2.2. Chemical and Sensory Analyses

The moisture content in cheese was assessed by the method IDF 4:1986, fat by the Soxhlet method, total protein by the Kjeldhal method, ash by muffle furnace incineration at 530 °C, and pH by the HI99165 pH Meter equipped with penetration electrode (Hanna Instruments, Woonsocket, RI, USA). Total fatty acids (TFA) analyses were performed on milk, whey and cheese by gas chromatography (GC), as reported by Faccia et al. [15], with some modifications. Briefly, the fatty acids methyl esters were obtained by transesterification of about 25 mg lipid sample obtained by Soxhlet extraction (for cheese) and centrifugation at 2.360 RCF at 4 °C (for milk and whey), previously dissolved in 2 mL of petroleum ether. Samples were injected into a Fisons model MFC 800 gas chromatograph (Fisons, Milan, Italy) equipped with a 60 m × 0.32 mm i.d. and 0.5 µm film thickness fused-silica capillary column (Stabilwax, Restek, Bellefonte, PA, USA). The operating conditions were: (a) oven temperature 70 °C for 5 min, then to 220 °C at 1 °C/min, and held at 220 °C for 30 min; (b) carrier gas helium 20 cm/s at 170 °C; (c) injector at 250 °C, 1 µL; split 40:1; (d) flame-ionisation detector, 250 °C. Volatile compounds were extracted by the HS-SPME technique using a DVB-CAR-PDMS fiber (Supelco, Bellefonte, Pa., USA), as reported by Gambacorta et al. [16], with some modifications. The sample (1.0 mL milk or 1 g cheese), 0.2 g sodium chloride and 10 L of 3-pentanone (internal standard, 81.9 ng/L in water) were placed in 20 mL screw-cap vial, tightly capped with a PTFE-silicon septum, and conditioned for 10 min at 40 °C. A TriPlus RSH™ Autosampler (Thermo Fisher Scientific, Rodano, Italy) was used to optimise and standardise the entire extraction procedure. The fiber then was introduced into the headspace of the vial for 40 min at 40 °C. Then, desorption of volatiles from fiber took place in a splitless mode for 3 min at 220 °C. The separation of volatile compounds was performed by a Trace 1300 gas chromatograph (Thermo Fisher Scientific) equipped with a VF-WAXms capillary column (Agilent, Santa Clara, CA, USA), 60 m length × 0.25 mm I.D. × 0.25 µm film, and coupled with an ISQ single quadrupole mass spectrometer (Thermo Fisher Scientific). The chromatographic conditions were: oven, 45 °C (5 min) to 210 °C at 4 °C/min, held for 3 min; detector, source temperature 250 °C; transfer line temperature 250 °C; carrier gas, helium at constant flow of 0.4 mL/min. The impact energy was 70 eV. Data were acquired using the full-scan mode in the range of 35 to 150 m/z at an acquisition rate of 7.2 Hz. Volatile compounds were tentatively identified by comparing the experimental spectra with those reported in the National Institute of Standards and Technology (NIST) Library, and with those obtained by the available pure standard compounds. Semi-quantitation was carried out by considering the relative areas of the compounds related to that of the internal standard, and assuming arbitrarily that their response factors were the same.

Acquisition and processing of peaks was carried out using Xcalibur v 4.1 software (Thermo Fisher Scientific). All analyses were done in triplicate, and means and standard deviations were calculated. Two different electrophoresis (EF) techniques were used for investigating the protein fraction in milk, whey and cheese. A sample of isoelectric whey, used as a standard, was also prepared. The milk and cheese samples were analysed by urea polyacrylamide gel electrophoresis (PAGE) as indicated by Andrews [17], whereas sodium dodecyl sulphate (SDS) PAGE as reported by Harper et al. [18] was applied to the whey samples. Identification of the protein bands was done by comparison with the data reported by Egito et al. [19,20] and Chianese et al. [12], and for SDS PAGE, the molecular weight was also taken into consideration.

For sensory analysis, the Flavour Profile Method (FPM) was applied, since the product was very innovative, and derived from an "uncommon" raw matter. In fact, due to lack of familiarity, during the training sessions the assessors familiarised with a sample of raw donkey milk before approaching the cheese. Even though FPM can suffer of possible influences among assessors during the final discussion, the analysis is easier to perform than the other descriptive methods [21,22]. It requires a restricted group of trained panellists (at least four) and the generation of a group consensus profile. A trained panel composed of five members (two males and three females aged 28–56 years with three years' experience in cheese analysis) coordinated by a panel leader, performed the evaluation. The identification and selection of descriptors was carried out according to the ISO 11035 protocol [23]. The panel had two open sessions for approaching the new product, during which the cheese was tested by smelling, touching and tasting. A first list of descriptive terms was developed, and then the list was reduced according to the value of frequency of citations × the perceived intensity [24]. For quantitation, a series of reference products were chosen, on which the panel performed three training sessions. Finally, two sessions (in duplicate) were used for developing the qualiquantitative profile by adopting a nonstructured scale. By collegial discussion, a final graph was developed in which the attributes appeared in temporal order (from left to right, from the first to the last perceived, respectively) and their intensity was expressed by a vector of different length.

3. Results and Discussion

The adopted protocol allowed rapid milk coagulation, since the time from rennet addition to curd cutting was 9.15 ± 3.10 min. However, as already reported for calf rennet [14], the coagulum was very soft, and gentle processing was needed to obtain a curd with sufficient firmness, suitable to be moulded (Figure 2). The process consisted of cutting the curd by knife, heating to 46 °C and keeping constant the temperature while progressively removing the whey. On the average, this phase required more than 30 min, and the total processing time (until moulding) was 42.10 min. The cheese texture remained soft and slightly compacted after 24 h. Differently (and as expected), when applied to cow milk, the same technological conditions caused immediate coagulation, very firm curd and intense syneresis, with a total processing time of about 13 min. The results of the present experimentation, together with those previously reported for the use of calf rennet [14], suggest that any type of rennet allows curd formation, if used under suitable conditions. These findings refute the hypothesis of Iannella [13], according to which only camel chymosin can be used for making donkey cheese. Differently, it was evident that the most relevant difficulty does not lie in the primary phase of coagulation, but in the secondary one, which is mainly driven by ionic calcium, heat and casein concentration. In this study, fortification with ionic calcium ($CaCl_2$) and heating were applied at the same level, as in the previous study on calf rennet (0.3 g/L^{-1} and to 46 °C, respectively; [14], since increasing them to 0.4 g/L^{-1} and 50 °C had no effect.

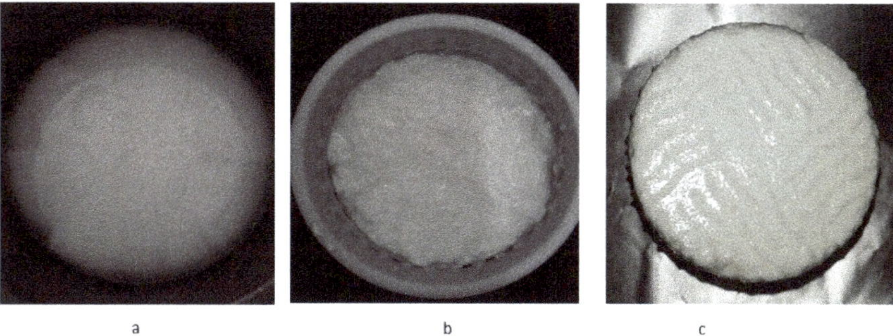

a   b   c

Figure 2. Donkey cheese manufacturing: curd at the cutting time (**a**); moulded curd (**b**); cheese after 24 h (**c**).

Table 1 shows the characteristics of the donkey cheese in comparison with the control one manufactured from cow milk. As expected, the processing time, yield and chemical composition were very different. For donkey, the processing time was more than three times longer and yield was more than three times lower. As to the chemical composition, it was very close to be a fat-free product, and the moisture content was typical of fresh, soft cheeses; the control was semi-hard.

Table 1. Chemical characteristics of donkey and cow milk and corresponding cheeses (mean values ± standard deviation (SD)).

Item	Donkey		Cow (Control)	
	Milk	Cheese	Milk	Cheese
pH	7.01 ± 0.03 [a]	5.85 ± 0.10 [c]	6.71 ± 0.02 [b]	5.61 ± 0.06 [d]
Total Solids%	n.a.	34.19 ± 2.64 [b]	n.a.	49.88 ± 1.16 [a]
Protein%	2.21 ± 0.11 [d]	25.87 ± 3.37 [a]	3.33 ± 0.07 [c]	17.29 ± 2.96 [b]
Fat%	0.33 ± 0.08 [d]	2.90 ± 0.11 [c]	3.62 ± 0.39 [b]	23.76 ± 3.11 [a]
Lactose%	6.76 ± 0.07 [a]	n.a.	4.91 ± 0.02 [b]	n.a.
Ash%	0.39 ± 0.06 [d]	3.31 ± 0.30 [b]	0.81 ± 0.05 [c]	4.83 ± 0.44 [a]
Total processing time *, min		42.10 ± 6.82 [a]		13.20 ± 0.20 [b]
Cheese yield%		7.21 ± 0.76 [b]		23.52 ± 1.50 [a]

Values in the same row bearing different superscripts ([a,b,c,d]) are significantly different, $p < 0.05$. n.a., not assessed; * from addition of rennet.

The total solids contentd of donkey cheese was close to those reported in the literature [13,14], whereas the yield was much higher. This latter result was probably connected to the high protein content of the milk, but the impact of microbial rennet should be investigated further. In fact, since all caseins contribute to the colloidal stability of the micelle (κ-fraction is poorly present in donkey milk), the typical, nonspecific, proteolytic activity of microbial coagulant could have some relevance [25,26]. As regards the total processing time, it was not very different from that reported when using calf rennet, but much shorter with respect to that when camel chymosin was applied. Such great difference could indicate that this latter product was not obtained by 'pure' enzymic coagulation, but by 'combined enzymic-acid' technology. In fact, the total processing time reported for camel chymosin (more than 5 h) very likely allowed acidification by the starter microflora added, which contributed to coagulation and curd formation.

TFA analyses of milk, whey and cheese samples were carried out in order to evaluate if the fatty acid profiles remained unchanged after cheesemaking. Table 2 shows the results obtained. As to donkey milk, the most abundant compound was oleic acid, followed by palmitic, linoleic, capric and lauric.

Table 2. Fat content (%) and fatty acid composition (% of total fatty acids) of donkey's and cow's milk, cheese and whey.

Item	Donkey			Cow (Control)		
	Milk	Cheese	Whey	Milk	Cheese	Whey
Total fat content	0.33 ± 0.08	2.90 ± 0.11	<0.1	3.62 ± 0.39	23.76 ± 3.11	0.28 ± 0.10
Butyric Acid	0.22 ± 0.11	0.19 ± 0.09	n.d.	4.99 ± 0.05	4.85 ± 0.11	5.19 ± 0.23
Caproic Acid	0.40 ± 0.21	0.72 ± 0.66	n.d.	2.73 ± 0.21	2.91 ± 0.19	2.03 ± 1.21
Caprylic Acid	3.69 ± 1.27 [a]	1.50 ± 0.19 [b]	2.71 ± 0.90 [a]	1.29 ± 0.24	1.44 ± 0.34	1.51 ± 0.54
Capric Acid	7.85 ± 1.22 [a]	4.08 ± 0.10 [b]	6.00 ± 1.73 [a]	2.41 ± 0.19	2.55 ± 0.49	2.92 ± 0.55
Undecanoic Acid	1.02 ± 0.22 [a]	0.30 ± 0.12 [b]	1.80 ± 0.64 [a]	0.81 ± 0.11	0.73 ± 0.15	0.33 ± 0.39
Lauric Acid	7.56 ± 0.40 [a]	4.07 ± 0.09 [b]	6.74 ± 1.78 [a]	4.21 ± 0.58	4.40 ± 0.27	4.33 ± 0.61
Tridecanoic Acid	0.03 ± 0.06	n.d.	n.d.	n.d.	n.d.	n.d.
Myristic Acid	6.07 ± 0.62 [b]	12.16 ± 0.33 [a]	7.77 ± 4.77 [a,b]	11.94 ± 0.31	12.41 ± 0.77	13.07 ± 1.00
Myristoleic Acid	0.29 ± 0.07 [b]	1.15 ± 0.46 [a]	n.d.	0.49 ± 0.11	0.32 ± 0.21	0.44 ± 0.33
Pentadecanoic Acid	0.29 ± 0.37 [b]	1.19 ± 0.46 [a]	n.d.	0.90 ± 0.23	0.70 ± 0.31	0.61 ± 0.42
Palmitic Acid	19.08 ± 0.34 [b]	33.08 ± 0.60 [a]	21.37 ± 1.47 [b]	28.45 ± 0.21	30.18 ± 0.37	29.41 ± 3.17
Palmitoleic Acid	2.27 ± 0.11 [a]	1.97 ± 0.08 [b]	2.48 ± 1.66 [a,b]	1.37 ± 0.05	1.01 ± 0.49	1.78 ± 0.52
Margaric Acid	0.29 ± 0.20	0.71 ± 0.26	n.d.	0.45 ± 0.02	0.38 ± 0.12	n.d.
Eptadecenoic Acid	0.34 ± 0.46	n.d.	n.d.	n.d.	n.d.	n.d.
Stearic Acid	1.87 ± 0.47 [c]	8.80 ± 0.15 [a]	5.18 ± 3.14 [b]	10.46 ± 0.17 [b]	13.09 ± 1.22 [a]	9.67 ± 0.94 [b]
Elaidic Acid	n.d.	0.95 ± 0.16	n.d.	0.84 ± 0.30	0.54 ± 0.23	0.89 ± 0.21
Oleic Acid	24.46 ± 1.96	21.84 ± 2.01	24.34 ± 0.84	23.61 ± 0.91	20.08 ± 1.88	22.57 ± 1.90
Linoleic Acid	18.27 ± 0.25 [a]	5.10 ± 0.13 [b]	16.54 ± 3.46 [a]	3.05 ± 0.56	2.80 ± 0.41	3.56 ± 0.68
Linolenic Acid	5.32 ± 0.16 [a]	1.45 ± 1.66 [b]	5.07 ± 2.56 [a,b]	1.21 ± 0.11	0.97 ± 0.05	1.69 ± 0.97
Arachidic acid	0.09 ± 0.05 [b]	0.21 ± 0.05 [a]	n.d.	0.11 ± 0.04	0.23 ± 0.09	n.d.
Gadoleic Acid	0.24 ± 0.16	0.52 ± 0.46	n.d.	0.30 ± 0.11	0.41 ± 0.15	n.d.
Eicosadienoic Acid	0.29 ± 0.07	n.d.	n.d.	0.38 ± 0.01	n.d.	n.d.
Mean saturated	48.46	67.01	51.57	70.35	73.87	68.27
Mean monounsaturated	27.60	26.43	26.82	26.61	21.95	25.68
Mean polyunsaturated	23.92	6.55	21.61	4.64	4.18	5.25

Values in the same row bearing different superscripts ([a,b,c,d]) are significantly different, $p < 0.05$. n.d., not detected.

The results agree with those obtained by Martemucci and D'Alessandro [27] in samples taken at 90 days of lactation. Some differences were observed with the data of Gastaldi et al. [28], who reported a different order of abundance. This discrepancy could be connected to a different lactation stage; however, the authors did not indicate the period in which they took the samples. The total amount of saturated fatty acids was very close to that found by Martini et al. [29] in milk from the Amiata donkey, whereas the concentration of polyunsaturated was much higher. The average TFA profile of the bovine milk used as control (bulk milk samples taken from an industrial dairy) was within the typical range of this type of milk, and the high saturated/polyunsaturated ratio suggests possible derivation from intensive farms. The TFA profile of the cheese showed some changes with respect to milk, due to an increase of some saturated compounds (in particular stearic acid) and decrease of the polyunsaturated ones (mainly linoleic and linolenic acids). This result was surprising, and was only partially reflected by the control, for which the differences were less pronounced and the significance was only found for stearic acid. At our knowledge, this argument has been poorly investigated, and more analyses are needed to confirm the result. A possible cause could lie in a different retention of the fat globules into the cheese, depending on their size. According to the literature, small sized and large sized globules of cow milk present some differences in the fatty acid composition of glycerides: the small globules contain less short chain fatty acids, less stearic and more oleic acid [30,31] than larger ones. For donkey milk, no information is available on TFA composition of fat globules glycerides, but Martini et al. [29] reported that globules have different size, ranging from less than 2 to more than 5 μm, and that the small ones are much more abundant (about 70% of total). If, as theoretically expected, the small globules are less retained into the curd, some differences in the TFA profile of milk, cheese and whey should arise. Unfortunately, the TFA analyses of the whey samples did not give further elements for supporting this hypothesis, since the standard deviation was too high. A specific study is required to in order to clarify the matter.

The volatile organic compounds (VOCs) analyses gave the results shown in Table 4, in which only the compounds detected in at least four trials (out of six) are shown. Information on VOCs in donkey milk is rare: Conte et al. [32], Tidona et al. [33] and Vincenzetti et al. [34] identified 7, 6 and 19 compounds, respectively. In our study, 11 individual volatile compounds were identified (versus 21 in the control), and the total peak area was much lower than in cow milk. The different fat content in the two types of milk should play a role in the formation and release of VOCs. In this regard, it must be underlined that the results on the donkey samples (both milk and cheese) had low standard deviations, differently from those obtained in the cow cheese. It could be connected to the different concentration at which VOCs are present in the two types of samples, or to the variability of the fat content (3.62 ± 0.39) in the cow milk samples used. The most abundant volatile compounds in donkey milk were acetic acid and acetone: they can both derive from animal metabolism and microbial activity. These two compounds were also detected in cheese, but at a lower level, probably because they were lost into the whey during cheesemaking. Among the other compounds, 6-methyl-5-hepten-2-one was the third in order of abundance: it is commonly called sulcatone, and presents an apple, bitter and citrus flavour. It is a powerful attractant for mosquitos, and is emitted by a variety of plants and animals, including humans, horses, cows and sheep. Sulcatone is also present in the faeces of mares in estrous and in the hair of horses [35,36]: this can explain the presence in donkey milk and consequently in cheese, where its concentration almost doubled. As regards cheese, 18 compounds were identified in the donkey product (versus 34 in the control one). The most abundant were acetoin, 2-methylbutanal and 2-ethylhexanol, followed by 2-ethylhexanal and hexanoic acid. Acetoin and methylbutanal are commonly found in any types of dairy products; they present butter and green odors and derive from lactic acid bacteria fermentation [37–39]. 2-Ethylhexanol has a fruity, green, cucumber odour [40] and likely derives from microbial metabolism, since it was not present in milk [41]. However, it has also been reported to be a pollutant deriving from plastic tools. In particular, it is a metabolic derivative of 2-ethylexylphtalate (DEHP), an additive commonly added to plastics to make them flexible [42]. Possible pollution from the cheesemaking environment (i.e., plastic electrode, tools for milking and plastic baskets used for cheese moulding) should be ascertained. All other newly formed compounds had been widely reported in the literature for cow's cheese.

Table 3. Volatile organic compounds (VOCs) detected in milk and cheese (μg/kg ± SD).

Compoun.d.	Donkey		Cow (Control)	
	Milk	Cheese	Milk	Cheese
Acids				
acetic	5.65 ± 0.03 [b]	0.19 ± 0.01 [d]	2.69 ± 0.17 [c]	66.03 ± 5.74 [a]
butanoic	n.d.	0.36 ± 0.03 [c]	5.09 ± 0.24 [b]	8.77 ± 1.05 [a]
hexanoic	n.d.	0.55 ± 0.10 [c]	7.55 ± 0.87 [b]	10.73 ± 0.82 [a]
heptanoic	n.d.	n.d.	n.d.	0.68 ± 0.09
octanoic	n.d.	0.29 ± 0.07 [c]	3.91 ± 0.33 [b]	5.02 ± 0.22 [a]
nonanoic	n.d.	n.d.	1..44 ± 0.03 [b]	2.10 ± 0.07 [a]
decanoic	n.d.	n.d.	3.22 ± 0.40	3.01 ± 0.44
Alcohols				
ethanol	n.d.	n.d.	n.d.	17.46 ± 2.07
2-ethylhexanol	n.d.	7.16 ± 0.12	n.d.	n.d.
isoamylalcohol	n.d.	0.42 ± 0.02	n.d.	n.d.
3-methyl,1-butanol	n.d.	n.d.	n.d.	71.84 ± 6.47
3-methyl-2-buten-1-ol	n.d.	n.d.	n.d.	1.19 ± 0.52
1-nonanol	n.d.	0.16 ± 0.01	n.d.	n.d.
2-octyloxy-ethanol	n.d.	n.d.	n.d.	16.26 ± 3.62
phenylethyl alcohol	n.d.	n.d.	n.d.	1.92 ± 0.09
2-propyl-1-pentanol	n.d.	n.d.	1.48 ± 0.39 [a]	0.47 ± 0.16 [b]

Table 4. Volatile organic compounds (VOCs) detected in milk and cheese (μg/kg $\pm$ SD).

Compoun.d.	Donkey		Cow (Control)	
Aldheydes				
diacetyl (2,3-butanedione)	n.d.	n.d.	n.d.	27.35 $\pm$ 2.70
hexanal	n.d.	n.d.	1.04 $\pm$ 0.05 [b]	5.33 $\pm$ 0.19 [a]
heptanal	n.d.	n.d.	0.27 $\pm$ 0.07 [b]	1.02 $\pm$ 0.10 [a]
octanal	n.d.	n.d.	0.48 $\pm$ 0.13 [b]	1.77 $\pm$ 0.51 [a]
nonanal	0.02 $\pm$ 0.01 [d]	0.39 $\pm$ 0.08 [c]	3.22 $\pm$ 1.26 [b]	17.97 $\pm$ 3.97 [a]
decanal	n.d.	n.d.	2.05 $\pm$ 1.11	2.54 $\pm$ 0.33
2-methylbutanal	0.08 $\pm$ 0.01	8.79 $\pm$ 0.37	n.d.	n.d.
3-methylbutanal	n.d.	0.15 $\pm$ 0.01	n.d.	63.66 $\pm$ 0.29
2-ethylhexanal	0.09 $\pm$ 0.11	0.71 $\pm$ 0.02	n.d.	n.d.
Esters				
ethylacetate	0.09 $\pm$ 0.01 [b]	0.08 $\pm$ 0.03 [b]	n.d.	17.83 $\pm$ 5.72 [a]
1-butanol-3-methyl-acetate	n.d.	n.d.	n.d.	1.03 $\pm$ 0.58
Ketones				
acetone	2.84 $\pm$ 0.03 [c]	0.65 $\pm$ 0.01 [d]	59.11 $\pm$ 9.11 [a]	31.55 $\pm$ 0.29 [b]
2-butanone	0.03 $\pm$ 0.01 [d]	0.14 $\pm$ 0.02 [c]	4.21 $\pm$ 1.21 [b]	13.78 $\pm$ 2.85 [a]
2-heptanone	0.09 $\pm$ 0.02 [d]	0.37 $\pm$ 0.01 [c]	4.81 $\pm$ 0.93 [b]	16.09 $\pm$ 3.74 [a]
3-heptanone	0.07 $\pm$ 0.01	n.d.	n.d.	n.d.
2-nonanone			n.d.	4.37 $\pm$ 0.47
2-un.d.ecanone			n.d.	0.85 $\pm$ 0.23
3-hydroxy 2-butanone (acetoin)	0.07 $\pm$ 0.02 [c]	11.93 $\pm$ 0.04 [b]	n.d.	91.71 $\pm$ 23.70 [a]
6-methyl-5-hepten-2-one	0.15 $\pm$ 0.04 [d]	0.38 $\pm$ 0.05 [c]	0.88 $\pm$ 0.29 [b]	1.83 $\pm$ 0.18 [a]
2,3 pentanedione	n.d.	n.d.	0.60 $\pm$ 0.38 [b]	23.56 $\pm$ 6.20 [a]
Others				
dimethyl sulfide	n.d.	n.d.	0.12 $\pm$ 0.07 [b]	1.33 $\pm$ 0.40 [a]
dimethyl sulfone	n.d.	0.26 $\pm$ 0.01 [b]	1.97 $\pm$ 0.28 [a]	1.83 $\pm$ 0.19 [a]
hexane	n.d.	n.d.	1.27 $\pm$ 0.53 [b]	2.98 $\pm$ 0.81 [a]
heptane	n.d.	n.d.	0.77 $\pm$ 0.12 [b]	7.20 $\pm$ 1.00 [a]

Values in the same row bearing different superscripts ([a,b,c,d]) are significantly different, $p < 0.05$. n.d., not detected.

Figures 3 and 4 show the results of the electrophoretic study. For interpreting the gel, the works of Chianese et al. [12] on donkey milk, and of Egito et al. [19,20] on equine caseins in buffered solutions, were considered. By comparing the urea-PAGE patterns of milk and cheese (Figure 3), it was evident that some new bands appeared after 24 h from cheesemaking. In the upper zone of the gel, the intensity of several bands strongly increased in cheese: they should correspond to γ-casein-like fragments deriving from β-caseins. In the central area, five new bands appeared: for those coded x_1 and x_2, no possible correspondence was found in the literature, whereas those coded x_3, x_4 and x_5 laid in the same zone of β-I- and β-II-caseins-like fragments. Finally, the diffuse band positioned in the lower zone (coded x_6 and probably composed of two or three overlapping bands), could correspond to proteose-peptone-like fragments deriving from β-casein, or to fragments with high electrophoretic mobility deriving from αs1-casein. Considering that the cheese was analysed at 24 h after production, it is likely that these fragments derive from the activity of microbial rennet, which is able to hydrolyse all casein fractions in very short time [43].

Due to scarce resolution of whey proteins in the urea-PAGE system, SDS-PAGE was used for analysing the whey samples [44]. The patterns of isoelectric and cheese whey were very similar, except for the total absence of residual casein in the latter (Figure 4).

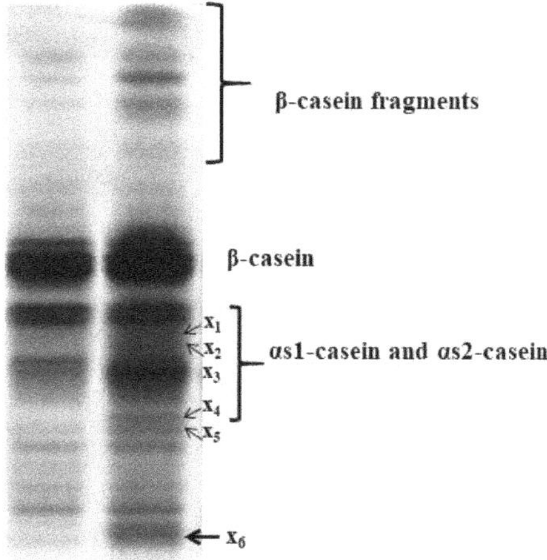

Figure 3. Urea polyacrylamide gel electrophoresis (urea-PAGE) patterns of donkey milk (left slot) and cheese after moulding (right slot).

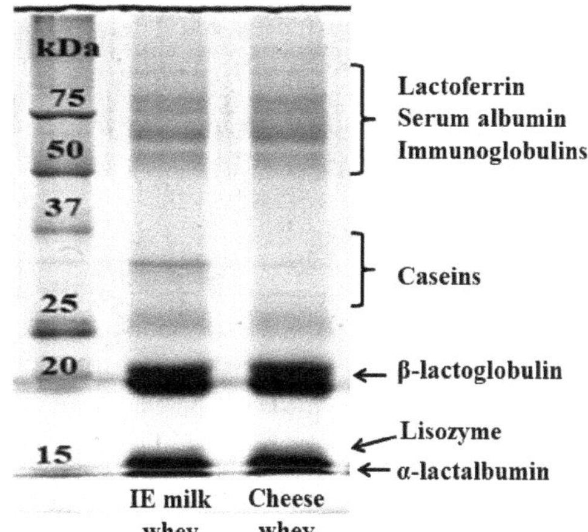

Figure 4. Sodium dodecyl sulphate (SDS)-PAGE patterns of isoelectric donkey milk whey (left slot) and donkey cheese whey after moulding (right slot).

Table 5 and Figure 5 show the results of the sensory analyses. The restricted list of descriptors and the corresponding reference products that were developed for the experimental and control cheeses are shown in Table 5. For the former, 9 descriptors were selected from a pool of 13 initial attributes, 3 of which regarded texture, 3 aroma, 2 taste and 1 colour. The term "gamy" was adopted to indicate the typical animal odour of donkey milk. For the control, the descriptors were mostly different (10 developed). The colour was white in both cases, but with different gradations. The texture was very close to that of fresh Feta cheese for donkey, and to that of Edam cheese for cow.

Table 5. List of descriptors and reference product developed for donkey cheese.

Descriptor	Reference Product	Donkey	Cow (Control)
Colour			
White		+	
Ivory white			+
Texture			
Corky	Rubber stopper		+
Crumbly	Feta cheese	+	
Semi-hard	Carrot		+
Soluble	Meringue	+	
Springy	Edam cheese		+
Juicy	Water melon	+	
Aroma			
Buttery	Butter		+
Caramel	Cooked sugar	+	
Egg yolk	Raw egg yolk	+	
Gamy	Fresh donkey milk	+	
Lactic	Fresh uncultured milk		+
Yogurt	Plain yogurt		+
Taste			
Salty	0.5% NaCl in water	+	+
Sweet	0.5% sucrose in water	+	+
Acid	0.5% citric acid		+

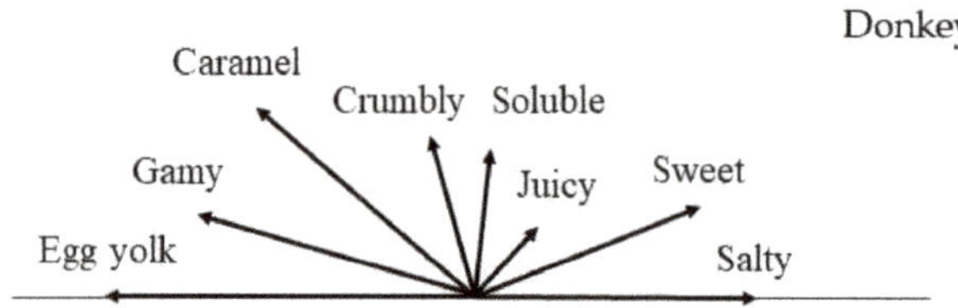

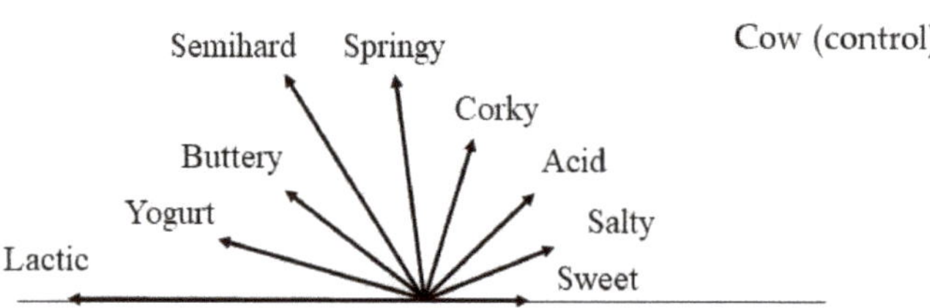

Figure 5. Flavour profile (aroma, texture and taste) of donkey and control cheese made with microbial rennet.

An extremely distinctive aroma was perceived in the former, for which a unique descriptor was developed, that is "egg yolk". As to taste, the most intense descriptor selected was "sweet", whereas the control also had a slight acid taste. This difference should be ascribed both to different pH values (it was slightly lower in the control cheese) and/or possible masking of the sour taste by the sweetness of donkey samples.

The differences in aroma, texture and taste can be clearly observed in Figure 5, where the flavour profiles are shown. The most intense odour in donkey cheese was "egg yolk", followed by "gamy" and "caramel". By comparing these results with those reported in the literature for egg yolk [45], a possible connection with the VOC profiles can be hypothesised. In fact, the two matrices present several volatile compounds in common, such as dimethyl sulphone, acetic, butanoic and hexanoic acids. As to texture, the cheese lacked elasticity, and became dry and chalky after a few days storage at 13 °C and 75% relative humidity, indicating that it was not suitable to ripening (results not shown). Overall, the sensory profile was rather similar to that previously reported for cheesemaking with calf rennet, and the flavour was judged as pleasant.

4. Conclusions

The present experimentation confirmed that donkey milk can be processed into fresh cheese by enzymatic coagulation under suitable technological conditions, by using microbial rennet. Such conditions result in being very severe when applied to cow milk, and lead to the formation of a semihard curd in a very short time. The chemical and sensory composition of fresh donkey cheese can be very interesting from a nutritional point view, due to very low fat levels and pleasant flavour. However, the scarce availability and high cost of the raw material currently limit practical application. Further studies are necessary by animal husbandry scientists to increase the milk yield of lactating donkeys, and by food scientists for improving the cheese texture and cheesemaking yield. Efforts for improving curd firmness should focus on increasing casein concentration: to this aim, protein enrichment could be pursued by water evaporation (at low temperature, in order to avoid deterioration of the already weak coagulation properties) or membrane filtration.

Author Contributions: Michele Faccia, Giuseppe Gambacorta, Giovanni Martemucci, Graziana Difonzo and Angela Gabriella D'Alessandro contributed equally to this work. All authors have read and agreed to the published version of the manuscript.

Funding: This research received no external funding.

References

1. Langlois, B. The history, ethnology and social importance of mare's milk consumption in Central Asia. *J. Life Sci.* **2011**, *5*, 863–872.
2. Cunsolo, V.; Muccilli, V.; Fasoli, E.; Saletti, R.; Righetti, P.G.; Foti, S. Poppea's bath liquor: The secret proteome of she-donkey's milk. *J. Proteom.* **2011**, *74*, 2083–2099. [CrossRef] [PubMed]
3. Salimei, E.; Fantuz, F. Equid milk for human consumption. *Int. Dairy J.* **2012**, *24*, 130–142. [CrossRef]
4. Camillo, F.; Rota, A.; Biagini, L.; Tesi, M.; Fanelli, D.; Panzani, D. The current situation and trend of donkey industry in Europe. *J. Equine Vet. Sci.* **2018**, *65*, 44–49. [CrossRef]
5. Giacometti, F.; Bardasi, L.; Merialdi, G.; Morbarigazzi, M.; Federici, S.; Piva, S.; Serraino, A. Shelf life of donkey milk subjected to different treatment and storage conditions. *J. Dairy Sci.* **2016**, *99*, 4291–4299. [CrossRef]
6. Chiavari, C.; Coloretti, F.; Nanni, M.; Sorrentino, E.; Grazia, L. Use of donkey's milk for a fermented beverage with lactobacilli. *Le Lait* **2005**, *85*, 481–490. [CrossRef]
7. Perna, A.; Intaglietta, I.; Simonetti, A.; Gambacorta, E. Donkey milk for manufacture of novel functional fermented beverages. *J. Food Sci.* **2005**, *80*, S1352–S1359. [CrossRef]
8. Turchi, B.; Pedonese, F.; Torracca, B.; Fratini, F.; Mancini, S.; Galiero, A.; Montalbano, B.; Cerri, D.; Nuvoloni, R. Lactobacillus plantarum and Streptococcus thermophilus as starter cultures for a donkey milk fermented beverage. *Int. J. Food Microbiol.* **2017**, *256*, 54–61. [CrossRef]
9. Uniacke-Lowe, T. Studies on Equine Milk and Comparative Studies on Equine and Bovine Milk Systems. Ph.D. Thesis, University College Cork, Eire, Ireland, 2011.
10. Tidona, F.; Criscione, A.; Devold, T.G.; Bordonaro, S.; Marletta, D.; Vegarud, G.E. Protein composition and micelle size of donkey milk with different protein patterns: Effects on digestibility. *Int. Dairy J.* **2014**, *35*, 57–62. [CrossRef]
11. Luo, J.; Jian, S.; Wang, P.; Ren, F.; Wang, F.; Chen, S.; Guo, H. Thermal instability and characteristics of donkey casein micelles. *Food Res. Int.* **2019**, *119*, 436–443. [CrossRef]

12. Chianese, L.; Calabrese, M.G.; Ferranti, P.; Mauriello, R.; Garro, G.; De Simone, C.; Quarto, M.; Addeo, F.; Cosenza, G.; Ramunno, L. Proteomic characterization of donkey milk "caseome". *J. Chromatogr. A* **2010**, *1217*, 4834–4840. [CrossRef] [PubMed]

13. Iannella, G. Donkey cheese made through pure camel chymosin. *Afr. J. Food Sci.* **2015**, *9*, 421–425.

14. Faccia, M.; Gambacorta, G.; Martemucci, G.; Natrella, G.; D'Alessandro, A.G. Technological attempts at producing cheese from donkey milk. *J. Dairy Res.* **2018**, *85*, 327–330. [CrossRef] [PubMed]

15. Faccia, M.; Trani, A.; Gambacorta, G.; Loizzo, P.; Cassone, A.; Caponio, F. Production technology and characterization of Fiordilatte cheese from ovine and caprine milk. *J. Dairy Sci.* **2015**, *98*, 1402–1410. [CrossRef] [PubMed]

16. Gambacorta, G.; Trani, A.; Fasciano, C.; Paradiso, V.M.; Faccia, M. Effects of prefermentative cold soak on polyphenols and volatiles of Aglianico, Primitivo and Nero di Troia red wines. *Food Sci. Nutr.* **2019**, *7*, 483–491. [CrossRef]

17. Andrews, A.T. Proteinases in normal bovine milk and their action on caseins. *J. Dairy Res.* **1983**, *50*, 45–55. [CrossRef]

18. Harper, J.; Meyer, M.; Knighton, D.; Lelievre, J. Effects of whey proteins on the proteolysis of Cheddar cheese slurries. (A model for the maturation of cheese made from ultrafiltered milk). *J. Dairy Sci.* **1989**, *72*, 333–341. [CrossRef]

19. Egito, A.S.; Miclo, L.; Lopez, C.; Adam, A.; Girardet, J.-M.; Gaillard, J.-L. Separation and characterization of mares' milk αs1-, β-, κ-caseins, γ-casein-like, and proteose peptone component 5-like peptides. *J. Dairy Sci.* **2002**, *85*, 697–706. [CrossRef]

20. Egito, A.S.; Girardet, J.-M.; Poirson, C.; Mollè, D.; Humbert, G.; Miclo, L.; Gaillard, J.-L. Action of plasmin on equine β-casein. *Int. Dairy J.* **2003**, *13*, 813–820. [CrossRef]

21. Piggott, J.R.; Simpson, S.J.; Williams, S.A.R. Sensory analysis. *Int. J. Food Sci. Technol.* **1998**, *33*, 7–18. [CrossRef]

22. Drake, M.A.; Civille, G.V. Flavor lexicon. *Compr. Rev. Food Sci. Food Saf.* **2002**, *2*, 33–40. [CrossRef]

23. ISO 11035. *Sensory Analysis—Identification and Selection of Descriptors for Establishing A Sensory Profile by A Multidimensional Approach*; International Organization for Standardization: Geneva, Switzerland, 1994.

24. Trani, A.; Gambacorta, G.; Loizzo, P.; Cassone, A.; Faccia, M. Short communication: Chemical and sensory characteristics of Canestrato di Moliterno cheese manufactured in spring. *J. Dairy Sci.* **2016**, *99*, 6080–6085. [CrossRef]

25. Jacob, M.; Jaros, D.; Rohm, H. Recent advances in milk clotting enzymes. *Int. J. Dairy Technol.* **2011**, *64*, 14–33. [CrossRef]

26. Uniacke-Lowe, T.; Chevalier, F.; Hem, S.; Fox, P.F.; Mulvihill, D.M. Proteomic comparison of equine and bovine milks on renneting. *J. Agric. Food Chem.* **2013**, *61*, 2839–2850. [CrossRef]

27. Martemucci, G.; D'Alessandro, A.G. Fat content, energy value and fatty acid profile of donkey milk during lactation and implications for human nutrition. *Lipids Health Dis.* **2012**, *11*, 113. [CrossRef]

28. Gastaldi, D.; Bertino, E.; Monti, G.; Baro, C.; Fabris, C.; Lezo, A.; Medana, C.; Baiocchi, C.; Mussap, M.; Galvano, F.; et al. Donkey's milk detailed lipid composition. *Front. Biosci.* **2010**, *E2*, 537–546.

29. Martini, M.; Altomonte, I.; Salari, F. Amiata donkeys: Fat globule characteristics, milk gross composition and fatty acids. *Ital. J. Anim. Sci.* **2014**, *13*, 123–126. [CrossRef]

30. Briard, V.; Leconte, N.; Michel, F.; Michalski, M.C. The fatty acid composition of small and large naturally occurring milk fat globules. *Eur. J. Lipid Sci. Technol.* **2003**, *105*, 677–682. [CrossRef]

31. Fauquant, C.; Briard, V.; Leconte, N.; Michalski, M.C. Differently sized native milk fat globules separated by microfiltration: Fatty acid composition of the milk fat globule membrane and triglyceride core. *Eur. J. Lipid Sci. Technol.* **2005**, *107*, 80–86. [CrossRef]

32. Conte, F.; Rapisarda, T.; Belvedere, G.; Carpino, S. Shelf-life del latte d'asina: Batteriologia e componente volatile. *Ital. J. Food Saf.* **2010**, *7*, 25–29. [CrossRef]

33. Tidona, F.; Charfi, I.; Povolo, M.; Pelizzola, V.; Carminati, D.; Contarini, G.; Giraffa, G. Fermented beverage emulsion based on donkey milk with sunflower oil. *Int. J. Food Sci. Technol.* **2015**, *50*, 2644–2652. [CrossRef]

34. Vincenzetti, S.; Cecchi, T.; Perinelli, D.R.; Pucciarelli, S.; Polzonetti, V.; Bonacucina, G.; Ariani, A.; Parrocchia, L.; Spera, D.M.; Ferretti, E.; et al. Effects of freeze-drying and spray-drying on donkey milk volatile compounds and whey proteins stability. *LWT* **2018**, *88*, 189–195. [CrossRef]

35. McBride, C.S.; Baier, F.; Omondi, A.B.; Spitzer, S.A.; Sang, J.L.R.; Ignell, R.; Vosshall, L.B. Evolution of mosquito preference for humans linked to an odorant receptor. *Nature* **2014**, *515*, 222–227. [CrossRef]

36. Nielsen, B.L.; Jerôme, N.; Saint-Albin, A.; Ouali, C.; Rochut, S.; Zins, E.L.; Briant, C.; Guettier, E.; Reigner, F.; Couty, I.; et al. Oestrus odours from rats and mares: Behavioural responses of sexually naive and experienced rats to natural odours and odorants. *Appl. Anim. Behav. Sci.* **2016**, *176*, 128–135. [CrossRef]

37. Chen, C.; Zhao, S.; Hao, G.; Yua, H.; Tiana, H.; Zhao, G. Role of lactic acid bacteria on the yogurt flavour: A review. *Int. J. Food Prop.* **2017**, *20*, S316–S330. [CrossRef]

38. Dan, T.; Wang, D.; Wu, S.; Jin, R.; Ren, W.; Sun, T. Profiles of Volatile flavor compounds in milk fermented with different proportional combinations of lactobacillus delbrueckii subsp. bulgaricus and streptococcus thermophiles. *Molecules* **2017**, *22*, 1633.

39. Picon, A.; López-Pérez, O.; Torres, E.; Garde, S.; Nuñez, M. Contribution of autochthonous lactic acid bacteria to the typical flavour of raw goat milk cheeses. *Int. J. Food Microbiol.* **2019**, *299*, 8–22. [CrossRef]

40. Niu, Y.; Zhang, X.; Xiao, Z.; Song, S.; Eric, K.; Jia, C.; Yu, H.; Zhu, J. Characterization of odor-active compounds of various cherry wines by gas chromatography–mass spectrometry, gas chromatography–olfactometry and their correlation with sensory attributes. *J. Chromatogr. B* **2011**, *879*, 2287–2293. [CrossRef]

41. Morales, P.; Calzada, J.; Juez, C.; Nunez, M. Volatile compounds in cheeses made with *Micrococcus* sp. INIA 528 milk culture or high enzymatic activity curd. *Int. J. Dairy Technol.* **2010**, *63*, 538–543. [CrossRef]

42. Fay, M.; Brattin, W.J.; Donohue, J.M. Public health statement. *Toxicol. Ind. Health* **1999**, *15*, 652–654. [CrossRef]

43. Awad, S.; Lthi-Peng, Q.-Q.; Puhan, Z. proteolytic activities of suparen and rennilase on buffalo, cow, and goat whole casein and β-casein. *J. Agric. Food Chem.* **1999**, *47*, 3632–3639. [CrossRef]

44. Van Hekken, D.L.; Thompson, M.P. Application of PhastSystem®to the resolution of bovine milk proteins on urea-polyacrylamide gel electrophoresis. *J. Dairy Sci.* **1992**, *75*, 1204–1210. [CrossRef]

45. Plagemann, I.; Zelena, K.; Krings, U.; Berger, R.G. Volatile flavours in raw egg yolk of hens fed on different diets. *J. Sci. Food Agric.* **2011**, *91*, 2061–2065. [CrossRef]

 *foods*

Article

Effect of Different Starches on the Rheological, Sensory and Storage Attributes of Non-fat Set Yogurt

Ali Saleh, Abdellatif A. Mohamed *, Mohammed S. Alamri, Shahzad Hussain, Akram A. Qasem and Mohamed A. Ibraheem

Department of Food Science and Nutrition, King Saud University, Riyadh 11456, Saudi Arabia;
Alisa@ksu.edu.sa (A.S.); msalamri@ksu.edu.sa (M.S.A.); shussain@ksu.edu.sa (S.H.);
akram@ksu.edu.sa (A.A.Q.); ibraheem@ksu.edu.sa (M.A.I.)
* Correspondence: abdmohamed@ksu.edu.sa; Tel.: +966-0530832995

Received: 20 November 2019; Accepted: 17 December 2019; Published: 7 January 2020

Abstract: This study was conducted to investigate the effect of various native starches on the rheological and textural properties of non-fat set yogurt. The yogurt samples were prepared while using five types of starches (potato, sweet potato, corn, chickpea, and Turkish beans). The physical properties of the prepared yogurt were analyzed while using shear viscosity, viscoelasticity, and texture analysis. The tests were performed after 0, 7, and 15 days storage. The effect of these starches on the yogurt viscoelastic properties, texture, syneresis, and sensory evaluation were determined under optimum conditions. The results showed that adding 1% starch could significantly ($p < 0.05$) reduce syneresis and improve yogurt firmness. Starches exhibited different effect on the overall quality of the yogurt due to their origin and amylose content. Regardless of the number of storage period duration, all of the samples, including the control behaved as pseudoplastic materials ($n < 1$) with various levels of pseudoplasticity. Yogurts with corn and tuber starches had the highest consistency coefficient (k), which indicated higher viscosity. The yogurt sample with chickpea starch exhibited the highest G´, making the gel more solid like. Therefore, the influence of tuber starches (potato and sweet potato) on G´ was different when compared to corn or legume starches. The behavior of the starches changed with storage time, where some starches performed better only at the beginning of the storage period duration. Wheying-off was significantly reduced, regardless of starch type. The pH of the yogurt remained unchanged through storage. Sensory evaluation showed a preference for starch-containing samples as compared to the control, regardless of the starch type. The variation in yogurt quality as a function of starch type could be attributed to the starch granule structure, gelatinization mechanism, or amylose content.

Keywords: starch; yogurt; rheology; sensory; texture

---

1. Introduction

Nowadays, yogurt is considered to be one of the most popular fermented milk products and it has gained widespread consumer acceptance as a healthy food [1]. It is prepared by fermenting milk with bacterial cultures consisting of a mixture of *Streptococcus thermophiles* and *Lactobacillus* [2]. Yogurt is as nutritious as many other fermented milk products, since it contains high level of milk solids in addition to the nutrients developed during the fermentation process. In the yogurt industry, the major concern is the production and maintenance of a product with optimal consistency and stability during transportation and storage. The texture, viscosity at high total solids, variation in the processing variables, and characteristics of the starter culture are the main components that determine yogurt's consistency [3].

The consumption of whole fat products (e.g., full fat yogurt) has declined due to the awareness of the probable harmful effect of fat on consumer's health, thus the dietary habits of consumers have changed and market interest moved in favor of low or nonfat dairy products [4]. The Code of Federal Regulations of FDA reported that the low fat yogurt and nonfat yogurt are similar in description to full fat yogurt, but low fat contains 0.5% to 2% and nonfat is less than 0.5% milk fat [5,6].

In yogurt products, milk fat plays a major role in the texture, flavor, and color development of the final products [7]. Therefore, the reduction in fat will subsequently reduce the total solids content (in low-fat and nonfat yogurt), leading to weak body, poor texture, and increased whey separation, unless various stabilizers are used [8].

Food hydrocolloids, such as starches, are usually used in the food industries as thickeners, stabilizer, gelling agents, syneresis controller, and emulsifiers [9,10]. On the other hand, they regulate flavor and aroma release [11]. The use of stabilizers in manufactured dairy products, such as yogurt, is very important for appropriate viscosity, sensory properties, and inhibiting/reducing wheying-off during storage and transportation, as well as boosting the ratio of total solids. There are many kinds of stabilizers, for instance, synthetic (carboxyl methyl cellulose) or natural. Plant origin stabilizers are considered to be the cheapest and they include the commonly used ones, such as corn starch. Starch is preferred in the yogurt industry, because it is a good thickener and its ability to reduce yogurt flaws by improving texture and make the product more appealing to consumers [12–14]. Starches, such as corn, sweet potato, potato, and chestnut, are commonly used by the yogurt industry at 0.25 to 1%". Although the use of starch in yogurt manufacturing is currently practiced, a comparison between the performance of starches form different sources and dissimilar amylose content, such as tubers, cereals, or legumes, is not done to the best of our knowledge. Therefore, the focus of this study was to explore the effect of corn, sweet potato, potato, Turkish beans, and chickpea starches on the physicochemical, rheological, and sensory properties of the none-fat set yogurt during and after storage.

2. Materials and Methods

2.1. Materials

Nonfat milk powder (34.5% protein, 3.5% moisture, 7.2% ash, 55% lactose) of Nino brand was purchased from a local supermarket. Fresh potato and sweet potato were obtained from the produce market (Riyadh, Saudi Arabia). Chickpea and Turkish beans were purchased from local store, whereas ARASCO (Riyadh, Saudi Arabia) donated the corn starch.

2.2. Starches Extraction

2.2.1. Potato and Sweet Potato Starches Isolation

The potato or sweet potato starch was extracted according to [15]. The tuber was thoroughly washed, peeled, and diced into small pieces. Slurry was prepared by blending the diced tubers in distilled water (50:50 *v/v*) for 3 min. while using kitchen aid blender at medium speed (B. Braun Melsungen, AG, Hessen, Germany). The slurry was filtered through a muslin cloth and the overs were re-suspended in distilled water (1:2 *v/v*), blended, and filtered in the same way, and finally the filtrate is sieved while using 200 mesh sieve. Starch was allowed to settle for 1 h at room temperature and the supernatant was discarded. The starch was re-suspended in distilled water and then centrifuged at 2000× *g* for 15 min. After centrifugation, the top dark layer was removed and the white material at the bottom of the bottle is reconstituted in distilled water and then centrifuged until pure white starch fraction is obtained. The isolated starch is dried, ground, and stored as in air tight jars at 5 °C.

2.2.2. Chickpea and Turkish Bean Starches Isolation

Whole meal of Turkish beans was prepared by crushing the dry beans in the blender at fast speed for 3 min. Slurry was prepared by blending the whole meal of chickpea or Turkish beans in distilled water (50/50; *w/w*) in heavy duty blender (B. Braun Melsungen, AG, Hessen, Germany) for 5 min. at medium speed. The slurry was passed through 200 mesh sieves and the filtrate was centrifuged at

2000× *g* for 15 min. [16,17]. After centrifugation, the top layer on the precipitate was removed and the white material at the bottom of the bottle (the pellet) was then re-suspended in distilled water and centrifuged while using the above-mentioned conditions. This procedure was repeated five times to get the pure white starch. The starch was air dried, ground in a coffee grinder, and stored in air-tight glass bottles at 4 °C for further analysis.

2.3. Amylose Content

Amylose content was determined according to the method of Williams, et al. [18]. The method is based on weighing 0.1 g of starch dry basis, 1 mL ethanol, 9 mL NaOH (I M). The mixture is boiled for 10 min. in water bath and then cooled to room temperature. To 5 mL mixture, 1 mL acetic acid (1 N), 2 mL iodine solution (2.0 g of potassium iodide and 0.2 g of iodine diluted to 100 mL with distilled water) and the absorbance (A) was red at 620 nm. The percent amylose content was calculated, as follows: 3.06 × A × 20.

2.4. Yogurt Preparation

Nonfat yogurt was prepared while using powdered skim milk (14.0% total solids) and starch. The blends were prepared by replacing 1.0 g of the skim milk with 1.0 g of starch. So as to ascertain complete solubility, dry starch and milk powder were mixed first. To the dry ingredients, water was added to finally maintain 14% total solids and the mixture was preheated to 60 °C for 30 min., cooled to 42 °C, and the yogurt starter was added at 3.0% of the dry ingredients. The microbial starter included Streptococcus thermophiles and Lactobacillus bulgaricus. The mixture (14 g) was divided into plastic cups (100 mL) and incubated at 42 °C until coagulation occurs or the pH reached 4.6 (Barrantes et al., 1994) [4]. The yogurt samples were stored at 5 ± 0.2 °C and then analyzed after 0, 7, and 15 days of storage. Three replicates were tested. The pH of yogurt samples was determined at 25 °C.

2.5. Yogurt Composition

2.5.1. Total Solids Contents

The total solids content of yogurt samples were determined according to the Association of Official Agricultural Chemists AOAC Method (940.09) [19]. Yogurt sample (10 g) was dried in air forced oven at 105 ± 5 °C for 1 h. The remaining weight was expressed as percent total solids content.

2.5.2. Total Ash Content

The ash content of the yogurt samples was determined according to the AOAC Method (942.05) [19], where 5 g of yogurt sample was heated in a muffle furnace at 550 °C for 5 h and the residue was expressed as % Ash content.

2.5.3. Crude Protein

The crude protein content was estimated according to the Kjeldahl method, as described in the AOAC method (992.15) [19]. The yogurt sample (2 g) was digested in concentrated sulfuric acid and the total nitrogen content was multiplied by 5.70.

2.5.4. Crude Fat

The crude fat content was determined while using the Gerber Method and the percent crude fat was determined by directly reading calibrated butyrometer (Badertscher et al., 2007) [20].

2.5.5. Total Carbohydrates

The carbohydrate contents were determined by the difference method given using the expression given below:

$$\text{Carbohydrates } (\%) = 100 - (\text{protein }\% + \text{fat }\% + \text{moisture }\% + \text{ash }\%). \tag{1}$$

2.6. Apparent Viscosity

The yogurt viscosity measurements were carried out at ambient temperature (25 °C) while using Brookfield viscometer (Brookfield Engineering Inc., Model RV-DV II Pro+, New York, NY, USA) with spindle number 63. The disk spindle was selected because of the nature of the yogurt sample, which allows for readings within the instrument sensitivity. According to the manufacturer (Brookfield), samples should be kept in a thermostatically controlled water bath at 25 °C for about 10 min. before measurements. The first measurements were taken 2 min. after the spindle was immersed in the sample to allow for thermal equilibrium to eliminate the effect of immediate time dependence effect. The data were collected every 40 sec and the measurement was duplicated and the average value was considered. The experimental data obtained was converted to shear rate/shear stress and fitted to the power law model.

$$\sigma = k \times \gamma^n \tag{2}$$

where σ is shear stress (Pas), γ is shear rate (s^{-1}), n is the flow behavior index, and k is the consistency index (Pas). The values for the flow behavior index n were obtained from plotting the log of shear stress versus log of shear rate and the slope of the line (if the dependence is sufficiently close to a linear one) is simply equal to the flow index (n).

2.7. Dynamic Rheology, Steady Flow Behavior

The dynamic viscoelastic properties of nonfat yogurt were determined while using TA Instrument Discovery Hybrid Rheometer (HR-1) that was installed with parallel plates system (40 mm in diameter and 50 μm gaps (TA Instruments, New Caste, PA, USA). The samples were transferred to the plate and rheometer was calibrated at 25°C for one minute and excess sample was wiped off with spatula. Dynamic shear data were obtained at frequency sweeps ranging from 0.1–100 rad/s and 0.5% constant strain at 25 °C. Experimental data were collected using the software that was provided by the manufacturer. The storage moduli (G′), loss moduli (G″), and viscosity are the parameters obtained for every run. A strain-sweep experiment was performed to establish that all measurements were done within the linear viscoelastic rage of the experiment. Linear viscoelasticity indicates that the measured parameters are independent of shear strains. To determine the liner viscoelastic range (LVR), a stress sweep was increased from 0.1 to 50.0 Pas at a constant frequency of 0.1 Hz (0.628 rad/s). Frequency sweeps between 0.1 to 10 (rad/s) were implemented within the LVR of the yogurt samples at a constant stress of 1.0 Pas. The frequency range used here is typically used for frequency sweep to ascertain that G′, G″, and η* were within the linear region. The behavior of all the measured materials in this study was in the linear range below 1% strain. Measurement was repeated three times with fresh samples. The relative errors were within the range of ±10%. All of the calculations were performed with the Rheology Advantage Data Analysis software (Version 5.7.0., TA Instruments, New Caste, PA, USA) that was provided by the manufacturer. The viscosity profile was used to provide information regarding the possibility of slippage, where the viscosity profile as a function of shear rate of duplicate runs was plotted in the same graph to ascertain the repeatability and slippage behavior of the material. No slippage was recorded.

2.8. Yogurt Texture Profile Analysis (TPA)

The double compression test was performed while using texture analyzer (TA-XT2 Texture Analyzer, Texture Technologies Crop, Scarsdale, NY, USA) equipped with a software. Plastic cylinder

(45 Perspex Cone, 432–081) is attached to crosshead moving at speed of 70 mm/min. in both upward and downward directions. The yogurt sample was placed on a flat holding plate and the plastic cylinder was inserted 20 mm below the surface of the yogurt sample. The firmness of the yogurt was calculated according to the method that was described by Steffe 1996 [21].

2.9. Whey Separation (Wheying-Off)

The volume of the separating whey (wheying-off) on the surface of the yogurt sample was collected as an indicator for wheying-off (g/100 g yogurt) according to the siphon method. The level of spontaneous whey separation of undisturbed set yogurt was determined while using a siphon drainage method [22]. In this study, a cup of set yogurt was taken out of the cold room (4 °C), weighed, and kept at an angle of approximately 45 °C to allow for whey collection on the side of the cup. A syringe was used to draw the whey from the surface of the sample every 10 s, and the cup of yogurt was weighed again. The syneresis was expressed as the percent weight of the whey over the initial weight of the yogurt samples.

2.10. Sensory Evaluation

The sensory evaluation was carried out by a group of 10 trained sensory assessors. The evaluation of the yogurt samples included the following sensory attributes; appearance, color, texture, aroma, taste, aftertaste, and overall acceptability. The scale used was nine points hedonic scale, where 1 represents dislike extremely, 5 for neither dislike or like, and 9 for extremely like.

2.11. Statistical Analysis

Statistical analysis was performed while using SPSS (SPSS Inc., Hong Kong, China). All of the measurements were done in triplicate and then subjected to analysis of variance (ANOVA) using factorial design. Duncan's multiple range tests at $p \leq 0.05$ was used to compare means while using PASW® Statistics 18 software (SPSS Inc., Hong Kong, China).

3. Results and Discussion

3.1. Shear Viscosity

Three different sets of temperatures were selected, 36–38 °C, 40–42 °C, and 44–46 °C, in order to determine the appropriate incubation temperature. The yogurt made at all three sets of temperature was tested for its texture, viscosity and wheying-off. Unlike the other temperatures, the yogurt that was prepared at 40–42 °C had good texture and viscosity as well as less wheying-off. Therefore, yogurt with or without starch was prepared at 42 °C. In addition, the pH was monitored throughout the incubation time and was found to decrease with the same rate in the presence of starch when compared to the control. Williams et al. (2003) [23] reported that when a modified waxy maize starch was added to yogurt prepared at 43 °C, the product had a grainy texture and high viscosity. The authors were able to improve the texture and viscosity by lowering the fermentation temperature to 35 °C and then increased the fermentation time. In this work, we used common native starch (non-waxy) and, by trying different temperatures, we found that 42 °C was the best.

The proximate composition of the yogurt showed no significant difference between the control and the yogurt/starch blends with respect to the total solids, lipids, and ash. However, lower carbohydrates and higher protein content were recorded for the control. The percent amylose content for potato, sweet potato, corn, chickpea, and Turkish bean starches were 21.8, 22.9, 20.4, 32.2, and 17.5, respectively. Figure 1 shows the effect of the starches on the shear rate and shear stress of the yogurt. By observing the shape of the curves in Figure 1, it is clear that yogurt exhibited shear thinning and yield stress behavior that indicates interactive structure [24], where the entangled molecules of the yogurt start to disentangle and become less resistant to flow which results in shear thinning as indicated by lower viscosity. Yield stress is the applied stress at which irreversible deformation is first observed across the sample. Yield stress studies can help to evaluate the product performance and process-ability and predict product's long-term stability and shelf life. A higher yield stress prevents the material to undergo phase separation or break down and it reduces flow under shipping vibrations. At the beginning of cold storage at, the yield stress of the control and the yogurt/starch blends can be ranked as: Turkish beans > corn > chickpea > potato > sweet potato > control (Figure 1a). The gaps between the profiles (curves) in Figure 1a point to similar effect of the tubers, where the chickpea was in the middle. The highest yield stress of corn and Turkish starches indicates structural stability of the gel. After seven days of storage, the yield stress of potato starch ranked first, Turkish beans and sweet potato were the lowest and the control became firmer as compared to zero storage days (Figure 1b). The gaps between the profiles were further apart than those of at the beginning of cold storage. The sweet potato and Turkish bean were similar, whereas the control was similar to the chickpea. After 15 days, the picture was not very different than seven days, except that chickpea starch exhibited a higher value (Figure 1c). The role of these starches appeared to be time dependent, because Turkish-beans starch exhibited the highest yield stress at zero storage days, while it has the least yield stress after seven days of storage. This could be due to the mechanism of starch gelatinization during the first steps of yogurt making. Another reason could be amylose retrogradation and syneresis (water separation due to amylose retrogradation). Therefore, the use of potato or corn starch for set yogurt to maintain good gel firmness is recommended for longer storage time. Yield stress was the highest at seven storage days than zero or 15 days, which could be attributed to maximum physical interaction between the starch and the casein. Shear viscosity analysis showed Turkish-beans as starch with the highest yield stress and stability at the beginning of cold storage, whereas potato starch with the highest yield stress and stability after 15 days of storage.

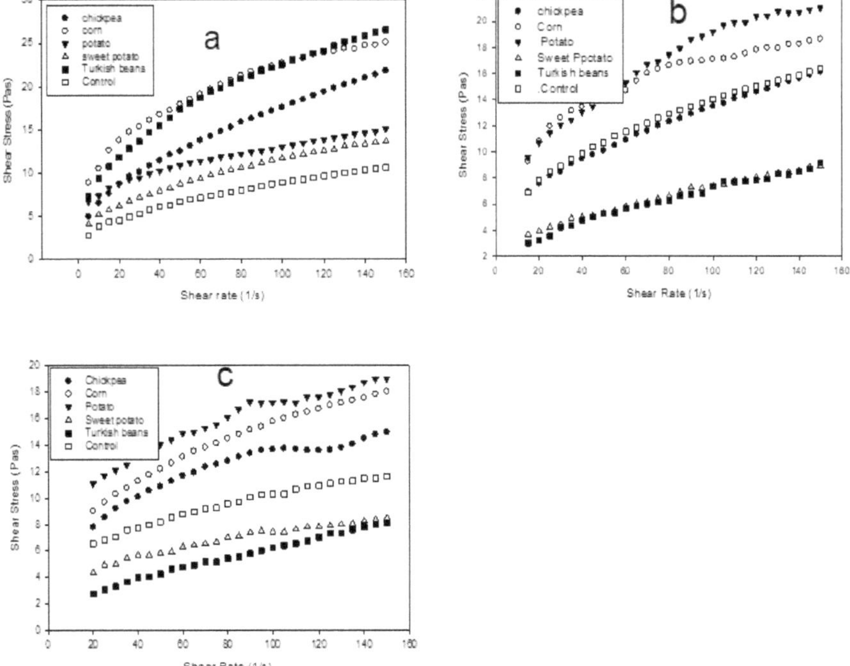

Figure 1. Shear rate shear stress relationship for fresh yoghurt fortified with deferent starches after (**a**) 0 storage days, (**b**) 7 storage days, and (**c**) 15 storage days.

The power law was applied to describe the rheological behavior of the yogurt with different types of starches (Equation (1)). Table 1 a,b present the effect of the starches on the power law parameters (k and n), where the data clearly fit the power law, because of the values of the high value of r^2 (>0.95). The Y-intercept represents the K and the slope represents the n. Regardless of the number of storage days, all the samples (Table 1), including the control behaved as pseudoplastic materials, since $n < 1$, but some are more pseudoplastic than others. Cruz et al. (2012) [25]; Yu et al. (2007) [26] noticed a similar observation. The difference in pseudoplasticity of yogurts with different types of starch as a function of storage time is obvious in Table 1. This is an indication of the structural changes during storage due to interactions between starch molecules, possibly amylose, and the casein network. Unlike the control and the corn starch samples, and by virtue of higher n values, yogurt sample containing Turkish beans starch was the least pseudoplastic, especially at longer storage time. This behavior was also reflected for the yield stress. Potato and sweet potato starches exhibited the most pseudoplasticity after seven days storage. This difference can be attributed to the different amylose content of these starches, where the lowest amount of amylose showed the smallest pseudoplasticity. Conversely, chickpea with the highest amylose content started as the least pseudoplastic, but it become more pseudoplastic as a function of storage time (Table 1). When compared to the control, all of the yogurt/starch blends exhibited higher k values, which indicated higher viscosity and a thicker structure (Table 1). The k values at the beginning of cold storage was similar for all starches, regardless of amylose content, but for longer storage time corn starch yogurt exhibited the thickest texture followed by the tubers. Once again, the Turkish bean starch yogurt was the thinnest of all samples, which could be accredited to the low amylose content. The thickest yogurt after 15 days storage was the one with potato starch, which was also true for the yield stress results. Other researchers attributed the high yogurt viscosity to the high milk solids [27].

Table 1. The consistency index (k) and flow behavior index (n) of yogurt prepared with different starches.

1a	0 Day		7 Days		15 Days	
	k	r^2	k	r^2	k	r^2
Control	0.381 ± 0.01 [b]	0.991 ± 0.02	0.401 ± 0.08 [c]	0.992 ± 0.02	0.471 ± 0.06 [c]	0.991 ± 0.01
Y. Ps	0.423 ± 0.02 [a]	0.989 ± 0.02	0.542 ± 0.02 [b]	0.994 ± 0.00	0.681 ± 0.01 [a]	0.992 ± 0.01
Y. SWPs	0.423 ± 0.03 [a]	0.993 ± 0.01	0.550 ± 0.05 [b]	0.989 ± 0.02	0.235 ± 0.05 [e]	0.995 ± 0.01
Y. Cs	0.423 ± 0.03 [a]	0.998 ± 0.02	0.682 ± 0.02 [a]	0.979 ± 0.02	0.496 ± 0.03 [b]	0.990 ± 0.00
Y. CPs	0.424 ± 0.06 [a]	0.996 ± 0.03	0.381 ± 0.03 [c]	0.990 ± 0.05	0.324 ± 0.02 [d]	0.975 ± 0.03
Y. TBs	0.431 ± 0.08 [a]	0.997 ± 0.07	0.271 ± 0.01 [d]	0.995 ± 0.03	0.101 ± 0.021 [f]	0.990 ± 0.02

1b	0 Day		7 Days		15 Days	
	n	r^2	n	r^2	n	r^2
Control	0.381 ± 0.11 [b,c]	0.991 ± 0.02	0.310 ± 0.01 [d]	0.992 ± 0.02	0.294 ± 0.04 [d]	0.991 ± 0.01
Y. Ps	0.246 ± 0.02 [c]	0.989 ± 0.02	0.365 ± 0.01 [c]	0.994 ± 0.01	0.275 ± 0.05 [e]	0.992 ± 0.01
Y. SWPs	0.370 ± 0.11 [c]	0.993 ± 0.01	0.410 ± 0.03 [b]	0.989 ± 0.02	0.317 ± 0.11 [c]	0.995 ± 0.01
Y. Cs	0.313 ± 0.03 [d]	0.998 ± 0.02	0.280 ± 0.02 [e]	0.979 ± 0.02	0.349 ± 0.22 [b]	0.990 ± 0.02
Y. CPs	0.440 ± 0.01 [a]	0.996 ± 0.03	0.380 ± 0.03 [c]	0.990 ± 0.05	0.304 ± 0.42 [d]	0.975 ± 0.03
Y. TBs	0.390 ± 0.01 [b]	0.997 ± 0.07	0.488 ± 0.22 [a]	0.995 ± 0.03	0.535 ± 0.23 [a]	0.990 ± 0.02

k consistency index (Pa). n = flow behavior index (dimensionless). C = Control, Y. Ps = potato starch, Y. SWPs = Sweet potato starch, Y. Cs = corn starch, Y. CPs = Chickpea starch, Y. TBs = Turkish bean starch. Values followed by different letters in each column are significantly different ($p \leq 0.05$).

3.2. Viscoelastic Properties

The dynamic rheological testing provides useful information regarding the internal structure of the yogurt, which is represented by storage modulus (G′) and loss modulus (G″) that denote the elastic and the viscous behavior, respectively, as well as the complex viscosity η* and the phase angle tan δ. The yogurt samples exhibited a weak gel behavior because G′ > G″ over the range of 0.1 and 50 Pas. A linear viscoelastic region was observed from 0.1 to 10 Pas. Therefore, 1.0 Pas stress was chosen for the frequency sweep test. These data are in agreement of other researchers who reported similar results for yogurt enriched with milk solids and inulin [28]. The longer LVR G′ region of the yogurt samples as a function of stress indicates stress independence (Figure 2). The extent of LVR of the yogurt samples can be ranked according to the type of starch, as follows: potato > corn > chickpea > sweet potato > Turkish beans > control (Figure 2). Therefore, the gel of the control yogurt was the weakest of all and potato starch produced yogurt with the firmest gel (structure), whereas yogurt with Turkish beans starch exhibited the least firm gel. These data are in agreement with the yield stress and the *k* and *n* values reported above.

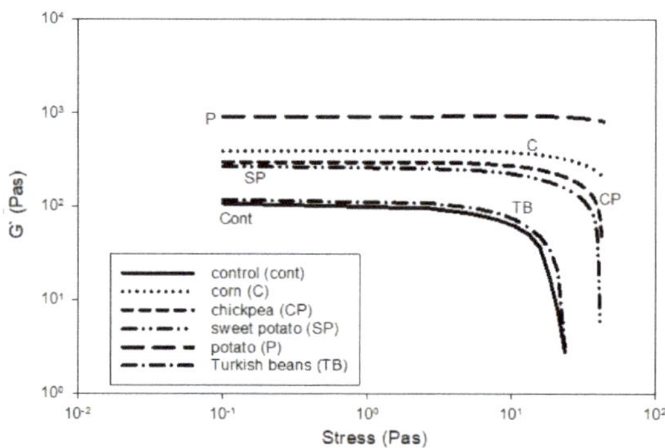

Figure 2. Linear viscoelastic region determination of the control and starch containing yogurt.

Figure 3 showed the oscillation dependency of G′. The G′ of the control stayed the same up to seven storage days, but it increased after that. Chickpea starch was more effective than other starches by maintaining the highest G′; nonetheless, G′ was almost the same after seven or 15 days (Figure 3b). Although on smaller scale, the G′ corn starch yogurt was similar to chickpea starch (Figure 3c), where the gel became firmer as a function of longer storage time. The firmest gel for potato and sweet potato was recorded after seven storage days i.e., high G′, but potato starch yogurt was softer after 15 days (Figure 3e,d). Generally, the G′ of sweet potato starch was higher than potato starch which can be attributed to the higher amylose content of the sweet potato starch. The remaining starches showed higher G′ after 15 days. In addition, the G′ appeared to be unchanged after seven and 15 days, except for the tuber starches (Figure 3b,c). Reports indicated that chemically modified starch can induce positive impact in syneresis and rheological properties as compared with a full-fat yogurt by forming a stable yogurt structure [13]. The texture of low fat yogurt that was prepared with acid treated crosslinked wheat starch was soft, but syneresis was lower [29]. Acetylated-crosslinked starch improved the properties of yogurt more effectively than native starch at 0.5% concentration, in terms of yield stress, consistency, apparent viscosity, thixotropy, and pseudoplasticity [30]. Generally, yogurt with chickpea starch exhibited the highest G′ after 15 days in storage (Figure 3f). Overall, the G′ of the tubers was similar and the legumes starch yogurt were alike, because the yogurt gels became firmer with time. In the presence of legume starches, a large G′ gap between zero and seven days storage was observed when compared to small G′ gap between the same storage days of the tubers. The G′ of the tested yogurts appeared to reach its maximum at seven days storage and remained unchanged, except for tuber starches.

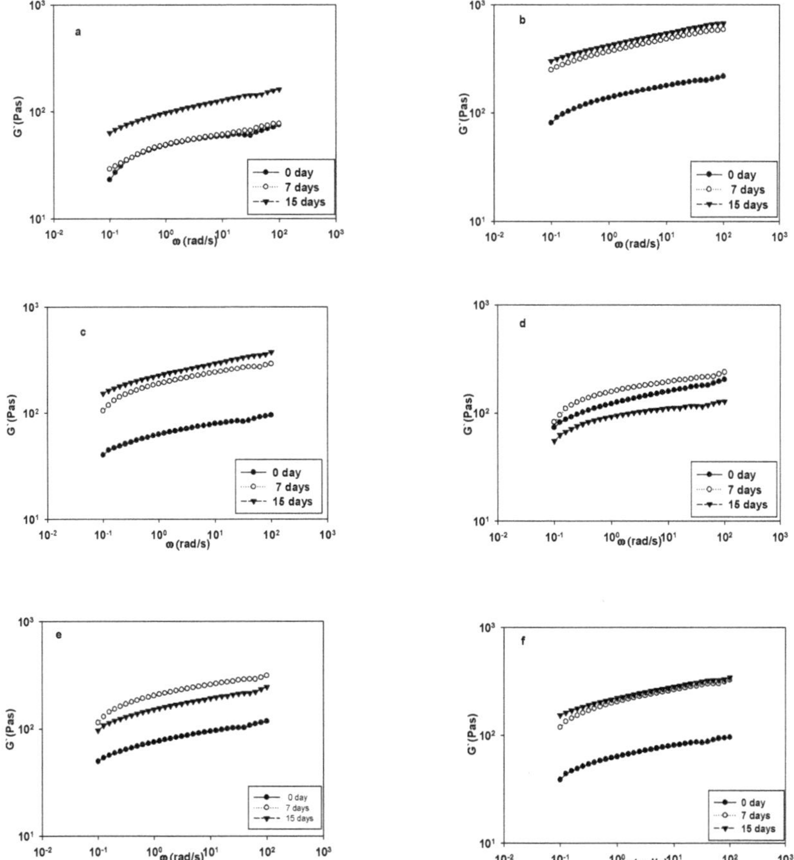

Figure 3. The G′ of yoghurt samples fortified with starch and stored for 0, 7, and 15 days: (**a**) control, (**b**) chickpea, (**c**) corn, (**d**) potato, (**e**) sweet potato, and (**f**) Turkish beans starches.

3.3. Yogurt Texture Profile Analysis

Generally, yogurt texture depends on the physical interaction between casein micelles [31]. Overall, yogurt that was prepared with or without starch exhibited harder texture with storage time, regardless of starch type (Table 2 a–c). Corn and sweet potato starches increased the hardness of the yogurt significantly ($p < 0.05$) after seven days, but the effect of corn starch dropped after 15 days. This could be attributed to weaker corn starch amylose network, which allows for the water to be free after it has been trapped in the gel network. Yogurt containing tuber starches (sweet potato and potato) exhibited the hardest gel through most of the storage time (Table 2). The highest yogurt hardness was recorded for chickpea starch after 15 storage days and corn starch after zero and seven days. Sweet potato starch was the only starch performed well at the three storage times followed by potato and corn. Chickpea starch appeared to significantly increase hardness after 15 days relative to the other starches (Table 2 c), which is in line with the highest G′ for chickpea starch mentioned earlier. This could be accredited to the high amylose content capable of forming stronger network. The cohesiveness is the degree of yogurt deformation during testing. Turkish beans yogurt exhibited the highest cohesiveness, which points to softer gel, as pointed by the hardness data. Adhesiveness is the attractive force between the food and the teeth that can predict the stickiness of the food. The stickiest yogurt is the one with chickpea starch and the control (Table 2). Generally, the higher value of adhesiveness suggests softer yogurt texture; this was not true for the Turkish beans starch, where soft gel faces high adhesiveness (Table 2), but this starch exhibited the highest cohesiveness value of all the starches. Once again, Turkish beans behaved differently from other starches. Gumminess is defined as the product of hardness and cohesiveness that is typical of semisolid foods with a low degree of hardness and a high degree of cohesiveness. Chickpea starch had the highest gumminess value in seven days and after which can be attributed to the high hardness. Texture analysis showed chickpea starch with the hardest yogurt through the storage period, whereas the tubers exhibited the highest hardness until seven days of storage. Therefore, it is recommended to use chickpea starch for longer storage time.

Table 2. Effect of starches addition on texture of yogurt after 0, 7, and 15 days of storage.

	Hardness (g)	Cohesiveness	Adhesiveness (mJ)	Gumminess (g)
2a	0 days storage			
Control	21.000 ± 1.500 [d]	0.351 ± 0.021 [c]	0.471 ± 0.110 [a]	7.351 ± 0.560 [c]
Y. Ps	22.230 ± 0.581 [c,d]	0.382 ± 0.011 [b]	0.300 ± 0.000 [b]	8.561 ± 0.200 [b]
Y. SWPs	27.670 ± 0.580 [b]	0.272 ± 0.022 [e]	0.330 ± 0.060 [b]	7.471 ± 0.300 [c]
Y. Cs	30.330 ± 0.588 [a]	0.321 ± 0.011 [d]	0.332 ± 0.060 [b]	9.810 ± 0.200 [a]
Y. CPs	22.670 ± 0.580 [c]	0.323 ± 0.011 [d]	0.532 ± 0.060 [a]	7.331 ± 0.060 [c]
Y. TBs	15.330 ± 1.530 [e]	0.433 ± 0.011 [a]	0.302 ± 0.002 [b]	6.541 ± 0.510 [d]
2b	7 days storage			
Control	24.000 ± 1.111 [d]	0.461 ± 0.011 [a]	0.330 ± 0.060 [a]	10.660 ± 0.500 [c]
Y. Ps	29.000 ± 1.000 [b]	0.212 ± 0.020 [d]	0.130 ± 0.062 [b]	6.190 ± 0.300 [e]
Y. SWPs	29.331 ± 0.581 [b]	0.320 ± 0.010 [c]	0.171 ± 0.061 [b]	9.391 ± 0.461 [d]
Y. Cs	31.330 ± 0.580 [a]	0.350 ± 0.010 [b]	0.172 ± 0.120 [b]	10.971 ± 0.51 [c]
Y. CPs	29.272 ± 0.462 [b]	0.471 ± 0.010 [a]	0.300 ± 0.000 [a]	13.702 ± 0.152 [a]
Y. TBs	25.330 ± 0.582 [c]	0.472 ± 0.012 [a]	0.330 ± 0.060 [a]	11.992 ± 0.22 [b]
2c	15 days storage			
Control	28.000 ± 1.000 [c]	0.420 ± 0.030 [b]	0.571 ± 0.061 [a]	11.850 ± 0.300 [b]
Y. Ps	30.000 ± 1.000 [a,b,c]	0.340 ± 0.010 [c]	0.220 ± 0.251 [e]	10.300 ± 0.211 [c]
Y. SWPs	30.331 ± 0.581 [a,b]	0.410 ± 0.010 [b]	0.230 ± 0.060 [d,e]	12.441 ± 0.150 [b]
Y. Cs	28.672 ± 0.582 [b,c]	0.360 ± 0.010 [c]	0.301 ± 0.001 [c]	10.222 ± 0.372 [c]
Y. CPs	32.232 ± 2.023 [a]	0.420 ± 0.010 [b]	0.431 ± 0.061 [b]	13.552 ± 1.222 [a]
Y. TBs	21.000 ± 1.111 [d]	0.460 ± 0.010 [a]	0.231 ± 0.061 [d,e]	9.661 ± 0.600 [c]

Mean of three replicates. Values followed by different letters in each column are significantly different ($p \leq 0.05$). C = Control, Y. Ps = potato starch, Y. SWPs = Sweet potato starch, Y. Cs = corn starch, Y. CPs = Chickpea starch, Y. TBs = Turkish bean starch.

3.4. Whey Separation

Wheying-off is a negative characteristic of set yogurt and it is defined as the expulsion of whey from the casein network. Spontaneous wheying-off is the separation of whey without the application of any external force that is associated with unstable gel network. This can be caused by increased rearrangements of the gel matrix or by mechanical damage to the weak gel network. Manufacturers used stabilizers, such as starch, pectin, and gelatin, to prevent wheying-off [11,32,33]. Common causes of wheying-off include high incubation time, disproportionate whey protein to casein ratio, low solid content, and physical mishandling of the product during storage and distribution. Yogurt treated with 1% crosslinked cassava, corn starch, and tapioca starch significantly reduced yogurt syneresis [32]. Altemimi 2018 [33] reported that the reduction in wheying-off by potato starch was significant. The reported data showed no significant difference between 0.25–0.50% potato starch and the control, but 0.75 and 1.00% significantly reduced wheying-off [33]. The results of this work showed a significant reduction ($p < 0.05$) in wheying-off, regardless of starch type for the same storage time (Figure 4). As a function of storage time, wheying-off significantly increased after seven or 15 days, but the rise was significantly less than that of the control. Therefore, wheying-off was more affected during storage and not by starch type. Significant syneresis reduction can be achieved by using modified starches when compared to native starches, as reported by other researchers [34]. With respect to wheying-off, all of the starches significantly reduced wheying-off the same way without any preference. Hence, starch selection should be based on other yogurt attributes, where starch performance was significantly different.

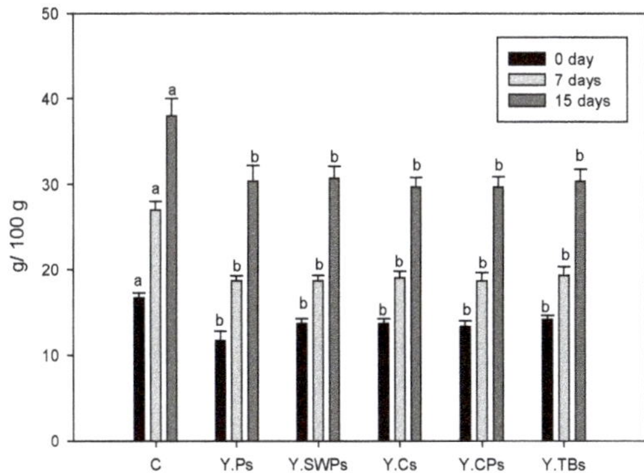

Figure 4. Effect of starches addition on whey separation of yogurt. C = Control, Y. Ps = potato starch, Y. SWPs = Sweet potato starch, Y. Cs = corn starch, Y. CPs = Chickpea starch, Y. TBs = Turkish bean starch Yogurt samples with the different types of starch with the same letter are not significantly different.

3.5. Sensory Evaluation

Yogurt texture is one of the main characteristics that define its sensory quality and affect appearance, mouth-feel, and overall consumer acceptability. In addition, yogurt consistency is perhaps as important as flavor. Acceptable firmness without syneresis is critical for excellent final product. Yogurt texture is usually measured in the cup by using a spoon or directly in the mouth, where the viscosity of the product can represent yogurt consistency on the tongue. The yogurt sample is considered to be thick (viscous) if it stays on the tongue or flows slowly and is swallowed with difficulty, while the visually thick sample can be measured by tilting the spoon and observing how slow the sample will flow. The statistical analysis of this work underlined that there was no significant difference ($p > 0.05$) between the starches with respect to sensory texture, but it was significantly better than the control (Table 3). The sensory viscosity of sweet potato and corn starches were significantly higher than the control, where other starches exhibited similar values. The best flavor was recorded for yogurt with corn starch and chickpea starch, whereas all the starches showed better flavor than the control. The panelist overall acceptability was not significantly different, but the starch containing samples were better accepted, especially yogurt with sweet potato starch. The panel preferred yogurt that was made with sweet potato starch the most.

Table 3. Effect of starches on the sensory evaluation of yogurt.

Treatments	Viscosity	Texture	Creaminess	Flavor	Overall Acceptability
Control	5.40 ± 1.07 [b]	5.11 ± 1.50 [b]	6.30 ± 1.34 [a,b]	6.80 ± 1.39 [b]	5.90 ± 0.99 [a,b]
Y. Ps	6.40 ± 1.18 [a,b]	6.21 ± 1.05 [a]	6.51 ± 0.82 [a]	7.50 ± 0.53 [a,b]	6.30 ± 0.67 [a,b]
Y. SWPs	6.50 ± 1.34 [a]	6.33 ± 0.99 [a]	6.40 ± 1.08 [a,b]	7.70 ± 1.05 [a,b]	6.50 ± 1.08 [a]
Y. Cs	6.70 ± 0.84 [a]	6.13 ± 0.63 [a]	5.91 ± 1.07 [a,b]	7.90 ± 0.32 [a]	6.10 ± 1.10 [a,b]
Y. CPs	6.40 ± 0.84 [a,b]	6.23 ± 0.63 [a]	6.12 ± 0.86 [a,b]	8.10 ± 1.07 [a]	6.20 ± 0.92 [a,b]
Y. TBs	6.40 ± 0.83 [a,b]	6.23 ± 0.63 [a]	6.14 ± 0.84 [a,b]	7.40 ± 0.60 [a,b]	6.21 ± 0.92 [a,b]

C = Control, Y. Ps = potato starch, Y. SWPs = Sweet potato starch, Y. Cs = corn starch, Y. CPs = Chickpea starch, Y. TBs = Turkish bean starch. The scale used was 1 represents dislike extremely, 5 for neither dislike or like and 9 for like extremely. Mean of three replicates. Values followed by different letters in each column are significantly different ($p \leq 0.05$).

4. Conclusions

Stabilizers are important ingredients in manufactured yogurt or other dairy products due to their capacity to improve viscosity and sensory properties, and to decrease wheying-off during storage. The results showed that adding 1% starch gave better results than the control for the syneresis, sensory evaluation, and viscoelastic properties. The outcome of this work showed that tuber starches can behave differently from cereal or legume starches. In addition, the data indicated that the storage time of yogurt could affect starches behavior. The variation in the effects of starches on yogurt quality could be due to the starch granule structure, gelatinization mechanism, and amylose content. The overall results obtained indicate tuber starches or chickpea starch as the most promising stabilizer of non-fat yogurt during storage.

Author Contributions: Conceptualization, A.A.M. and M.S.A.; methodology, S.H. and A.A.Q.; formal analysis, A.S. and M.A.I.; investigation, S.H.; resources, M.A.; data curation, A.A.Q. and A.S.; writing—original draft preparation, A.S.; writing—review and editing, A.A.M. and S.H.; supervision, A.A.M.; project administration, A.A.M.; funding acquisition, A.A.Q. All authors have read and agreed to the published version of the manuscript.

Acknowledgments: The authors extend their appreciation to the Deanship of Scientific Research at King Saud University for funding this work through research group no RGP-114.

Conflicts of Interest: Authors declare no conflict of interest.

References

1. McKinley, M.C. The nutrition and health benefits of yoghurt. *Int. J. Dairy Technol.* **2005**, *58*, 1–12. [CrossRef]

2. Tamime, A.; Robinson, R. *Yoghurt Science and Technology*, 2nd ed.; CRC Press: Boca Raton, FL, USA; New York, NY, USA, 1999.

3. Omer, S. Chemical and Physical Properties of Yoghurt from Khartoum Dairy Product Company (KDPC). Master's Thesis, Faculty of Animal Production, University Khartoum, Khartoum, Sudan, 2003.

4. Barrantes, E.; Tamime, A.Y.; Sword, A.M. Production of low-calorie yogurt using skim milk powder and fatsubstitute. 3. Microbiological and organoleptic qualities. *Milchwissenschaft* **1994**, *49*, 205–208.

5. Brennan, C.S.; Tudorica, C.M. Carbohydrate-based fat replacers in the modification of the rheological, textural and sensory quality of yoghurt: Comparative study of the utilisation of barley beta-glucan, guar gum and inulin. *Int. J. Food Sci. Technol.* **2008**, *43*, 824–833. [CrossRef]

6. FDA. *Low Fat Yogurt, 21 CFR 131.203, Code of Federal Regulations*; US Department of Health and Human Services: Washington, DC, USA, 1996.

7. FDA. *Nonfat Yogurt, 21 CFR 131.206, Code of Federal Regulations*; US Department of Health and Human Services: Washington, DC, USA, 1996.

8. Haque, Z.U.; Ji, T. Cheddar whey processing and source: II. Effect on non-fat ice cream and yoghurt1. *Int. J. Food Sci. Technol.* **2003**, *38*, 463–473. [CrossRef]

9. Mistry, V.; Hassan, H. Manufacture of Nonfat Yogurt from a High Milk Protein Powder. *J. Dairy Sci.* **1992**, *75*, 947–957. [CrossRef]

10. Lucey, J. Formation and Physical Properties of Milk Protein Gels. *J. Dairy Sci.* **2002**, *85*, 281–294. [CrossRef]

11. Nikoofar, E.; Hojjatoleslami, M.; Shariaty, M.A. Surveying the effect of quince seed mucilage as a fat replacer on texture and physicochemical properties of semi fat set yoghurt. *Int. J. Farm. Alli. Sci.* **2013**, *2*, 861–865.

12. Zhao, Q.; Zhao, M.; Yang, B.; Cui, C. Effect of xanthan gum on the physical properties and textural characteristics of whipped cream. *Food Chem.* **2009**, *116*, 624–628. [CrossRef]

13. Alakali, J.; Okonkwo, T.; Iordye, E. Effect of stabilizers on the physico-chemical and sensory attributes of thermized yoghurt. *Afr. J. Biotechnol.* **2008**, *7*, 7–15.

14. Sameen, A.; Khan, M.I.; Sattar, M.U.; Javid, A.; Ayub, A. Quality evaluation of yoghurt stabilized with sweet potato (*Ipomoea batatas*) and taro (*Colocassia esculenta*) starch. *Int. J. Food Allied Sci.* **2016**, *2*, 23–33. [CrossRef]

15. Malik, A.; Faqir, M.; Ayesh, S.; Muhammad, I.; Muhammad, S. Extraction of starch from Water Chestnut (*Trapa bispinosa Roxb*) and its application in yogurt as a stabilizer. *Pak. J. Food Sci.* **2012**, *22*, 209–218.

16. Sit, N.; Misra, S.; Deka, S.C. Physicochemical, functional, textural and colour characteristics of starches isolated from four taro cultivars of North-East India. *Starch Stärke* **2013**, *65*, 1011–1021. [CrossRef]

17. Singh, N.; Sandhu, K.S.; Kaur, M. Characterization of starches separated from Indian chickpea (*Cicer arietinum* L.) cultivars. *J. Food Eng.* **2004**, *63*, 441–449. [CrossRef]

18. Williams, V.R.; Wu, W.-T.; Tsai, H.Y.; Bates, H.G. Rice Starch, Varietal Differences in Amylose Content of Rice Starch. *J. Agric. Food Chem.* **1958**, *6*, 47–48. [CrossRef]

19. AOAC. *Official Methods of Analysis*, 18th ed.; Association of Official Analytical Chemists: Washington, DC, USA, 2007.

20. Badertscher, R.; Berger, T.; Kühn, R. Densitometric determination of the fat content of milk and milk products. *Int. Dairy J.* **2007**, *17*, 20–23. [CrossRef]

21. Steffe, J.F. *Rheological Methods in Food Process Engineering*; Freeman Press: East Lansing, MI, USA, 1996.

22. Amatayakul, T.; Halmos, A.; Sherkat, F.; Shah, N. Physical characteristics of yoghurts made using exopolysaccharide-producing starter cultures and varying casein to whey protein ratios. *Int. Dairy J.* **2006**, *16*, 40–51. [CrossRef]

23. Williams, R.P.; Glagovskaia, O.; Augustin, M.A. Properties of stirred yogurts with added starch: Effects of alterations in fermentation conditions. *Aust. J. Dairy Technol.* **2003**, *58*, 228–232.

24. Butler, F.; McNulty, P. Time dependent rheological characterisation of buttermilk at 5 °C. *J. Food Eng.* **1995**, *25*, 569–580. [CrossRef]

25. Cruz, A.; Castro, W.; Faria, J.; Lollo, P.; Amaya-Farfan, J.; Freitas, M.; Rodrigues, D.; Oliveira, C.; Godoy, H. Probiotic yogurts manufactured with increased glucose oxidase levels: Postacidification, proteolytic patterns, survival of probiotic microorganisms, production of organic acid and aroma compounds. *J. Dairy Sci.* **2012**, *95*, 2261–2269. [CrossRef]

26. Yu, H.-Y.; Wang, L.; McCarthy, K.L. Characterization of yogurts made with milk solids nonfat by rheological behavior and nuclear magnetic resonance spectroscopy. *J. Food Drug Anal.* **2016**, *24*, 804–812. [CrossRef]

27. Özlem, G.; Işıklı, N.D. Effect of fat and non-fat dry matter of milk, and starter type, on the rheological properties of set during the coagulation process. *Int. J. Food Sci. Technol.* **2007**, *42*, 352–358.

28. Paseephol, T.; Small, D.M.; Sherkat, F. Rheology and texture of set yogurt as affected by inulin addition. *J. Texture Stud.* **2008**, *39*, 617–634. [CrossRef]

29. Vercet, A.; Oria, R.; Marquina, P.; Crelier, S.; Lopez-Buesa, P. Rheological Properties of Yoghurt Made with Milk Submitted to Manothermosonication. *J. Agric. Food Chem.* **2002**, *50*, 6165–6171. [CrossRef] [PubMed]

30. Isleten, M.; Karagül-Yüceer, Y. Effects of Dried Dairy Ingredients on Physical and Sensory Properties of Nonfat Yogurt. *J. Dairy Sci.* **2006**, *89*, 2865–2872. [CrossRef]

31. Guven, M.; Yasar, K.; Karaca, O.B.; Hayaloglu, A.A. The effect of inulin as a fat replacer on the quality of set-type low-fat yogurt manufacture. *Int. J. Dairy Technol.* **2005**, *58*, 180–184. [CrossRef]

32. Mwizerwa, H.; Abong, G.O.; Okoth, M.; Ongol, M.; Onyango, C.; Thavarajah, P. Effect of Resistant Cassava Starch on Quality Parameters and Sensory Attributes of Yoghurt. *Curr. Res. Nutr. Food Sci. J.* **2017**, *5*, 353–367. [CrossRef]

33. Altemimi, A.B. Extraction and Optimization of Potato Starch and Its Application as a Stabilizer in Yogurt Manufacturing. *Foods* **2018**, *7*, 14. [CrossRef]

34. Singh, M.; Byars, J.A. Starch-lipid composites in plain set yogurt. *Int. J. Food Sci. Technol.* **2009**, *44*, 106–110. [CrossRef]

Article

Characterization of Defatted Products Obtained from the Parmigiano–Reggiano Manufacturing Chain: Determination of Peptides and Amino Acids Content and Study of the Digestibility and Bioactive Properties

Sofie Buhler [1], **Ylenia Riciputi** [2], **Giuseppe Perretti** [3], **Maria Fiorenza Caboni** [2], **Arnaldo Dossena** [1], **Stefano Sforza** [1] and **Tullia Tedeschi** [1,*]

[1] Department of Food and Drug, University of Parma, Parco Area delle Scienze 27/A, I-43124 Parma, Italy; sofie.buhler@yahoo.it (S.B.); arnaldo.dossena@unipr.it (A.D.); stefano.sforza@unipr.it (S.S.)

[2] Department of Agro-Food Sciences and Technologies, Alma Mater Studiorum-University of Bologna, Piazza Goidanich 60, I-47521 Cesena (FC), Italy; ylenia.riciputi@yahoo.it (Y.R.); maria.caboni@unibo.it (M.F.C.)

[3] Department of Agricultural, Food and Environmental Sciences, University of Perugia, Via S. Costanzo, n.c.n., 06126 Perugia, Italy; giuseppe.perretti@unipg.it

* Correspondence: tullia.tedeschi@unipr.it

Received: 9 January 2020; Accepted: 6 March 2020; Published: 9 March 2020

Abstract: Parmigiano–Reggiano (PR) is a worldwide known Italian, long ripened, hard cheese. Its inclusion in the list of cheeses bearing the protected designation of origin (PDO, EU regulation 510/2006) poses restrictions to its geographic area of production and its technological characteristics. To innovate the Parmigiano–Reggiano (PR) cheese manufacturing chain from the health and nutritional point of view, the output of defatted PR is addressed. Two defatting procedures (Soxhlet, and supercritical CO_2 extraction) were tested, and the obtained products were compared in the composition of their nitrogen fraction, responsible for their nutritional, organoleptic, and bioactive functions. Free amino acids were quantified, and other nitrogen compounds (peptides, proteins, and non-proteolytic aminoacyl derivatives) were identified in the extracts and the mixtures obtained after simulated gastrointestinal digestion. Moreover, antioxidant and angiotensin converting enzyme (ACE) inhibition capacities of the digests were tested. Results obtained from the molecular and biofunctional characterization of the nitrogen fraction, show that both the defatted products keep the same nutritional properties of the whole cheese.

Keywords: defatted cheese; peptides; amino acids; bioactivity; digestibility

---

1. Introduction

Parmigiano–Reggiano (PR) is a worldwide known Italian hard-cooked and slowly-matured cheese. Its inclusion in the list of cheeses bearing the protected designation of origin (PDO, EU regulation 510/2006) poses restrictions to its geographic area of production and its technological characteristics. Parmigiano Reggiano P.D.O. is made from raw cow's milk, partially skimmed by natural surface skimming and produced by cows whose feed consists mainly of forage grown in the area of origin. The milk may not undergo any heat treatment, and no additives may be used [1]. It is a highly concentrated cheese and contains only 30% water and 70% nutrients: 30% protein, 30% fat. The remaining 10% is composed of vitamins, mineral salts, and free amino acids. Free amino acid fraction, which generally increases during cheese maturation, gives an essential contribution to the original taste of the product [2]. The fat content of dairy products is a crucial feature since many studies have been reported concerning their possible adverse effects on human health. Cow milk is, in fact, rich in saturated fatty acids (approximately 70% of total milk fat [3], the intake of which has been associated with an elevated risk of cardiovascular disease (CVD) [4] through increased plasma cholesterol and low-density lipoprotein (LDL) [5]. Moreover, the increased consumption of dietary fat in industrialized nations has been related to some types of cancer and obesity [6]. For those reasons, consumer demand for low-fat food products is steadily growing [7,8].

Lowering fat in cheese manufacturing can be quite easily performed by using totally skimmed milk for production. Still, it is commonly reported that low-fat cheese often suffers from undesirable flavor and texture [9]. Therefore, different approaches have been developed, such as processing techniques, adjunct cultures, use of additives such as fat replacers, and fat removing methods. It has been reported by Whetstine et al. [10] that the flavor release is different in the mouth with reduced-fat products than in full-fat products because hydrophobic flavor molecules have a higher sensory threshold in oil than they do in the water. When fats are extracted from milk before the cheese is made, there are fewer fat molecules for the sensory compounds to bind to, resulting in a lack of flavor reduced-fat cheese. Developing an extract method for defatting cheese is also useful to keep the same manufacturing chain of the new cheeses, following indications for the PDO rules. The use of supercritical fluid extraction technology for the production of low-fat parmesan and cheddar cheese has been previously reported [11].

The objective of this work is the complete characterization of the nitrogen fraction of two samples of defatted Parmigiano–Reggiano obtained with CO_2 supercritical extraction technology and Soxhlet procedure. In particular, the nutritional content of the products was evaluated in terms of peptide and amino acid content. Moreover, digestibility, antioxidant, and ACE-inhibitor properties were tested in vitro.

2. Materials and Methods

2.1. Cheese Defatting

A cheese sample aged 24 months was obtained from the Parmigiano–Reggiano Consortium (Reggio Emilia, Italy). Defatting was performed according to the following procedures. Soxhlet extraction procedure: The fat was extracted from 10 g of grated cheese, according to a slightly modified method of Manirakiza et al. [12], with 60 mL hexane for 2 h in a Soxhlet apparatus. Each extraction was performed in duplicate. Supercritical CO_2 extraction procedure: The SFE (Spe-ed SFETM; Applied Separation Allentown, PA, USA) conditions for total fat removal, according to a slightly modified method of Perretti et al. [13], were 24 g (50-mL vessel) grated cheese; pressure: 60 MPa; extractor temperature: 353 K; CO_2 flow rate: 0.990 g/L; static extraction time: 5 min; dynamic extraction time: 60 min. The extraction was performed five times.

2.2. Fatty Acids Determination by GasChromatrography(GC-FID)

The fatty acids (FA) composition of Parmigiano–Reggiano cheese was determined from the lipid fraction as FAMEs by capillary gas chromatography analysis, as reported by Verardo et al. [14].

To convert fatty acids to their corresponding methyl esters(FAMEs), the method of Christie [15] was used. The phospholipid fraction obtained from TLC separation was transmethylated and the FAMEs were analyzed by capillary gas chromatography using a BPX70 fused silica capillary column (10 m × 0.1 mm i.d., 0.2 μm film thickness; SGE Analytical Science, Ringwood, VIC, Australia). The column was fitted on a GC-2010 Plus gas chromatograph (Shimadzu, Tokyo, Japan). The injector and flame ionization detector temperatures were set at 240 °C. Hydrogen was used as carrier gas at a flow rate of 0.8 mL min^{-1}. The oven temperature was held at 50 °C for 0.2 min, increased from 50 to 175 °C at 120 °C min^{-1}, held at 175 °C for 2 min, increased from 175 to 220 °C at 20 °C min^{-1}, and finally risen from 220 to 250 °C at 50 °C min^{-1}. Samples were injected in split mode (0.4 μL) with a split ratio set at 1:10. Peak identification was accomplished by comparing peak retention times with GLC-463 and FAME 189-19 standard mixtures.

2.3. Cholesterol Analysis by GC-FID

The lipid fraction was saponified at room temperature using 10 mL of methanolic 0.5 M KOH for 18 h in the dark under constant stirring. After saponification, the organic fraction was washed with deionized water, and the unsaponifiable matter was extracted three times with diethyl ether. The organic fractions were pooled together and the solvent was removed under vacuum. The unsaponifiable matter was stored in *n*-hexane/2-propanol (4/1 *v/v*) at −18 °C until GC analysis. GC analyzed the previous extract after silylation. The trimethylsilyl derivatives (TMS) of sterols were analyzed as reported by Guerra et al. [16] by GC-FID. The analysis was carried out in duplicate for each sample.

2.4. Total Nitrogen and Fat Content Determination

For total nitrogen determination, and consequently, the determination of the protein content, the Kjeldahl instrument was used following the standard protocol, according to the European Regulation EC 152/20096. Analyses were performed in duplicate on the whole cheese and on the two defatted products.

The fat content determination was performed by using the Soxhlet method following the standard protocol, according to AOAC Official Method 948.22. [17] Analyses were performed in duplicate on the control sample and on the two defatted products.

2.5. Isolation of the Peptide and Amino Acid Fractions

The defatted samples, in an amount corresponding to 1 g of the treated cheese, were suspended in 4.5 mL of HCl 0.1 M; the dipeptide Phe-Phe was added as an internal standard to a final concentration of 50 µM. Control samples were prepared in the same way using 1 g of untreated cheese. The produced suspensions were homogenized by Ultraturrax (90″, 14,000 rpm) and centrifuged for 30′ at 5 °C; the samples were then filtered on 0.45 µm membranes to obtain clear extracts that were analyzed directly by LC-MS. All samples were prepared in triplicate.

2.6. Simulated Gastrointestinal Digestion

The defatted cheese samples and a whole cheese sample were digested according to the procedure described by Minekus et al. [18], starting with an amount of defatted sample corresponding to 1 g of untreated cheese. Briefly, 1 mL of salivary buffer (12.08 mM KCl, 2.96 mM KH_2PO_4, 10.88 mM $NaHCO_3$, 0.12 mM $MgCl_2$, and 0.048 mM $(NH4)_2CO_3$, 0.6 mM $CaCl_2$), containing 75 U/mL amylase, was added to an amount of sample corresponding to 1 g of untreated cheese. The sample was vortexed and incubated for 2 min at 37 °C. Then, 2 mL of gastric buffer (5.52 mM KCl, 0.72 mM KH_2PO_4, 20 mM $NaHCO_3$, 37.76 mM NaCl, 0.08 mM $MgCl_2$, 0.4 mM $(NH4)_2CO_3$ and 0.06 mM $CaCl_2$), containing 6250 U/mL pepsin, adjusted to pH = 3 with HCl were added. The mixture was vortexed and incubated for 2 h at 37 °C. Finally, 4 mL of intestinal buffer (5.44 mM KCl, 0.64 mM KH_2PO_4, 68 mM $NaHCO_3$, 30.72 mM NaCl, 0.264 mM $MgCl_2$ and 0.24 mM $CaCl_2$), containing 100 U/mL pancreatin and 37.5 mg/mL bile, adjusted to pH = 7 with NaOH, were added (final ratio cheese:digestive fluids 1:7, *w:v*). The sample was vortexed and incubated for 2 h at 37 °C. The digestion was stopped heating the sample at 95 °C for 15 min. After cooling, 125 the samples were centrifuged for 45 min at 4 °C at 3220× *g*.

A digestion blank was obtained by treating 1 mL of H_2O to the same procedure. Each sample was prepared in triplicate. Samples for LC–MS analyses were prepared by adding Phe-Phe as an internal standard to a final concentration of 50 µM, to the clear supernatants.

2.7. LC-MS Amino Acids Quantification

First, 50 μL of the previously obtained extracts or the clear supernatants produced after the simulated digestion process were mixed with 50 μl of 2.5 mM norleucine in 0.1 M HCl; 10 μL of the produced solutions were derivatized with the Waters AccQ-Fluor reagent kit, according to the instructions of the manufacturer. The derivatized samples were analyzed on a UPLC/ESI–MS system (UPLC Acquity Waters with a single quadrupole mass spectrometer Waters Acquity Ultra performance, Waters, Milford, MA, USA) using a RP column (ACQUITY UPLC BEH 300 C18 1.7 μm 2.1 × 150 mm, Waters, Milford, MA, USA) and a gradient elution. Eluent A was H_2O with 0.1% formic acid, eluent B was acetonitrile with 0.1% formic acid; gradient: 0–7 min 100% A, 7–30 min from 100% A to 73.3% A; flow: 0.2 mL/min; column temperature: 35 °C; sample temperature: 18 °C; injection volume: 2 μL. The samples were analyzed in the SIR Scan mode (monitored ions are reported in Table 1); ionization type: positive ions; capillary voltage: 3.2 kV; cone voltage: 30 V; source temperature: 150 °C; desolvation temperature: 300 °C; cone gas flow: 100 l/h; desolvation gas flow: 650 l/h.

Table 1. Total nitrogen and fat determination.

Sample	Protein (%)	Fat (%)
Whole Cheese	35.7 ± 0.1	37.7 ± 0.4
Soxhlet-defatted	50.7 ± 2.2	2.6 ± 0.2
CO_2-defatted	46.2 ± 0.9	4.6 ± 0.1

Calibration curves for the amino acids were obtained as follows: amino acid standard H solution (Thermo Scientific, Rockford, IL, USA) was mixed with an equal volume of norleucine 2.5 mM in HCl 0.1 M; 1:2, 1:4, 1:8, and 1:16 dilutions in H_2O were prepared, and the produced solutions were derivatized and analyzed as previously mentioned for the samples. The calibration curve for tryptophan, asparagine, and glutamine was obtained in the same way, after mixing equal volumes of a solution containing these three amino acids at a concentration of 2.5 mM each, in HCl 0.1 M and a solution containing norleucine 2.5 mM in HCl 0.1 M. In the case of the digests, also the blank digestion was analyzed according to the procedure outlined above. The produced results were subtracted from the values obtained for the digested cheese samples.

2.8. LC–MS Characterization of Peptides and Proteins

The extracts and the digested samples were analyzed on a UPLC/ESI–MS system (UPLC Acquity Waters with a single quadrupole mass spectrometer Waters Acquity Ultraperformance, Waters, Milford, MA, USA) using a RP column (ACQUITY UPLC BEH 300 C18 1.7 μm 2.1 × 150 mm, Waters, Milford, MA, USA) and a gradient elution. Eluent A was H_2O with 0.1% formic acid, eluent B was acetonitrile with 0.1% formic acid; gradient: 0–7 min 100% A, 7–47 min from 100% A to 53.5% A; flow: 0.2 mL/min; column temperature: 35 °C; sample temperature: 18 °C; injection volume: 6 μL. The samples were analyzed in the Full Scan mode; ionization type: positive ions; scan range: 100–2000 m/z; capillary voltage: 3.2 kV; cone voltage: 30 V; source temperature: 150 °C; desolvation temperature: 300 °C; cone gas flow: 100 l/h; desolvation gas flow: 650 l/h. Each produced chromatogram was elaborated, determining characteristic ions, molecular weights, possible in-source collision-induced dissociation (CID) fragments of the main peaks and retention times. In the case of molecular weights higher than 2000 Da, the value was confirmed by the MaxEnt application of MassLynx Software (Waters, Milford, MA, USA). The integration of the area of each compound was performed by QuanLynx software (Waters, Milford, MA, USA), after the extraction of the characteristic XICs. The integrated areas of each species in each extract were determined, and every compound was semi-quantified by dividing by the area of the internal standard (Phe-Phe) in the same sample (Semi-quantification value = Area compound/Area Phe-Phe).

2.9. Antioxidant Capacity of the Digested Samples

The antioxidant capacity was measured by ABTS assay, according to the method proposed by Re et al. [19] with slight modifications. ABTS (2,20-azinobis(3-ethylbenzothiazoline-6-sulfonic acid) and potassium persulfate ($K_2S_2O_8$) were dissolved in H_2O to obtain concentrations of 70 and 2 mM, respectively. The stock solution containing the ABTS radical cation was generated by adding 1% (*v/v*) of K_2S2O8 to the ABTS solution and incubating the mixture overnight in the dark. The ABTS working solution was prepared by diluting the ABTS stock solution 1:25 in phosphate buffer solution (PBS; pH = 7.4), to obtain an absorbance of about 0.7 ± 0.02. The digested samples and the digestion blank were diluted 1:500, and 200 µL of the obtained solutions were added to 1.8 mL of ABTS working solution. The control solution was made by mixing 0.2 mL of PBS with 1.8 mL of work solution. All the samples were incubated for 1 h in the dark before the measurement of the absorbance at 734 nm, performed using a Jasco V-530 UV–vis Spectrophotometer (Jasco Inc, Easton, USA). To express the antioxidant capacity in terms of TEAC (Trolox equivalent antioxidant capacity—mmol of Trolox for mL of digested sample), aqueous solutions containing variable amounts of Trolox were analyzed with the same procedure outlined for the samples. All samples and standards were analyzed in duplicate.

2.10. ACE Inhibition Capacity of the Digested Samples

The percentage of ACE inhibitory activity for the digested samples was determined using the methods of Cushman et al. and Nakamura et al. [20,21] with slight modifications. The following solutions were prepared: sodium borate buffer (0.1 M, NaBB) with NaCl (300 mM), pH 8.3; potassium phosphate buffer (0.01 M, KPB) with NaCl (500 mM), pH 7; 5 mM hippuryl-histidyl-leucine (HHL) in NaBB buffer; and ACE 0.1 U/mL in KPB + 5% glycerol (g/mL). The following samples were prepared: ACEmax = 100 µL of HHL + 40 µL of NaBB + 10 µL of ACE; ACEmin = 100 µL of HHL + 40 µL of digested sample (diluted 1:100) + 10 µL of ACE. After 60 min of incubation at 37 °C, the reaction was quenched adding 125 µL of HCl 1 M. The analysis was performed by HPLC–UV (Alliance 2695 separation, Waters, Milford, MA, USA) with dual λ absorbance detector model 2487 (Waters), using an RP JUPITER C18 column (5 µm, 300 Å, 250 × 2 mm, Phenomenex, Torrance, CA, USA); eluent A: H2O with 0.1% trifluoroacetic acid; eluent B: acetonitrile with 0.1% trifluoroacetic acid; gradient elution: 0–10 min, 100% A; 10–25, linear from 100 to 33% A; flow, 0.2 mL/min; column temperature, 35 °C; injection volume: 10 µL; UV detection, λ = 228. Data analysis was performed with Empower software (Waters, Milford, MA, USA). The integration values for the peak eluting at 22.5′, corresponding to hippuric acid (HA), were used to calculate the inhibition values, according to the following equation: I% = (Area HA(ACEmax) − Area HA (ACEmin))/Area HA (ACEmax) × 100.

2.11. Data Analysis

Statistical analyses were performed using the Statistical Package for the Social Sciences (SPSS for Windows version 11.0; SPSS Inc., Chicago, IL, USA). One-way analyses of variance (ANOVA) were run using Duncan's test.

3. Results and Discussion

3.1. Defatted Products

PR aged 24 months, which is the most consumed aging, was obtained from the Parmigiano–Reggiano Consortium and defatted according to 2 different procedures: Soxhlet and supercritical CO_2 extraction. The obtained yields in the defatted product were comparable, respectively: 65% and 66%.

Defatted products were analyzed for their total nitrogen and fat content, as reported in Table 1.

As shown in Table 1, the total protein amount is higher in the two defatted products, due to an effect of concentration. Considering the residual fat content, both the methods are very efficient if compared with the literature [10,11], since the reduction of fat far exceeds 50%.

At the same time, the quality of the fat extracted with the two extraction methods was assessed by fatty acids gas chromatographic evaluation. The composition did not show significant differences between the two extracts: the primary fatty acid was palmitic acid (28.6%) and oleic acid (22.8%); in both cases, cholesterol represented 2.7% of the fat. This content was comparable to what declared from the nutritional composition by Parmigiano–Reggiano Consortium.

Thus, these results confirm that the methods chosen for lipid extraction were able to recover, efficiently, both total fatty acids and cholesterol.

Concerning both vegetal and animal foodstuffs, these two methodologies are used in different applications [22–24]. Soxhlet is usually more diffuse in lab-scale and economical; on the other hand, supercritical CO_2 is quite expensive but more compatible with food-grade and sustainable procedures.

The nitrogen compounds, peptides, and amino acids were obtained from the defatted products and the whole cheese by acidic extraction, centrifugation, and filtration. The use of HCl 0.1 M as extracting solvent allowed us to achieve the denaturation and precipitation of entire caseins, subsequently eliminated by the filtration step. Thus, the lower molecular weight (MW) nitrogen compounds (species generated by proteolytic processes and whey proteins) remain in solution.

To evaluate possible variations in the composition of the nitrogen fraction due to the defatting procedures, an amount corresponding to 1 g of starting material (untreated cheese, control sample) was used. For each extraction, 1.00, 0.65, and 0.66 g were used, respectively, to obtain the extracts from the whole cheese, and the samples defatted according to the different yields in defatted products.

3.2. Free Amino Acids in the PR Samples

Free amino acids account for up to 25% of the total nitrogen in PR [25], and they were reported to be, among the non-volatile nitrogen molecules, that ones mostly influence its taste [26,27]. Therefore, their preservation during the defatting process was considered as a critical factor for the comparison of the two methodologies used. The content of free amino acids was quantified in the extracts obtained by UPLC–ESI–MS analysis, after derivatization with the Waters AccQ-Fluor reagent kit, to enhance the chromatographic separation. Single ions were monitored for each amino acid (Table S1). The produced quantification results are reported in Figure 1.

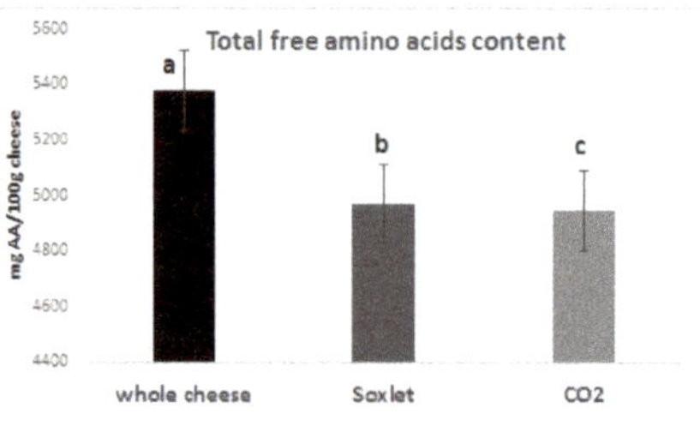

Figure 1. Total amount of free amino acids in the differently treated PR (mg amino acid/100 g PR; means ± SD). Bars carrying different letters are significantly different ($p < 0.05$) from each other.

The composition of the free amino acid fraction was found to be comparable with the data already reported by Careri et al. [28] for PR aged 24 months: with the most abundant amino acids being glutamic acid, lysine, proline, and leucine. However, little loss in the free amino acid content has been observed for both the defatted products, mainly due to the possible co-extraction of them with fat moiety.

3.3. Characterization of Peptides and Proteins in the Aqueous Extracts

A UPLC–MS analysis of the acidic extracts was performed to evaluate their content in nitrogen compounds.

The mass spectra associated with the most intense chromatographic signals were analyzed to obtain the molecular masses of the most abundant nitrogen compounds. Non-proteolytic aminoacyl derivatives (NPADS), peptides, and proteins were identified starting from their deduced MW, by using the software "FindPept" tool [29] and "Proteomics Toolkit" [30] fragment ion calculator of the four bovine milk casein sequences from which the peptides originate. The in-source fragmentation signals were used to discriminate between possible identifications.

The characterized compounds are reported in the supporting information (Table S1). The molecular weights of the found species were spread over a wide range; 12 lactoyl- and γ-glutamyl- derivatives of amino acids, already extensively characterized in cheese and other foods, and casein fragments of up to 81 amino acids were recognized. Moreover, α-lactalbumin and the isoforms A and B of β-lactoglobulin were detectable between 40 and 43 min.

Additionally, 19 of the 28 identified compounds have already been described by Sforza et al. (2012) [31] in a PR sample aged 24 months. It should be underlined that in the previous work, the species having an MW lower than 10 kDa were isolated and concentrated through an ultrafiltration process, resulting in a slightly different peptide mixture composition.

Each identified compound was then semi-quantified against the internal standard Phe-Phe, added during the extraction procedure, according to a method previously reported. [32] The obtained results are displayed in Figure 2.

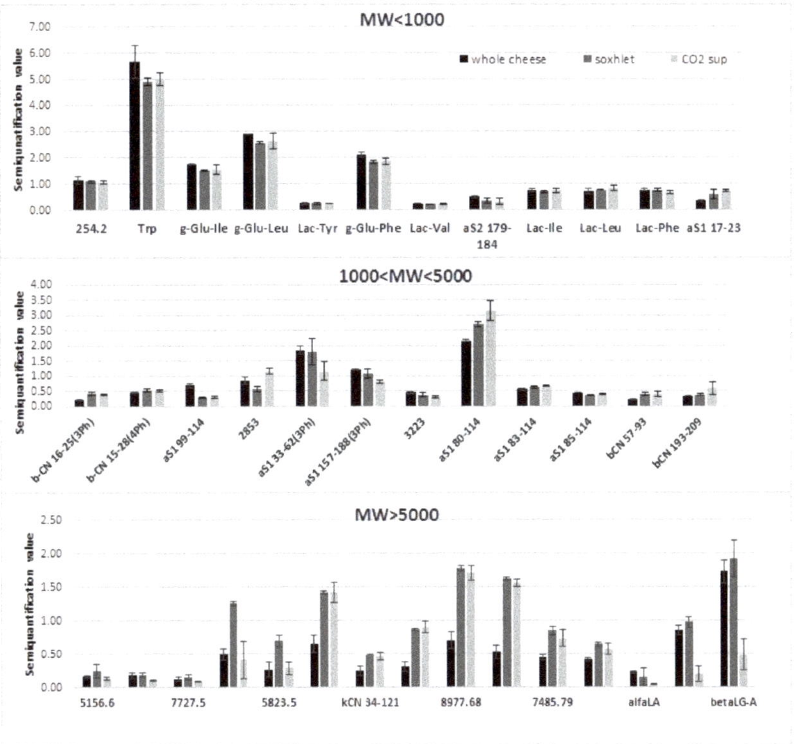

Figure 2. Semi-quantification of the main nitrogen compounds in the different PR extracts (means ± SD).

The data presented in Figure 2 show that the whole cheese, Soxhlet, and supercritical CO_2 samples seem to keep the same trend, except for the peptides group with MW >5000. In this case, higher amounts are observed for both the defatted samples compared with the control one. Maybe the presence of fat in the whole cheese prevented the extraction of high molecular weight peptides from the aqueous medium.

Altogether, data concerning the molecular characterization of the nitrogen fraction of the defatting products are comparable with the ones of the whole cheese, thus indicating that these new products keep the same nutritional features.

3.4. Simulated Gastrointestinal Digestion

Simulated gastrointestinal digestion, according to the procedure described by Minekus et al. [18] was performed on the defatted cheese samples and on the whole cheese, to detect if the digestibility of PR cheese is affected by defatting processes. As already done for the extraction, an amount corresponding to 1 g of starting material (untreated cheese, control sample) was used for each digestion.

Free amino acids were quantified in the digests following the same procedure previously outlined for the extracts.

All the samples show an increase of free amino acid content after digestion, as compared with Figure 1. Interestingly, after digestion, the total free amino acid amount of the whole cheese was lower than that observed for the digested sample obtained from Soxhlet and quite similar to the one of supercritical CO_2 extraction (Figure 3). This behavior suggests that the defatting processes could enhance the digestibility of PR: probably the lipids present in the whole cheese interfere with the digestive mixtures, lowering the efficiency of amino acid release. Defatted samples, instead, allow the peptides and proteins present to be more thoroughly digested. The ability of lipids present in milk to hamper the digestibility of proteins has already been reported in the literature. [33]

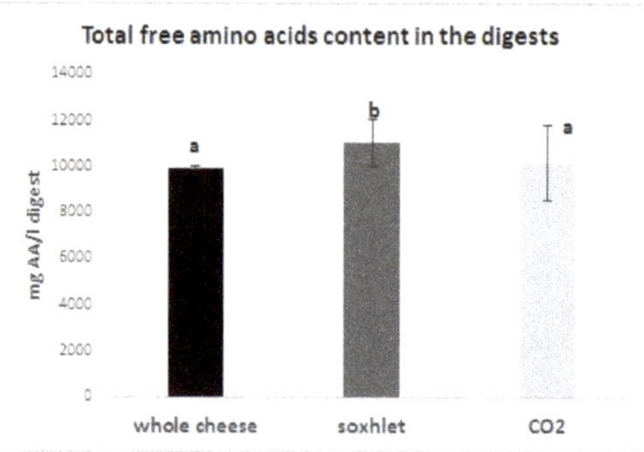

Figure 3. Total amount of free amino acids in the digested samples from differently treated PR (mg amino acid/100 g PR; means ± SD). Bars carrying different letters are significantly different ($p < 0.05$) from each other.

Furthermore, the digestion products were characterized by LC–MS analysis to evaluate their content in peptides. This evaluation is particularly important since the digestion of dairy products has been reported to release many peptides exerting biological activities besides of their nutritional functions [34].

The most abundant nitrogen compounds were identified in the digested samples applying the same procedure outlined previously for the acidic extracts; the characterized species are reported in the supporting information (Table S2). Twenty-eight of the 64 identified compounds were already described by Bottari et al. (2017) [35] in the digests obtained from PR aged 16, 24, and 36 months.

Interestingly, none of the nitrogen compounds previously characterized in the acidic extracts were detected in significant amounts in the digested samples, except for NPADS. Compared with the species found in the acidic extracts, the molecular weight distribution was, as expected, considerably shifted towards lower values, with the most extended peptide being constituted by 23 amino acids.

3.5. Antioxidant Activity of the Digested Samples

To evaluate the effect of the defatting process on the antioxidant properties of the PR samples after ingestion, the antioxidant capacity of the digested cheese samples was measured by a standard ABTS test [19]. The results were expressed in terms of Trolox equivalents (TEAC) related to the amount of digested PR (mmols of Trolox per digested sample in an amount corresponding to 100 g of the whole PR) and are reported in Figure 4.

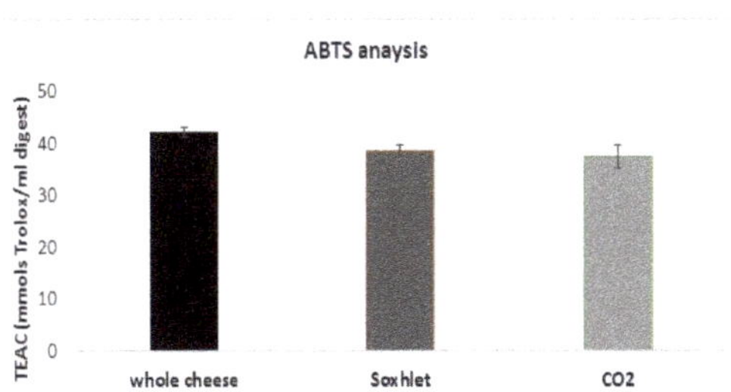

Figure 4. Antioxidant capacity (TEAC; mmols Trolox/mL digested sample) of the digested cheese samples. No significant differences are observed between the bars.

As highlighted by the results reported in Figure 4, the data confirmed that, even after a digestion process, the biases introduced in the nitrogen fraction by the extraction procedure were still very visible. No statistically significant difference was found, although the antioxidant capacity of the defatted cheeses was slightly higher if compared to the one observed for the whole cheese.

From the detailed molecular composition of the nitrogen fraction determined, it is possible to speculate which species can be responsible for the scavenging activity observed. Among the total free amino acids, tyrosine is known to be the main one responsible for the antioxidant capacity of PR [36].

Known antioxidant descriptors for peptides include the presence of redox-active amino acids (Tyr, Trp, Met, Cys, and His) [37] and a limit on their length (the most active peptides are 3–10 amino acids long) [38]. Seven peptides identified fulfilled the first requirement.

Moreover, research in the BIOPEP [39] database also revealed peptides d13, d24, d27, and d43 to be included in (or completely covering) sequences known as antioxidants.

3.6. ACE inhibition Capacity of the Digested Samples

The angiotensin I-converting enzyme belongs to the renin–angiotensin system. It acts in the regulation of blood pressure, by cleaving the C-terminal dipeptide portion of angiotensin I and producing the vasoconstrictor angiotensin II. Some peptides, which possess a specific C-terminal sequence, can bind to ACE, and they represent competitive substrates for the enzyme mentioned above. Peptides released during proteolytic processes in food, such as cheese ripening, share these features and, therefore, they have been studied and assessed for their in vitro ACE-inhibition potential [40,41].

To also evaluate the influence of the defatting processes on the presence of ACE inhibiting compounds in the digested samples, the ACE inhibition capacity was measured according to a procedure reported in the literature [20,21]. The results were expressed as percentages of ACE inhibition and are reported in Figure 5.

Again, the effect of the defatting procedures was very evident also after the digestion. The samples treated with Soxhlet and supercritical CO_2 show the most intense ACE inhibiting activity. This is likely a consequence of the selective enrichment of the fraction in short peptides, which are known to be the most effective, generally speaking, in ACE inhibition.

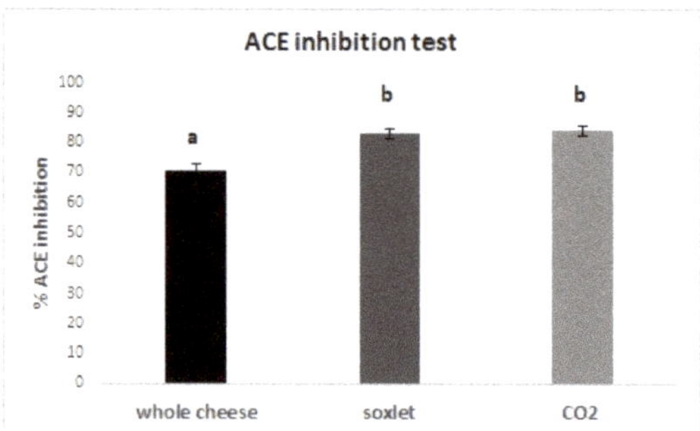

Figure 5. Percentages of ACE inhibition of the digested cheese samples. Bars carrying different letters are significantly different ($p < 0.05$) from each other.

Exploring the peptide sequences identified in the digested samples (Table S2), it was found that some residues (MPFPK from peptide d36, FVAP from peptide d46, and YPFGPIPN from peptide d55) were already found to be part of potential ACE inhibitory peptides in Cheddar and Gouda cheeses [40,41].

It should be noted that most of the studies reported in literature directly investigated the ACE inhibitory activity of peptides in aqueous extracts of cheeses and not of peptides generated by their gastrointestinal digestion.

Further research in the BIOPEP database [35] also revealed peptides d4, d13, d24, d36, d46, and d55 to be included in (or completely covering) sequences known as ACE inhibitors.

4. Conclusions

Based on the total protein content, free amino acids, and other nitrogen compounds (NPADs, peptides, and proteins) identified in the acidic extracts and in the gastrointestinal digestion products, it is evident how Soxhlet and supercritical CO_2 extraction can be applied to obtain defatted products from PR cheese since they allow the minimization of the loss of nitrogen compounds. Moreover, the tested samples also revealed that the defatting procedures seem to enhance the digestibility of PR. More studies will also be performed to test other PR cheeses with different aging, and the consumer attitudes towards flavor, texture, and general acceptance of the defatting product would also be considered as part of future studies.

Testing some bio-functionalities after simulated digestion indicated that the biases introduced by the defatting procedures remain evident also in the biological activity, even if the digestion itself completely changes the composition of the peptide fraction. The measurement of bio-functional properties of the digests highlighted that Soxhlet and supercritical CO_2 extraction allowed us to obtain products with antioxidant capacities, which were comparable to the ones of whole cheese. At the same time, the ACE inhibitory activities were found to be even enhanced by these two defatting procedures. The measured activities were related to the nitrogen species contained in the digested samples. More studies will be performed to better elucidate and identify the peptide sequences involved in these bioactive mechanisms.

Supplementary Materials: The following are available online at http://www.mdpi.com/2304-8158/9/3/310/s1. Table S1: Characteristic ions, chromatographic retention time and potential identification for the main nitrogen compounds found in the PR extracts; Table S2: Characteristic ions, chromatographic retention time and potential identification for the main nitrogen compounds found in the PR extracts in PR digests.

Author Contributions: All authors have read and agree to the published version of the manuscript. Conceptualization, T.T., S.S. and A.D.; methodology, S.B. G.P.; formal analysis, S.B., Y.R., G.P.; data curation, S.B.; writing—original draft preparation, S.B., T.T.; writing—review and editing, T.T. S.S.; supervision, T.T., A.D. M.F.C.; funding acquisition, M.F.C., A.D.

Funding: This research was funded by the Emilia-Romagna Region, Italy (PARENT project POR-FESR 2014–2020).

Acknowledgments: Parmigiano-Reggiano Consurtium is thanked for providing PR cheese samples.

Conflicts of Interest: The authors declare no conflict of interest.

References

1. Parmigiano-Reggiano Single Document. Available online: https://www.parmigianoreggiano.com/ (accessed on 20 December 2019).
2. Parmigiano Reggiano Official Website. Available online: https://www.parmigianoreggiano.com/made/nutritional_features/default.aspx (accessed on 20 December 2019).
3. MacGibbon, A.H.K.; Taylor, M.W. Composition and structure of bovine milk lipids. In *Advanced Dairy Chemistry*; Fox, P.F., McSweeney, P.L.H., Eds.; Springer: New York, NY, USA, 2006; pp. 1–42.
4. Gordon, T. The diet-heart idea: Outline of a history. *Am. J. Epidemiol.* **1988**, *127*, 220–225. [CrossRef] [PubMed]
5. Mensink, R.P.; Katan, M.B. Effect of dietary fatty acids on serum lipids and lipoproteins. A meta-analysis of 27 trials. *Arterioscler. Thromb.* **1992**, *12*, 911–919. [CrossRef] [PubMed]
6. Lin, C.-T.; Yen, S.T. Knowledge of Dietary Fats among US Consumers. *J. Am. Diet. Assoc.* **2010**, *110*, 613–618.
7. Undeland, I.; Härröd, M.; Lingnert, H. Comparison between methods using low-toxicity solvents for the extraction of lipids from herring (*Clupea harengus*). *Food Chem.* **1998**, *61*, 355–365. [CrossRef]
8. Biondo, P.B.F.; dos Santos, V.J.; Montanher, P.F.; de OS Junior, O.; Matsushita, M.; Almeida, V.C.; Visentainer, J.V. A new method for lipid extraction using low toxicity solvents developed for canola (*Brassica napus* L.) and soybean (*Glycine max* L. Merrill) seeds. *Anal. Methods* **2015**, *7*, 9773–9778. [CrossRef]
9. Ashraf Gaber, M. Low-Fat Cheese: A Modern Demand. *Int. J. Dairy Sci.* **2015**, *10*, 249–265.
10. Carunchia Whetstine, M.E.; Drake, M.A.; Nelson, B.K.; Barbano, D.M. Flavor profiles of full-fat and reduced-fat cheese and cheese fat made from aged cheddar with the fat removed using a novel process. *J. Dairy Sci.* **2006**, *89*, 505–517. [CrossRef]
11. Yee, J.L.; Khalil, H.; Jimenez -Flores, R. Flavor partition and fat reduction in cheese by supercritical fluid extraction: processing variables. *Lait* **2007**, *87*, 269–285. [CrossRef]
12. Manirakiza, P.; Covaci, A.; Schepens, P. Comparative Study on Total Lipid Determination using Soxhlet, Roese-Gottlieb, Bligh & Dyer, and Modified Bligh & Dyer Extraction Methods. *J. Food Compost. Anal.* **2001**, *14*, 93–100.
13. Perretti, G.; Marconi, O.; Montanari, L.; Fantozzi, P. Rapid determination of total fats and fat-soluble vitamins in Parmigiano cheese and salami by SFE. *LWT* **2004**, *37*, 87–92. [CrossRef]
14. Verardo, V.; Gómez-Caravaca, A.M.; Gori, A.; Losi, G.; Caboni, M.F. Bioactive lipids in the butter production chain from Parmigiano Reggiano cheese area. *J. Sci. Food Agric.* **2013**, *93*, 3625–3633. [CrossRef] [PubMed]
15. Christie, W.W. A simple procedure for rapid transmethylation of glycerolipids and cholesteryl esters. *J. Lipid Res.* **1982**, *23*, 1072–1075. [PubMed]
16. Guerra, E.; Downey, E.; O'Mahony, J.A.; Caboni, M.F.; O'Shea, C.A.; Ryan, A.C.; Kelly, A.L. Influence of duration of gestation on fatty acid profiles of human milk. *Eur. J. Lipid Sci. Technol.* **2016**, *118*, 1775–1787. [CrossRef]
17. AOAC International Website. Available online: https://www.aoac.org (accessed on 3 March 2020).
18. Minekus, M.; Alminger, M.; Alvito, P.; Ballance, S.; Bohn; Bourlieu, C.; Carriere, F.; Boutrou, R.; Corredig, M.; Dupont, D.; et al. A standardized static in vitro digestion method suitable for food—An international consensus. *Food Funct.* **2014**, *5*, 1113–1124. [CrossRef] [PubMed]
19. Re, R.; Pellegrini, N.; Proteggente, A.; Pannala, A.; Yang, M.; Rice-Evans, C. Antioxidant capacity applying an improved ABTS radical cation decolorization assay. *Free Radic. Biol. Med.* **1999**, *26*, 1231–1237. [CrossRef]
20. Cushman, D.W.; Cheung, H.S. Spectrophotometric assay properties of the angiotensin-converting enzyme of rabbit lung. *Biochem. Pharmacol.* **1971**, *20*, 1637–1648. [CrossRef]
21. Nakamura, Y.; Yamamoto, N.; Sakai, K.; Okubo, A.; Yamazaki, S.; Takano, T. Purification and characterization of angiotensin I-converting enzyme inhibitors from sour milk. *J. Dairy Sci.* **1995**, *78*, 777–783. [CrossRef]
22. Vibha, D.; Shabina, K. Comparative study of different extraction processes for hemp (*Cannabis sativa*) seed oil considering physical, chemical and industrial-scale economic aspects. *J. Clean. Prod.* **2019**, *207*, 645–657.

23. Kyo-Yeon, L.M.; Shafiur, R.; Ah-Na, K.; Khalid, G.; Sung-Won, K.; Jiyeon, C.; William, L.K.; Sung-Gil, C. Quality characteristics and storage stability of low-fat tofu prepared with defatted soy flours treated by supercritical-CO2 and hexane. *LWT* **2019**, *100*, 237–243.

24. Rahman, M.S.; Seo, J.K.; Choi, S.G.; Gul, K.; Yang, H.S. Physicochemical characteristics and microbial safety of defatted bovine heart and its lipid extracted with supercritical-CO_2 and solvent extraction. *LWT* **2018**, *95*, 355–371. [CrossRef]

25. *Istituto Nazionale di Ricerca per gli Alimenti e la Nutrizione (INRAN) Dossier Il Parmigiano Reggiano un Prodotto Naturalmente Funzionale*; National Research Institute on Food and Nutrition: Rome, Italy, 2008.

26. Hillmann, H.; Hofmann, T. Quantitation of Key Tastants and Reengineering the Taste of Parmesan Cheese. *J. Agric. Food Chem.* **2016**, *64*, 1794–1805. [CrossRef] [PubMed]

27. Virgili, R.; Parolari, G.; Bolzoni, L.; Barbieri, G.; Mangia, A.; Careri, M.; Spagnoli, S.; Panari, G.; Zannoni, M. Sensory-Chemical Relationships in Parmigiano-Reggiano Cheese. *Food Sci. Technol.* **1994**, *27*, 491–495. [CrossRef]

28. Careri, M.; Spagnoli, S.; Panari, G.; Zannoni, M.; Barbieri, G. Chemical parameters of the non-volatile fraction of ripened Parmigiano-Reggiano cheese. *Int. Dairy J.* **1996**, *6*, 147–155. [CrossRef]

29. Indpep Tool. Available online: http://web.expasy.org/findpept/?_ga=1.4745075.1565211071.1480347985 (accessed on 20 December 2019).

30. Proteomics Toolkit. Available online: http://db.systemsbiology.net:8080/proteomicsToolkit/FragIonServlet.html (accessed on 20 December 2019).

31. Sforza, S.; Cavatorta, V.; Lambertini, F.; Galaverna, G.; Dossena, A.; Marchelli, R. Cheese peptidomics: a detailed study on the evolution of the oligopeptide fraction in Parmigiano-Reggiano cheese from curd to 24 months of aging. *J. Dairy Sci.* **2012**, *95*, 3514–3526. [CrossRef]

32. Sforza, S.; Ferroni, L.; Galaverna, G.; Dossena, A.; Marchelli, R. Extraction, semi-quantification, and fast on-line identification of oligopeptides in Grana Padano cheese by HPLC-MS. *J. Agric. Food Chem.* **2003**, *51*, 2130–2135. [CrossRef]

33. Mackie, A.; Macierzanka, A. Colloidal aspects of protein digestion. *Curr. Opin. Colloid. Interface Sci.* **2010**, *15*, 102–108. [CrossRef]

34. Korhonen, H. Milk-derived bioactive peptides: from science to applications. *J. Funct. Foods* **2009**, *1*, 177–187. [CrossRef]

35. Bottari, B.; Quartieri, A.; Prandi, B.; Raimondi, S.; Leonardi, A.; Rossi, M.; Ulrici, A.; Gatti, M.; Sforza, S.; Nocetti, M.; et al. Characterization of the peptide fraction from digested Parmigiano Reggiano cheese and its effect on growth of lactobacilli and bifidobacteria. *Int. J. Food Microbiol.* **2017**, *16*, 32–41. [CrossRef]

36. Bottesini, C.; Paolella, S.; Lambertini, F.; Galaverna, G.; Tedeschi, T.; Dossena, A.; Marchelli, R.; Sforza, S. Antioxidant capacity of water soluble extracts from Parmigiano-Reggiano cheese. *Int. J. Food Sci. Nutr.* **2013**, *64*, 953–958. [CrossRef]

37. Sarmadi, B.H.; Ismail, A. Antioxidative peptides from food proteins: A review. *Peptides* **2010**, *31*, 1949–1956. [CrossRef]

38. Kitts, D.D.; Weiler, K. Bioactive proteins and peptides from food sources. Applications of bioprocesses used in isolation and recovery. *Curr. Pharm. Des.* **2003**, *9*, 1309–1323. [CrossRef] [PubMed]

39. BIOPEP Database. Available online: http://www.uwm.edu.pl/biochemia/index.php/pl/biopep, (accessed on 20 December 2019).

40. Lu, Y.; Govindasamy-Lucey, S.; Lucey, J.A. Angiotensin-I-converting enzyme-inhibitory peptides in commercial Wisconsin Cheddar cheeses of different ages. *J. Dairy Sci.* **2016**, *99*, 41–52. [CrossRef] [PubMed]

41. Saito, T.; Nakamura, T.; Kitazawa, H.; Kawai, Y.; Itoh, T. Isolation and Structural Analysis of Antihypertensive Peptides That Exist Naturally in Gouda Cheese. *J. Dairy Sci.* **2000**, *83*, 1434–1440. [CrossRef]

Article

Characterisation of Formaggella della Valle di Scalve Cheese Produced from Cows Reared in Valley Floor Stall or in Mountain Pasture: Fatty Acids Profile and Sensory Properties

Paolo Formaggioni [1], Massimo Malacarne [1,*], Piero Franceschi [1,*], Valentina Zucchelli [2], Michele Faccia [3], Giovanna Battelli [4], Milena Brasca [4] and Andrea Summer [1]

[1] Department of Veterinary Science, University of Parma, Via del Taglio 10, I-43126 Parma, Italy; paolo.formaggioni@unipr.it (P.F.); andrea.summer@unipr.it (A.S.)

[2] Veterinary Freelance, Via Monte Grappa 7, I-24020 Vilminore di Scalve (BG), Italy; valentina.zucchelli78@gmail.com

[3] Department of Soil, Plant and Food Sciences, University of Bari, Via Amendola 165/A, 70125 Bari, Italy; michele.faccia@uniba.it

[4] Consiglio Nazionale delle Ricerche, Istituto di Scienze delle Produzioni Alimentari, UT di Milano, Via Celoria 2, IT-20133 Milano, Italy; giovanna.battelli@ispa.cnr.it (G.B.); milena.brasca@ispa.cnr.it (M.B.)

* Correspondence: massimo.malacarne@unipr.it (M.M.); piero.franceschi@unipr.it (P.F.); Tel.: +39-0521032617 (P.F.)

Received: 5 February 2020; Accepted: 16 March 2020; Published: 26 March 2020

Abstract: An important problem in mountain areas is the abandonment of pasture. This trend can be combated by the valorisation of typical dairy products, such as "Formaggella della Valle di Scalve", a semi-cooked traditional cheese made from whole milk in a mountain area in Italy. The aim of the present research was to compare the fatty acid (FA) profile and the sensory properties of this cheese as manufactured under different conditions: i) from the milk of cows grazing on mountain or valley pasture or fed indoors; ii) from the milk of cows fed hay or fed silage. In the first case, five cheesemaking trials were conducted during two years for each of the following situations: mountain pasture (A); pasture at the bottom of the valley (P) (about 1000m asl); stall (S). In the second case, three cheesemaking trials were conducted for each of the following situations: cows fed silage (I); cows fed hay (F). S cheese was richer in medium-chain FAs, while long-chain FAs were higher in P and A cheeses. On the other hand, long chain fatty acids (LCFA) were more abundant in P and A cheeses than in S. In general, MUFA, PUFA and, consequently, total unsaturated FA (UFA), were significantly higher in the P and A cheeses than S (UFA: 36.55 and 38.34, respectively, vs. 31.13; $p < 0.001$), while SFA showed higher values in S (68.85 vs. 63.41 and 61.68 in P and A, respectively; $p < 0.001$). Conjugated linoleic acid isomers (CLA) were more represented in the P and A samples (1.86 in P and 1.52 in A, vs. 0.80 in S; $p < 0.001$); Omega 3 fatty acids, and in particular α-linolenic acid, were more abundant in P than in S cheese. In winter, the I sample (silage) presented higher percentages of myristic (C14), myristoleic (C14:1) and omega 6 acids, whereas F cheese (hay) contained higher concentrations of CLA. The triangular test of sensory analysis showed that, in general, F cheeses were judged as "sweeter" than I, with aromatic profiles characterized by higher content of 2- butanol and ethyl capronate.

Keywords: cheese quality; mountain cheese; fatty acid profile; volatile organic compounds; sensory properties

---

1. Introduction

One of the main problems in the mountain areas in Italy is the abandonment of marginal portions of the territory, widely utilised in the past for traditional activities (agriculture, livestock, forestry). In the specific case of Valle di Scalve (Lombardy Region, province of Bergamo), the causes of the phenomenon are to be found in the social, cultural and economic changes that affected mountain areas in the second post-war period. Agricultural and forestry activities, which were previously the basis of the self-consumption economy characterizing the existence of each family, were progressively supplanted first by industrial activity and successively by that of the tertiary sector. The national and local government continue to put in place interventions aiming to encourage the resumption of active management of the mountain pastures in this area. A common strategy is focusing on the valorisation of typical dairy products, such as Formaggella della Valle di Scalve cheese. This product is made from raw whole milk added to with a natural milk-starter (milk of the day before, left to ferment overnight) and coagulated with commercial rennet. It is a semi-cooked cheese weighing about 2 kg, ripened from 20 days up to 3 months. It is cylindrical in shape, with about a 20 cm diameter and a 6 cm heel, and bears the impressed brand (a stylised bear, the symbol of the Valley) of the cooperative company at which it is manufactured (Latteria Sociale Montana). The average gross composition at 30 days' ripening is shown in Table 1.

Table 1. Average gross composition of Formaggella della Valle di Scalve cheese at 30 days' ripening produced during the winter and summer periods.

		Gross Composition on 100 of Cheese		Gross Composition on 100g of Dry Matter	
		Mean	SD	Mean	SD
Winter:					
Moisture	g/100g	42.48	3.56	-	-
Protein	g/100g	25.60	0.80	44.64	3.16
Fat	g/100g	25.58	4.26	44.25	4.98
Ash	g/100g	4.04	0.16	7.05	0.71
NaCl	g/100g	1.61	0.21	2.82	0.50
Summer:					
Moisture	g/100g	44.63	2.88	-	-
Protein	g/100g	24.35	1.70	44.08	3.91
Fat	g/100g	27.24	2.95	49.12	1.27
Ash	g/100g	3.59	0.63	6.49	1.08
NaCl	g/100g	1.47	0.34	2,65	0.59

Formaggella della Valle di Scalve is manufactured according to traditional technology, descending from the multigenerational experience of the local cheese makers. Milk is produced in valley floor stall (indoor) during winter and on mountain pastures during summer, even though some farmers remain at the valley stall for the whole year. Even though the traditional breeding system of dairy cattle involves the use of high-altitude pastures during summer, the share of farms keeping animals in the stable all year long with a hay-based diet is increasing [1]. Pastures are still widespread in mountain environments and, apart from their valuable contribution to livestock production, they contribute the promotion of local tourism, biodiversity conservation, maintenance of landscapes and mitigation of pollution [2]. In this context, the role of mountain farms in preserving ecological equilibrium and historic traditions, and in maintaining the landscape for protection against hydro-geological disorder, has been widely recognized [2].

Unfortunately, mountain farms are poorly competitive and have high production costs due to the unfavourable natural conditions. Therefore, promoting mountain products is an appropriate strategy for generating wealth and preventing the abandoning of mountain farming. Consumers' perception of the mountain food products is good since they are commonly linked to concepts such as tastiness, healthiness, wholesomeness, animal welfare, history and local culture [2,3]. Mountain cheeses often have unique sensory properties that are deeply connected to the environmental conditions of milk production [4]. In particular, milk native microbiota deriving from the farm and cheesemaking environment are known to play a key role in determining the organoleptic characteristics [5]; for this reason, raw milk cheeses tend to develop a more intense flavour than pasteurised milk cheeses [3,6].

Besides the role of microbiota, animal feeding can modify the composition and rennet-coagulation properties of the milk [7,8], and the characteristics of the cheese [9]. In particular, the nature of the pasture is responsible for changes in cheese colour and aroma according to the type of forage fed to animals, as frequently reported by farmers and cheese makers. Recent studies conducted on various cheese varieties [10] have demonstrated the influence of forage preservation (e.g., grass versus hay [11], silage versus hay [12]) and the botanical composition of dry forage [13]. Other studies have been conducted to identify, quantify, and understand the effects of forage type (e.g., maize silage, hay, grass silage, pasture) [14,15]. According to several studies, the sensory properties of the cheeses made from "pasture milk" can reflect the characteristics of the fresh pasture plants [16,17] and of the natural microbiota from both the animal and the environment [18]. For certain, the diet of dairy cattle influences the colour [19], in connection with seasonal variations of the concentration of β-carotene in milk [20]. In general, consumption of green forage increases the β-carotene content of milk and cheese [21]. There is also evidence about the desirable effect of grasslands on the distinctive flavour of dairy products [19]. Kilcawley et al. [22], in a review, have recently analysed the factors influencing the sensory characteristics of bovine milk and cheese from grass-based versus non-grass-based milk production systems. All the sensory properties—odour, aroma, taste and texture—of cheeses made with pasture-derived milk may be different from those made with dry forage milk [1,23]. Compared with hay, pasture is known to provide a less firm, and more creamy, texture [14,15,24]. These characteristics are also modified during ripening because of the different enzymatic processes, including proteolysis and lipolysis, which also play a key role in aroma development [1]. Carpino et al. [19], found that the cheeses from pasture-fed cows had a significantly more floral and greener odour, as measured by quantitative descriptive analysis.

The type of forage deeply influences the milk fatty acid (FA) composition, both as to the type and the proportion of the compounds [25]. Fresh green forage (having higher levels of polyunsaturated fatty acids relative to silage) allows the production of milk with higher content of these type of FA [26] (especially *cis*-9-C18:1, *trans*-11-C18:1, *cis*-9,*trans*-11-CLA, and C18:3 n-3; CLA: Conjugated Linoleic Acids) and poorer in saturated FA (in particular C16:0, C14:0, and C12:0 [27,28]) compared with cows fed with preserved forages and concentrates [14]. Many studies have been carried out on this topic, particularly with regard to the effects of different diets on the content of unsaturated long-chain fatty acids, such as linolenic acid and conjugated linoleic acid [26,29–31]. These acids are claimed to have positive effects on human health, and a number of reports are available in the literature about ways of naturally increasing, through cow feeding, their content in milk [31]. In this regard, attention has been paid to the level of polyunsaturated fatty acids (PUFA), especially conjugated linoleic acid (CLA), in milk fat [29]. Although most of the CLA of milk fat are synthesized in the mammary gland, some of them represent an intermediate of ruminal biohydrogenation of linoleic acid [32]. For this reason, pasture feeding, yielding higher levels of PUFA, causes a higher CLA content of milk fat compared to feeding conserved forage [29].

As a consequence of this, consumers have a growing interest in mountain dairy products, which can be considered as functional foods in the proper sense. In fact, from a nutritional point of view, pasture-derived dairy products seem particularly interesting [1]: the FA profile is favourable to human health, being characterised by a higher content of PUFA. In effect, the haymaking process, i.e., mechanical damage to plant tissues combined with air access, causes extensive oxidation of PUFA [1]. In addition, CLA can beneficially modulate several important physiological functions [33,34]. Moreover, the traditional feeding system can also transmit biologically active molecules (such as β-carotene), which have beneficial effects on human health as powerful antioxidants protecting against oxidative stress [35].

The aim of the study was to compare the quality characteristics of Formaggella della Valle di Scalve produced under different conditions: a) in summer, from milk of cows grazing on mountain pasture, in valley floor pasture (1000 m quota) or fed indoors with a hay-based diet; b) in winter, from milk of cows fed hay or fed silage.

2. Materials and Methods

2.1. Experimental Design and Sampling Procedure

Summer phase. Five cheesemaking trials were conducted in summer over two years for each of the following experimental situations: stall (S) (indoor, at valley floor, about 1000 m asl; cows were fed permanent meadow hay and grass); pasture (P) (at the bottom of the valley, about 1000 m asl); mountain pasture (A). This latter is defined as the grazing of cattle in the high mountains, from altitudes greater than 1000 m up to 2300–2500 m, carried out from late May to mid-September.

Winter phase. Three cheesemaking trials were conducted for each of the following situations: cows reared in stall fed permanent meadow hay silage (I); cows reared in stall fed permanent meadow hay without silage (F). The winter tests began after the adaptation phase, when cows returned to the barn from the mountain pasture or pasture at 1000 m.

In both cases, the groups of cattle were, as far as possible, homogeneous by breed, lactation stage and deliveries of the cows. Just before cheesemaking, a milk sample was taken from the vat of each experimental case and, successively, the technological parameters of processing were monitored and noted on specially prepared technical sheets. This made it possible to verify that the cheesemaking operations performed during the trials were the same.

At the end of processing, the cheese wheels produced in each experiment were marked and left to ripen in the ripening cell of the social dairy, as normally done for the commercial product. The wheels were taken for analyses at 30 days of ripening.

2.2. Analyses

2.2.1. Gross Composition and Fatty Acids

For each cheese sample, fat content was determined by the volumetric Gerber method [36], as described by Formaggioni et al. [37]. Moreover, according to Malacarne et al. [38], crude protein by Kjeldahl, [39], salt (NaCl) by potentiometric titration method [40], ash after muffle calcination at 530 °C, and dry matter after oven drying at 102 °C [41], were determined, from which moisture was calculated.

The determination of total fatty acids and CLA was carried out by gas chromatography, by means of two Association of Official Analytical Chemists (AOAC) standard methods [42,43], after derivatisation of the sample according to AOAC standard [44]. Briefly, lipids were extracted from ground cheese (approximately 1 g) with ether–heptane mixture (rate in volume 1:1) after the addition of sodium sulphate and of 2.5 M sulphuric acid 2.5 M. The fatty acids were separated and determined by capillary chromatography with a Carlo Erba GC 6000 Vega Series gas chromatograph (Carlo Erba Instruments, Milan, Italy), equipped with a fused silica capillary column coated with Polyethylene Glycol (30 m × 0.25 mm) Supelco, SP TM 2330 (Sigma-Aldrich Corporation, Saint Louis, MO, USA). The operating conditions were as follows: programmed column temperature from 45 up to 175 °C (13 °C/minutes); then up to 215 °C (4 °C/minutes); stationary at 215 °C for 35 min; injector and detector temperature: 250 °C; carrier gas: hydrogen; column pressure: 175 kPa. The compounds were identified by standard co-injection and relative retention time to FAME 13:0 (internal standard).

Short chain fatty acids (SCFA) were calculated by adding the compounds from C4 to C11; medium chain fatty acids (MCFA) were calculated by adding the compounds from C12 to C16; long chain fatty acids (LCFA) were calculated by adding the compounds from C17 to C24. The other classes were odd fatty acids (OCFA), saturated fatty acids (SFA), monounsaturated fatty acids (MUFA) and polyunsaturated fatty acids (PUFA). Unsaturated fatty acids (UFA) derived from the sum of MUFA and PUFA.

2.2.2. Sensory Analysis

The sensory analysis was only performed on the samples from the winter experimentation. The Triangular test method was adopted [45], which is useful to compare two samples with even small differences. It consisted in presenting three samples to the tasters, two of which were identical: the taster was asked to identify the different samples, and the choice was forced. The comparison between the number of correct and incorrect choices provided the test result. The panel was composed of 40 members who were experts in mountain dairy products.

In order to avoid errors due to the tasting sequence, the sampling plan provided for random distribution of any of the six possible combinations to each taster. The samples were identified by different codes for each judge, using three-digit numbers generated by an algorithm randomly. The chance that the taster has to guess the different sample is 33%, regardless of the perceivability of the difference. The data used were the number of total judgments, the number of correct choices, and the level of significance required for the test. The number of correct choices was compared with the significant theoretical minimum number in two-entry probability tables. If the number of correct choices is greater than or equal to the theoretical one, then we can conclude that there is a significant difference between the two types of samples tested at a certain level of significance.

2.2.3. Volatile Organic Compound (VOC) analysis

Cheese volatile organic compound (VOC) analysis was determined by means of a Head-Space Solid Phase Micro Extraction module (Combi-Pal automated sampler CTC Analytics, Zwingen, Switzerland) equipped with DVB/CAR/PDMS 50/30 μm fiber (Supelco, Bellefonte in Centre, 16823, Pennsylvania, USA) and coupled to a gas chromatograph-mass spectrometer (6890N/5973N Agilent Technologies, Inc., Wilmington, DE, USA). Two and a half grams of cheese were put in a 20 mL head-space glass bottle sealed with a PTFE-silicone septum. Operating conditions: 10 min at 50 °C at 250 rpm; fiber exposition, at 50 °C for 40 min; desorption directly in the injection port of the GC at 260 °C for 10 min. GC column: Zebron ZB-WAX plus (60m × 0.25mm × 10.25 μm, Phenomenex, Torrance, CA, USA) with the following separation conditions: carrier gas helium, in constant flow mode at 1.2 mL/min; oven temperature at 45 °C (10 min), then rising to 150 °C at 5 °C/min, then to 222 °C at 12 °C/min (13 min). Acquisition was performed in electronic impact mode. Transfer line at 280 °C, ion source at 230 °C, quadrupole at 150 °C. Further details are described in Battelli et al. [46]. The mass range used was 39–220 amu. The volatile compounds were identified using the Wiley 7n-1 MS library of Agilent MSD ChemStation®software (Agilent Technologies Inc.). Confirmation of the identity of the volatile compounds was achieved by comparing the GC retention indices and mass spectra of individual components with those of authentic reference compounds injected under the same operating conditions. Data are expressed as arbitrary units of the area of the quant ion of each compound.

2.3. Statistical Analysis

The significance of the differences between seasons and between cheese-factories was tested by analysis of variance, using the general linear model procedure of SPSS (IBM SPSS Statics 23, Armonk, New York, NY, USA), according to the following univariate model:

$$Y_i = \mu + \beta_i + \varepsilon_i, \tag{1}$$

where: Y_i = dependent variable; μ = overall mean; β_i = effect considered in each comparison: housing (summer period only), three levels: stall (S), valley pasture (P), mountain pasture (A); cow feeding (winter only), two levels: silage (I), hay (F); ε_i = residual error.

3. Results and Discussion

3.1. Gross Composition of Formaggella della Valle di Scalve

Table 1 shows the gross composition of Formaggella della Valle di Scalve cheese produced during the winter and summer periods. Data are expressed both for 100 g of cheese and 100 g of dry matter. The cheese was produced with full cream milk, and had a fat-to-protein ratio of approximately 1:1. Since, during the summer period, milk contains more fat (data not shown in the table), the cheese produced in this season showed a higher fat content relative to the winter period, expressed both in 100 g of cheese and 100 g of dry matter. Moisture was just below 45%, and the NaCl content was rather low (around 1.5 g/100g of the cheese).

3.2. Fatty Acids

3.2.1. Comparison among Cheese Produced during Summer Season

The results for cheese fatty acid profile reflected the data already found for the corresponding milks (data not shown), confirming the differences in the fatty acid profile. These are not only due to the seasonal variation [47,48], but are closely linked to feed factors. In fact, various authors [30,49] have reported a complete transfer of FA from milk to cheese; hence, their profile largely reflected the raw milk from which they were made. Additionally, Dhiman et al. [50] reported that the FA profile of cow milk was not altered in cheese processing, even when cows were fed different diets.

Table 2 shows the percentage distribution of the fatty acids of the cheese at 30 days. In general, the cheese fat produced from stall milk (S) was richer in medium chain fatty acids (MCFA) relative to that produced from valley (P) and mountain pasture (A). In particular, the most abundant MCFA was palmitic acid (C16), which was higher in S relative to P and A; the second most represented was myristic acid (C14), which was higher in S relative to P, and in P relative to A; finally, C12 (lauric), C12:1 (lauroleic) and C13 (tridecanoic) were higher in S than in the P and A samples. On the other hand, long chain fatty acids (LCFA) were higher in P and A cheeses than in S ones. Oleic acid, the most represented LCFA, had a lower concentration in S and P samples relative to A; stearic acid, was lower in S than in P and A cheeses. Cheese from mountain pasture (A) also had more arachidic (C20) acid, while linolenic acid (C18:3) was higher in valley pasture (P).

In general, MUFA, PUFA and, consequently, total unsaturated (UFA), were significantly higher in the cheeses produced from both valley (P) and mountain pasture (A) milk (UFA: 31.13 S vs. 36.55 P and 38.34 A; $p < 0.001$), while SFA showed higher values in stall milk cheeses.

Carafa et al. [51], for traditional mountain cheeses, reported similar values for myristic acid (9.4 g/100g FA), but lower values for palmitic acid (22.6 g/100g FA), and stearic acid (8.1 g/100g FA), while oleic acid (22.5 g/100g FA) is consistent with value registered for P, but lower than of A, in the present research. In the literature, differences in milk FA profiles from cows fed pasture or hay and concentrates are well known. Various authors [49,52] have found higher amounts of short-, and medium-chain FAs, C16:0 and total saturated FAs (SFA) in indoor cows' milk.

In contrast, all the mono-unsaturated FAs (MUFA), all the conjugated linoleic acid (CLA) isomers, and the total poly-unsaturated FAs (PUFAs), were higher in pasture-based milk and cheese. The mechanism is well known: compared with indoor cows' milk, the content of de novo (<16 C) FA slightly decreased in pasture-based milk. A general reduction of de novo FA occurs in milk from grazing cows because high levels of dietary PUFA from pasture can compete with de novo fatty acids for esterification in the mammary gland, and thus determine a decrease in the synthesis of short- and medium-chain fatty acids [49].

Moreover, a negative energy balance may occur in lactating cows on pasture, thus reducing the synthesis of short- and medium-chain FA in the mammary gland [27,53]. In agreement with several studies, i.e., those reported by Collomb et al. [54] in a review, FA of mixed origin (C16:0 and C16:1) were lower in pasture-based milk, while C18 was higher in pasture milk.

Table 2. Percent distribution of FA of 30 days ripening cheese (comparison between milk produced in stall, valley pasture and mountain pasture). Mean ± SD. a,b,c, differ for $p < 0.05$. NS, $p > 0.05$; * $p \leq 0.05$; ** $p \leq 0.01$; *** $p \leq 0.001$.

		Stall		Valley Pasture		Mountain Pasture		p
Number of Observations		**5**		**5**		**5**		
C4—Butiric	%	2.28 ± 1.15		2.25± 0.48		2.77 ± 0.92		NS
C6—Capronic	%	1.74 ± 0.86		1.55 ± 0.26		1.49 ± 0.32		NS
C8—Caprilic	%	1.26 ± 0.52		1.04 ± 0.14		0.86 ± 0.14		NS
C10—Caprinic	%	3.28 ± 0.97	b	2.43 ± 0.30	ab	1.83 ± 0.17	a	*
C10:1—Decenoic	%	0.32 ± 0.12		0.26 ± 0.05		0.21 ± 0.04		NS
C12—Lauric	%	3.92 ± 0.69	b	2.93 ± 0.33	a	2.30 ± 0.14	a	**
C12:1—Lauroleic	%	0.12 ± 0.02	b	0.09 ± 0.02	a	0.07 ± 0.02	a	*
C13—Tridecanoic	%	0.14 ± 0.02	b	0.10 ± 0.02	a	0.10 ± 0.03	ab	*
C14—Myristic	%	12.79 ± 0.95	c	10.91 ± 0.73	b	9.57 ± 0.33	a	***
C14:1—Myristoleic	%	1.02 ± 0.12	b	0.90 ± 0.14	ab	0.80 ± 0.06	a	*
C15—Pentadecanoic	%	1.31 ± 0.06		1.37 ± 0.03		1.28 ± 0.12		NS
C16—Palmitic	%	30.58 ± 1.86	b	27.44 ± 1.12	a	26.64 ± 2.29	a	*
C16:1—Palmitoleic	%	1.29 ± 0.04		1.28 ± 0.07		1.26 ± 0.11		NS
C17—Eptadecanoic	%	0.82 ± 0.04		0.85 ± 0.07		0.87 ± 0.08		NS
C17:1—Eptadecenoic	%	0.25 ± 0.05		0.25 ± 0.04		0.28 ± 0.03		NS
C18—Stearic	%	10.31 ± 1.21	a	12.15 ± 0.98	b	13.57 ± 0.53	b	**
C18:1—Elaidic t-9	%	0.50 ± 0.13		0.70 ± 0.26		0.73 ± 0.33		NS
C18:1—Vaccenic t-11	%	1.89 ± 0.40	a	4.29 ± 0.48	b	3.49 ± 0.78	b	***
C18:1—Oleic	%	20.63 ± 2.54	a	22.70 ± 0.80	a	26.03 ± 1.25	b	**
C18:1—Vaccenic c-11	%	0.41 ± 0.07		0.41 ± 0.05		0.46 ± 0.05		NS
C18:2—Linoelaidic t-6	%	0.20 ± 0.02		0.23 ± 0.01		0.21 ± 0.03		NS
C18:2—Linoleic	%	2.62 ± 0.43		2.26 ± 0.17		2.19 ± 0.69		NS
C20—Arachidic	%	0.18 ± 0.03	a	0.20 ± 0.02	a	0.25 ± 0.05	b	*
C20:1—Eicosenoic	%	0.06 ± 0.01	ab	0.04 ± 0.01	a	0.07 ± 0.01	b	*
C18:3—α-linolenic	%	0.78 ± 0.07	a	1.02 ± 0.07	b	0.83 ± 0.21	a	*
C18:2—Rumenic c9,t11 CLA	%	0.78 ± 0.26	a	1.82 ± 0.18	b	1.49 ± 0.26	b	***
C18:2—t10-c12 CLA	%	0.02 ± 0.03		0.04 ± 0.06		0.03 ± 0.03		NS
SCFA—short chain FA	%	8.97 ± 3.49		7.58 ± 1.04		7.18 ± 1.44		NS
MCFA—middle chain FA	%	51.17 ± 2.26	b	45.01 ± 2.02	a	42.02 ± 2.52	a	***
LCFA—long chain FA	%	39.84 ± 4.10	a	47.37 ± 2.29	b	50.82 ± 1.30	b	***
OCFA—odd FA	%	2.69 ± 0.20		2.67 ± 0.23		2.61 ± 0.23		NS
MUFA—monounsaturated FA	%	26.47 ± 2.52	a	30.93 ± 0.99	b	33.38 ± 1.57	b	***
PUFA—polyunsaturated FA	%	4.66 ± 0.42	a	5.62 ± 0.25	b	4.96 ± 0.78	ab	*
UFA—unsaturated FA	%	31.13 ± 2.86	a	36.55 ± 1.11	b	38.34 ± 0.94	b	***
SFA—saturated FA	%	68.85 ± 2.84	b	63.41 ± 1.10	a	61.68 ± 0.91	a	***
CLA (conjug. linoleic acids)	%	0.80 ± 0.28	a	1.86 ± 0.21	b	1.52 ± 0.27	b	***
Omega 3 FA	%	0.83 ± 0.12	a	1.09 ± 0.14	b	0.86 ± 0.22	ab	*
Omega 6 FA	%	2.83 ± 0.49		2.44 ± 0.21		2.37 ± 0.77		NS

Esposito et al. [49] observed that, consequently, unsaturated FA content increased to the detriment of SFA content in the cheese. This can result in a more favourable FA composition in cheese from grazing cows. Therefore, a wider use of pasture may be promoted in order to accentuate this positive feature; this is important from the point of view of promoting mountain dairy products. From a nutritional and health point of view, it is important to reduce the level of saturated fatty acids relative to unsaturated ones in the diet [55].

In the present study, conjugated linoleic acids, and in particular rumenic acid (and its precursor vaccenic acid t-11) were more represented in the P and A samples. Values for total CLAs were in agreement with the data reported by Carafa et al. [51] for traditional mountain cheeses (but the concentration of vaccenic acid of 2.4 g/100g FA was higher than that found in the present experiment). A partial agreement with the results of Revello Chion et al. [30] was also found: these authors registered, in Toma Piemontese cheese (a semi-hard cheese like Formaggella della Valle di Scalve), average values of 2.09% and 0.81% total FA in summer and winter, respectively. Seasonal variation is mainly due to animal feeding: in fact, as in the present study, also in the research of Revello Chion et al. [30] cows in winter were housed indoors and fed hay and concentrates, whereas in summer were fed natural pasture. Relative to the values found by these authors, the CLA values registered in the present study for valley pasture, and particularly those of mountain pasture (A), were perceptibly lower, and this is probably due to the different compositions of the pastures. Our values are in agreement with those reported by Lobos-Ortega et al. [56] in a study on the different CLA value of cheeses from three different species.

The variation of CLA content in milk has been associated with several factors, such as diet of cows, breed and stage of lactation, but diet is the most important variation factor. The difference between cows reared indoor and cows fed pasture has been reported by various authors: all studies indicated higher CLA values in the case of pasture [52,57–60]. Di Grigoli et al. [61], in their study on Caciocavallo, an Italian ripened pasta filata cheese, reported that the utilization of pasture, relative to hay-based feeding, almost doubled the level of C18:2 *trans*-10,*cis*-12 (CLA). Coppa et al. [57] also reported that all the conjugated linoleic acid (CLA) isomers were lower in the milk of cows reared indoors. Esposito et al. [49] and Chilliard et al. [27] found an increase in the unsaturated fraction and CLA contents in dairy products derived from grazing systems. Prandini et al. [62], in Grana Padano cheese, reported that CLA concentration was higher in mountain cheese (mainly Trentin Grana, a particular Grana Padano for which animals are bred in the mountains with altitude > 800 m, and silages are forbidden) (6.52 and 9.47 mg/g fat in spring and summer respectively) compared with lowland cheese (5.29 and 5.75 mg/g fat in spring and summer respectively), with values increasing from spring to summer in all analyzed samples, but especially in mountain Grana Padano.

As described by many authors, [62–64], in milk from ruminants there is an amount of rumenic acid (*cis*-9, *trans*-11 CLA) due to incomplete biohydrogenation of polyunsaturated fatty acids (PUFA), specially linoleic and α-linolenic acids, in the rumen, and then from desaturation of vaccenic acid in the mammary gland via Δ9-desaturase [65]. In particular, up to 99% of α-linolenic acid and linoleic acid consumed by cows is biohydrogenated in the rumen, with vaccenic acid being a main derivative [66]. Vaccenic acid is then partly desaturated to CLA in the mammary gland, explaining the elevated CLA content in milk from predominantly-grass-fed cows [63]. In fact, when fed indoors, cows have no access to grass, which is rich in the linoleic and α-linolenic acids involved in the synthesis of CLA [59]; a higher dietary intake of α-linolenic acid should consequently lead to a higher amount of CLA in milk.

In fact, the value of α-linolenic acid in the present study is higher in P cheese, also relative to A cheese; moreover, although the difference is not significant, our CLA value for valley pasture tends to be higher than that for mountain (alpine) pasture. This result is supported by Leiber et al. [63] who reported that lowland pasture contained almost twice as much α-linolenic acid, resulting in 25% more CLA in milk relative to cows grazing alpine pasture. Various authors have suggested that fresh grass promotes the synthesis of CLA through a greater activity of Δ^9-desaturase in the udder [67,68]. Moreover, the high concentrations of soluble fiber and fermentable sugars in fresh grass can create an environment in the gastrointestinal tract of ruminants, without lowering the pH, that is favorable for the growth of the bacteria responsible for synthesizing CLA and the production of vaccenic acid [50].

Omega 3 fatty acids, and in particular α-linolenic acid, were in the present research more represented in cheese from valley pasture (P) milk relative to cheese from stall (S) milk (A milk being in an intermediate position). In contrast, Omega 6 fatty acid percentages were not statistically different among the three experimental situations. The higher content of Omega 3 FA in the cheese from pasture is confirmed by Cozzi et al. [69], who in mountain pasture cheese found a lower content of short chain fatty acids and C16:0, and an increase in unsaturated fatty acids and Omega 3 fatty acids. It is well-known that milk fatty acid composition is affected by the cow's feeding plan and particularly by the amount and the quality of the forage included in the diet [70]. Additionally, Zeppa et al. [71], in the production of Ossolano cheese (an Italian semi-hard cheese), in cows fed exclusively green forage in mountain pastures, found an increase in Omega 3 fatty acids and a decrease in the n-6/n-3 ratio, with a very important nutritional effect. In fact, Omega 3 polyunsaturated fatty acids are recognized as playing an essential role in human health and are particularly important for the proper functioning of the brain, the heart and the retina of the eye [72].

3.2.2. Comparison between Cheeses Produced in Winter Season

The fatty acid profile (Table 3) showed numerous differences between the two experimental groups. In particular, the cheeses in case I (silage) presented higher percentages of myristic (C14) and myristoleic (C14:1) acids and a lower percentage of C15, C17, C17:1, stearic acid and arachidic acid. A statistically significant difference was found between these two cases for SFA and MUFA, as well as for MCFA and LCFA, while total PUFA were higher for I than for F. Moreover, for I, the percentages of omega 6 were higher (in particular for linoleic acid) and the percentages of CLA (rumenic acid) and also of its metabolic precursor, vaccenic acid, were lower.

Table 3. Percentage distribution of the fatty acids of the cheese at 30 days ripening relative to the comparison of the winter period between milk produced from cows fed silage and milk produced from cows fed hay. Mean ± SD. a, b, differ for $p < 0.05$. NS, $p > 0.05$; * $p \leq 0.05$; ** $p \leq 0.01$.

		Silage (I)		Hay (F)		p
Number of Observations		3		3		
C4—Butiric	%	2.85 ± 1.33		2.27 ± 0.20		NS
C6—Capronic	%	2.19 ± 1.14		1.58 ± 0.36		NS
C8—Caprilic	%	1.52 ± 0.72		1.04 ± 0.25		NS
C10—Caprinic	%	3.75 ± 1.35		2.53 ± 0.44		NS
C10:1—Decenoic	%	0.41 ± 0.14	b	0.26 ± 0.01	a	*
C12—Lauric	%	4.33 ± 0.90		3.21 ± 0.24		NS
C12:1—Lauroleic	%	0.14 ± 0.01		0.11 ± 0.03		NS
C13—Tridecanoic	%	0.14 ± 0.01		0.12 ± 0.02		NS
C14—Myristic	%	13.22 ± 0.82	b	12.18 ± 0.31	a	*
C14:1—Myristoleic	%	1.15 ± 0.09	b	0.94 ± 0.14	a	*
C15—Pentadecanoic	%	1.30 ± 0.05	a	1.42 ± 0.08	b	*
C16—Palmitic	%	31.29 ± 2.64		32.99 ± 2.99		NS
C16:1—Palmitoleic	%	1.31 ± 0.02		1.52 ± 0.24		NS
C17—Eptadecanoic	%	0.79 ± 0.10	a	0.96 ± 0.02	b	*
C17:1—Eptadecenoic	%	0.23 ± 0.08	a	0.35 ± 0.06	b	*
C18—Stearic	%	9.02 ± 1.15	a	10.81 ± 0.96	b	*
C18:1—Elaidic t-9	%	0.68 ± 0.06	b	0.48 ± 0.03	a	**
C18:1—Vaccenic t-11	%	1.28 ± 0.24	a	1.83 ± 0.17	b	**
C18:1—Oleic	%	19.45 ± 1.74		20.87 ± 0.35		NS
C18:1—Vaccenic c-11	%	0.40 ± 0.08		0.46 ± 0.02		NS
C18:2—Linoelaidic t-6	%	0.20 ± 0.01	a	0.25 ± 0.02	b	**
C18:2—Linoleic	%	2.55 ± 0.20	b	1.87 ± 0.17	a	**
C20—Arachidic	%	0.16 ± 0.03	a	0.21 ± 0.01	b	*
C18:3—α-linolenic	%	0.87 ± 0.07		0.76 ± 0.08		NS
C18:2—Rumenic c9,t11 CLA	%	0.65 ± 0.08	a	0.87 ± 0.08	b	**
C18:2—t10-c12 CLA	%	0.01 ± 0.00		0.02 ± 0.01		NS
SCFA—short chain FA	%	10.70 ± 4.68		7.67 ± 1.26		NS
MCFA—middle chain FA	%	52.87 ± 0.89		52.47 ± 2.97		NS
LCFA—long chain FA	%	36.39 ± 3.72		39.84 ± 1.69		NS
OCFA—odd FA	%	2.45 ± 0.23	a	2.85 ± 0.06	b	*
MUFA—monounsaturated FA	%	25.03 ± 1.84		26.80 ± 0.17		NS
PUFA—polyunsaturated FA	%	4.40 ± 0.37	b	3.88 ± 0.17	a	*
UFA—unsaturated FA	%	29.43 ± 2.21		30.68 ± 0.35		NS
SFA—saturated FA	%	70.53 ± 2.27		69.29 ± 0.36		NS
CLA (conjug. linoleic acids)	%	0.66 ± 0.08	a	0.89 ± 0.07	b	**
Omega 3 FA	%	0.87 ± 0.07		0.76 ± 0.08		NS
Omega 6 FA	%	2.68 ± 0.23	b	1.99 ± 0.14	a	**

There are few studies on the effect of hay or silage cow feeding on the cheese fatty acid profile, and most of them are related to maize silage [73], while another considers vetch hay vs. vetch silage in sheep milk and cheese [74]. The result that PUFA are higher in hay silage (haylage) than in hay is consistent with the study of Schingoethe et al. [75] that found the same result, explaining it as a trend toward increased unsaturation for feeding fermented forages [75]. On the contrary, other authors [74] found higher values of PUFA in the hay group than in the silage group, but the silage is not the same (in this case it is vetch silage, and the research is on sheep milk and cheese).

The result for SFA, which had no significant variations, is consistent with results of Renes et al. [74]. The higher values of CLA and vaccenic acid found in hay vs. silage in the present research is confirmed by Segato et al. [73], although these authors consider maize silage instead of hay silage.

In the present research, the concentration of elaidic (trans) acid was higher in type I cheeses. Elaidic acid is a trans fatty acid associated with an increased risk of coronary heart disease, which can deleteriously affect lipoproteins by increasing LDL and decreasing HDL [76].

Odd fatty acids, in the present research, were more represented in cheese F. They are an index of a higher ruminal activity, as attested also by the higher contents of vaccenic and rumenic acids in the F thesis. Odd fatty acids have, by some authors [60,73], been associated with beneficial effects on human health [77]; in synthesis, except for the lower contents of PUFA and Omega 6 fatty acids, hay cow feeding without silage results in an overall more favourable fatty acid profile for human health.

3.3. Sensory and VOC Analyses: Comparison between Cheeses Obtained from Cows Fed Hay and Silage at 30 Days of Ripening

In order to evaluate any possible differences in taste and flavour between hay and silage production, cheeses at 30 days ripening were submitted to both sensorial and VOC analyses. The results of the triangular test carried out on the cheeses produced on 5 different days (two different winters), show that the cheese samples were judged significantly different in only 2 out of 5 days of production (Table 4). Such "insignificance" can also be due to the difficult standardisation of cheesemaking procedures in mountain environments, which in some cases can prevail over differences in the raw material used.

Table 4. Results of the triangular test made on cheese samples at 30 days' ripening.

Day of Cheesemaking	Difference [1]
1 February	*
16 March	NS
6 January	***
24 February	NS
26 March	NS

[1] NS, not significantly different; * $p \leq 0.05$; *** $p \leq 0.001$.

After choosing the "different" sample, the tasters were also asked to make a preference judgment, and to motivate it, and both hay and silage cheeses were equally preferred. It should be noted that in this kind of cheese (artisanal mountain cheese), personal tastes can differ dramatically, and what is a pleasant note for one taster, can be a defect for another. Apart from the appreciation, the recurring descriptors were "sweet", "spicy", "savoury", and "fruity". The first three attributes are manly perceived by taste and are connected to water or fat-soluble compounds released during chewing, such as free amino acids and short chain free fatty acids. On the other hand, fruity perception is due to volatile substances, mainly esters, but also some alcohols, aldehydes and ketones [78,79]. In this regard, the evaluation of the volatile profile of the cheeses obtained by the SPME-GC-MS analysis of the sample headspace (data not reported) confirms sensorial analysis. Even though further investigations must be made, the total ion chromatograms reported in Figure 1 suggest a possible consistency between testers' assessments and the instrumental data. The chromatogram for a "hay" sample judged as "sweet" and "fruity", compared to the corresponding "silage", shows higher levels of 2-butanol (fruity aroma) and ethyl capronate (banana, pineapple smell), which can both explain the higher fruity and sweet perceptions [80].

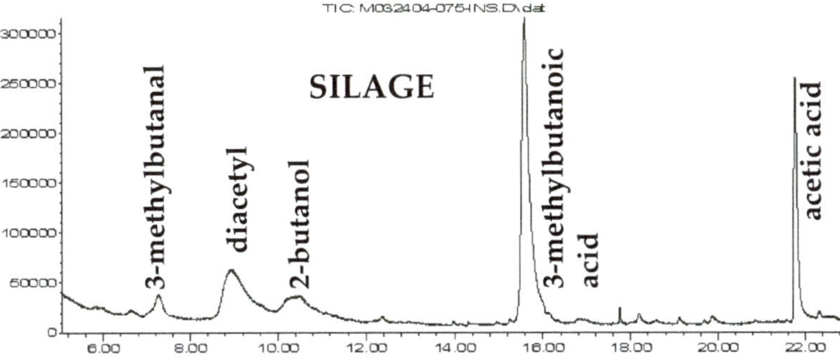

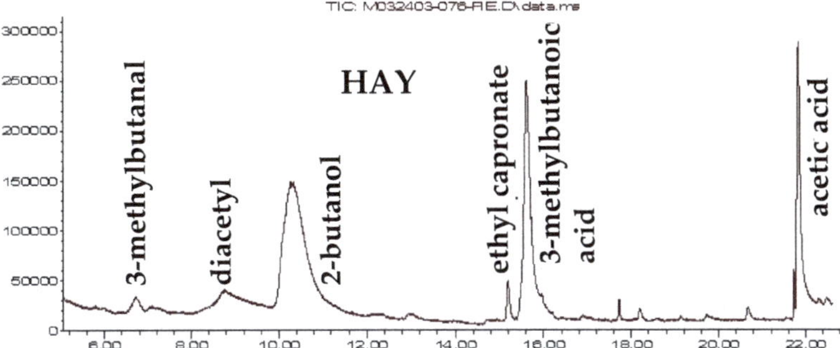

Figure 1. Chromatogram of total ionic current from dynamic headspace coupled for the two cheese typologies.

Verdier-Metz et al. [12], even though without statistically significant differences, also found that the cheeses produced with milk from cows fed silage seem to be less sticky and more bitter than the cheeses produced with milk from cows fed hay. The more pronounced bitterness of "silage" cheeses could be due to quicker ripening [81] and to proteolysis tending to produce hydrophobic peptides, which are generally bitter [82].

4. Conclusions

In conclusion, the experimentation demonstrated that the fat of the cheese obtained from stall milk was richer in medium-chain fatty acids. On the other hand, long-chain fatty acids were higher in the cheeses produced from pasture milk. Oleic acid was higher in the case of mountain pasture milk, whereas omega 3 fatty acids were more present in the cheeses produced on low altitude pasture. The content of unsaturated fatty acids and CLAs was always positively influenced by pasture.

The results of the sensory analysis, conducted on winter samples, show that the tasters were not always able to find significant differences. In general, cheeses produced with hay feeding were judged as "sweeter" than cheeses produced with silage feeding. The volatile profile of the F cheese presented higher levels of 2-butanol (fruity aroma) and ethyl capronate (smell of banana, pineapple), which can explain both the higher fruity and sweet perceptions. The results of a triangular test were compared with "cheese volatile organic compound (VOC) analysis"; the obtained chromatogram suggests a possible consistency between the assessments of the tasters and the instrumental data.

Author Contributions: Conceptualization, A.S., M.B. and V.Z.; methodology, A.S., M.B. and V.Z.; software, P.F. (Piero Franceschi), P.F. (Paolo Formaggioni), M.M. and M.F.; validation, A.S., M.M., P.F. (Paolo Formaggioni), M.B., G.B. and P.F. (Piero Franceschi); formal analysis, P.F. (Piero Franceschi), P.F. (Paolo Formaggioni), M.M., M.B. and G.B.; investigation, P.F. (Paolo Formaggioni), P.F. (Piero Franceschi), M.M., M.B., A.S., M.F. and G.B..; resources, A.S. and M.B.; data curation, P.F. (Paolo Formaggioni), P.F. (Piero Franceschi), A.S., G.B., V.Z. and M.B.; writing—original draft preparation, P.F. (Paolo Formaggioni), P.F. (Piero Franceschi), M.M., M.B., A.S., M.F., V.Z. and G.B.; writing—review and editing, P.F. (Paolo Formaggioni), P.F. (Piero Franceschi), M.M., M.B., A.S., M.F., V.Z. and G.B.; visualization, P.F. (Paolo Formaggioni), P.F. (Piero Franceschi), M.M., M.B., A.S., M.F., V.Z. and G.B.; supervision, A.S. and M.B.; project administration, A.S., M.B. and V.Z.; funding acquisition, A.S. and M.B. All authors have read and agreed to the published version of the manuscript.

Funding: Work carried out with the contribution of the Lombardy Region (Regional Agricultural Program 2007–2009)—TLead institution: Municipality of Vilminore di Scalve (BG); Partner: Scalve Valley Mountain Community (Comunità Montana della Valle di Scalve) and Scalve Social Mountain Dairy (Latteria Sociale Montana di Scalve).

Conflicts of Interest: The authors declare that there are no conflicts of interest in this research article.

References

1. Romanzin, A.; Corazzin, M.; Piasentier, E.; Bovolenta, S. Effect of rearing system (mountain pasture vs. indoor) of Simmental cows on milk composition and Montasio cheese characteristics. *J. Dairy Res.* **2013**, *80*, 390–399. [CrossRef] [PubMed]

2. Bovolenta, S.; Corazzin, M.; Saccà, E.; Gasperi, F.; Biasioli, F.; Ventura, W. Performance and cheese quality of Brown cows grazing on mountain pasture fed two different levels of supplementation. *Livest. Sci.* **2009**, *124*, 58–65. [CrossRef]

3. Fallico, V.; Chianese, L.; Carpino, S.; Licitra, G. Preliminary evaluation of the influence of pasture feeding on proteolysis of Ragusano cheese. *J. Food Technol.* **2006**, *4*, 128–134.

4. Urbach, G. Effect of feed on flavor in dairy foods. *J. Dairy Sci.* **1990**, *73*, 3639–3650. [CrossRef]

5. Grappin, R.; Beuvier, E. Possible implications of milk pasteurization on the manufacture and sensory quality of ripened cheese. *Int. Dairy J.* **1997**, *7*, 751–761. [CrossRef]

6. Buchin, S.; Delague, V.; Duboz, G.; Berdagué, J.L.; Beuvier, E.; Pochet, S.; Grappin, R. Influence of pasteurization and fat composition of milk on the volatile compounds and flavour characteristics of a semi-hard cheese. *J. Dairy Sci.* **1998**, *81*, 3097–3108. [CrossRef]

7. Malossini, F.; Bovolenta, S.; Piras, C.; Ventura, W. Effect of concentrate supplementation on herbage intake and milk yield of dairy cows grazing an alpine pasture. *Livest. Prod. Sci.* **1995**, *43*, 119–128. [CrossRef]

8. Berry, N.R.; Bueler, T.; Jewell, P.L.; Sutter, F.; Kreuzer, M. The effect of supplementary feeding on composition and renneting properties of milk from cows rotationally grazed at high altitude. *Milchwissenschaft* **2001**, *56*, 123–126.

9. Bugaud, C.; Buchin, S.; Coulon, J.B.; Hauwuy, A.; Dupont, D. Influence of the nature of alpine pastures on plasmin activity, fatty acid and volatile compound composition of milk. *Lait* **2001**, *81*, 401–414. [CrossRef]

10. Panthi, R.R.; Kelly, A.L.; Hennessy, D.; O'Sullivan, M.G.; Kilcawley, K.N.; Mannion, D.T.; Fenelon, M.A.; Sheehan, J.J. Effect of pasture versus indoor feeding regimes on the yield, composition, ripening and sensory characteristics of Maasdam cheese. *Int. J. Dairy Technol.* **2019**, *72*, 435–446. [CrossRef]

11. Coulon, J.-B.; Verdier, I.; Pradel, P. Effect of forage type (hay or grazing) on milk cheesemaking ability. *Lait* **1996**, *76*, 479–486. [CrossRef]

12. Verdier-Metz, I.; Coulon, J.-B.; Pradel, P.; Viallon, C.; Berdagué, J.-L. Effect of forage conservation (hay or silage) and cow breed on the coagulation properties of milks and on the characteristics of ripened cheeses. *J. Dairy Res.* **1998**, *65*, 9–21. [CrossRef]

13. Viallon, C.; Verdier-Metz, I.; Denoyer, C.; Pradel, P.; Coulon, J.B.; Berdagué, J.-L. Desorbed terpenes and sesquiterpenes from forages and cheeses. *J. Dairy Res.* **1999**, *66*, 319–326. [CrossRef]

14. Coppa, M.; Ferlay, A.; Monsallier, F.; Verdier-Metz, I.; Pradel, P.; Didienne, R.; Farruggia, A.; Montel, M.C.; Martin, B. Milk fatty acid composition and cheese texture and appearance from cows fed hay or different grazing systems on upland pastures. *J. Dairy Sci.* **2011**, *94*, 1132–1145. [CrossRef]

15. Martin, B.; Verdier-Metz, I.; Buchin, S.; Hurtaud, C.; Coulon, J.B. How does the nature of forages and pastures diversity influence the sensory quality of dairy livestock products? *Anim. Sci.* **2005**, *81*, 205–212. [CrossRef]

16. Dumont, J.P.; Adda, J. Occurrence of sesqiterpenes in mountain cheese volatiles. *J. Agric. Food Chem.* **1978**, *26*, 364–367. [CrossRef]

17. Bugaud, C.; Buchin, S.; Hauwuy, A.; Coulon, J.B. Relationships between flavour and chemical composition of Abundance cheese derived from different types of pastures. *Lait* **2001**, *81*, 757–773. [CrossRef]

18. Carpino, S.; Mallia, S.; La Terra, S.; Melilli, C.; Licitra, G.; Acree, T.E.; Barbano, D.M.; Van Soest, P.J. Composition and aroma compounds of Ragusano cheese: Native pasture and total mixed rations. *J. Dairy Sci.* **2004**, *87*, 816–830. [CrossRef]

19. Carpino, S.; Horne, J.; Melilli, C.; Licitra, G.; Barbano, D.M.; Van Soest, P.J. Contribution of native pasture to the sensory properties of Ragusano cheese. *J. Dairy Sci.* **2004**, *86*, 308–315. [CrossRef]

20. Kosikowski, F.V.; Mistry, V.V. *Cheese and Fermented Milk Foods. Volume 1: Origins and Principles*; Kosikowski, F.V., Ed.; LLC: Westport, CT, USA, 1997; Volume 1, p. 379.

21. Panfili, G.; Manzi, P.; Pizzoferrato, L. High-performance liquid chromatographic method for the simultaneous determination of tocopherols, carotenes, and retinol and its geometric Isomers in Italian cheeses. *Analyst* **1994**, *119*, 1161–1165. [CrossRef]

22. Kilcawley, K.N.; Faulkner, H.; Clarke, H.J.; O'Sullivan, M.G.; Kerry, J.P. Factors influencing the flavour of bovine milk and cheese from grass based versus non-grass based milk production systems. *Foods* **2018**, *7*, 37. [CrossRef]

23. Nozière, P.; Graulet, B.; Lucas, A.; Martin, B.; Grolier, P.; Doreau, M. Carotenoids for ruminants: From forages to dairy products. *Anim. Feed Sci. Technol.* **2006**, *131*, 418–450. [CrossRef]

24. Frétin, M.; Martin, B.; Buchin, S.; Desserre, B.; Lavigne, R.; Tixier, E.; Cirié, C.; Cord, C.; Montel, M.-C.; Delbés, C.; et al. Milk fat composition modifies the texture and appearance of Cantal-type cheeses but not their flavor. *J. Dairy Sci.* **2019**, *102*, 1131–1143. [CrossRef]

25. Corazzin, M.; Romanzin, A.; Sepulcri, A.; Pinosa, M.; Piasentier, E.; Bovolenta, S. Fatty acid profiles of cow's milk and cheese as affected by mountain pasture type and concentrate supplementation. *Animals* **2019**, *9*, 68. [CrossRef]

26. White, S.L.; Bertrand, J.A.; Wade, M.R.; Washburn, S.P.; Green, J.T.; Jenkins, T.C. Comparison of fatty acid content of milk from Jersey and Holstein cows consuming pasture or a total mixed ration. *J. Dairy Sci.* **2001**, *84*, 2295–2301. [CrossRef]

27. Chilliard, Y.; Glasser, F.; Ferlay, A.; Bernard, L.; Rouel, J.; Doreau, M. Diet, rumen biohydrogenation and nutritional quality of cow and goat milk fat. *Eur. J. Lipid Sci. Technol.* **2007**, *109*, 828–855. [CrossRef]

28. Slots, T.; Butler, G.; Leifert, C.; Kristensen, T.; Skibsted, L.H.; Nielsen, J.N. Potential to differentiate milk composition by different feeding strategies. *J. Dairy Sci.* **2009**, *92*, 2057–2066. [CrossRef]

29. De Noni, I.; Battelli, G. Terpenes and fatty acid profiles of milk fat and "Bitto" cheese as affected by transhumance of cows on different mountain pastures. *Food Chem.* **2008**, *109*, 299–309. [CrossRef]

30. Revello Chion, A.; Tabacco, E.; Giaccone, D.; Peiretti, P.G.; Battelli, G.; Borreani, G. Variation of fatty acid and terpene profiles in mountain milk and "Toma piemontese" cheese as affected by diet composition in different seasons. *Food Chem.* **2010**, *121*, 393–399. [CrossRef]

31. Povolo, M.; Pelizzola, V.; Lombardi, G.; Tava, A.; Contarini, G. Hydrocarbon and fatty acid composition of cheese as affected by the pasture vegetation type. *J. Agric. Food Chem.* **2012**, *60*, 299–308. [CrossRef]

32. Tanaka, K. Occurrence of conjugated linoleic acid in ruminant products and its physiological functions. *Anim. Sci. J.* **2005**, *76*, 291–303. [CrossRef]

33. Dewhurst, R.J.; Shingfield, K.J.; Lee, M.R.F.; Scollan, N.D. Increasing the concentrations of beneficial polyunsaturated fatty acids in milk produced by dairy cows in high-forage systems. *Anim. Feed Sci. Technol.* **2006**, *131*, 168–206. [CrossRef]

34. Summer, A.; Formaggioni, P.; Franceschi, P.; Di Frangia, F.; Righi, F.; Malacarne, M. Cheese as functional food: The example of Parmigiano Reggiano and Grana Padano. *Food Technol. Biotechnol.* **2017**, *55*, 277–289. [CrossRef]

35. Copani, V.; Guarnaccia, P.; Biondi, L.; Lesina, S.C.; Longo, S.; Testa, G.; Cosentino, S.L. Pasture quality and cheese traceability index of Ragusano PDO cheese. *Ital. J. Agron.* **2015**, *10*, 667. [CrossRef]

36. IDF Standard. *Milk and Milk Products, Determination of Fat Content, General Guidance on the Use of Butyrometric Methods*; International Dairy Federation Standard 152/ISO11870: Brussels, Belgium, 2009.

37. Formaggioni, P.; Summer, A.; Malacarne, M.; Franceschi, P.; Mucchetti, G. Italian and Italian-style hard cooked cheeses: Predictive formulas for Parmigiano-Reggiano 24 h cheese yield. *Int. Dairy J.* **2015**, *51*, 52–58. [CrossRef]

38. Malacarne, M.; Summer, A.; Franceschi, P.; Formaggioni, P.; Pecorari, M.; Panari, G.; Mariani, P. Free fatty acid profile of Parmigiano-Reggiano cheese throughout ripening: Comparison between the inner and outer regions of the wheel. *Int. Dairy J.* **2009**, *19*, 637–641. [CrossRef]

39. IDF Standard. *Milk, Determination of Nitrogen Content, Part 1: Kjeldahl Method*; International Dairy Federation Standard 20-1/ISO8968-1: Brussels, Belgium, 2001.

40. IDF Standard. *Cheese and Processed Cheese Products, Determination of Chloride Content, Potentiometric Titration Method*; International Dairy Federation Standard 88/ISO5943: Brussels, Belgium, 2006.

41. IDF Standard. *Cheese and Processed Cheese, Determination of the Total Solids Content (Reference Method)*; International Dairy Federation Standard 4/ISO5534: Brussels, Belgium, 2004.

42. Association of Official Analytical Chemists [AOAC]. AOAC Official Method no. 996.06. Fat (total, saturated, and unsaturated) in foods; hydrolytic extraction gas chromatographic method. In *Official Methods of Analysis of AOAC International*, 17th ed.; Horwitz, W., Ed.; AOAC International: Gaithersburg, MD, USA, 2000; pp. 20–24.

43. Association of Official Analytical Chemists [AOAC]. Fat in milk: Modified Mojonnier ether extraction method. In *Official Methods of Analysis of AOAC International*; Heldrich, K., Ed.; AOAC International: Arlington, VA, USA, 1990; pp. 811–812.

44. AOCS. Official Method CE 2-66. Preparation of methyl esters of fatty acids. In *Official Methods and Recommended Practices of the AOCS*, 6th ed.; Firestone, D., Ed.; AOCS Press: Urbana, IL, USA, 2013.

45. ISO Standard. *Sensory Analysis, Methodology, Triangle Test*; International Organization for Standardization Standard ISO4120: Genève, Switzerland, 2004; 15p.

46. Battelli, G.; Scano, P.; Albano, C.; Cagliani, L.R.; Brasca, M.; Consonni, R. Modifications of the volatile and nonvolatile metabolome of goat cheese due to adjunct of non-starter lactic acid bacteria. *LWT* **2019**, *116*, 108576. [CrossRef]

47. Summer, A.; Franceschi, P.; Formaggioni, P.; Malacarne, M. Characteristics of raw milk produced by free-stall or tie-stall cattle herds in the Parmigiano-Reggiano cheese production area. *Dairy Sci. Technol.* **2014**, *94*, 581–590. [CrossRef]

48. Summer, A.; Lora, I.; Formaggioni, P.; Gottardo, F. Impact of heat stress on milk and meat production. *Anim. Front.* **2019**, *9*, 39–46. [CrossRef]

49. Esposito, G.; Masucci, F.; Napolitano, F.; Braghieri, A.; Romano, R.; Manzo, N.; Di Francia, A. Fatty acid and sensory profiles of Caciocavallo cheese as affected by management system. *J. Dairy Sci.* **2014**, *97*, 1918–1928. [CrossRef]

50. Dhiman, T.R.; Helmink, E.D.; McMahon, D.J.; Fife, R.L.; Pariza, M.W. Conjugated linoleic acid content of milk and cheese from cows fed extruded oilseeds. *J. Dairy Sci.* **1999**, *82*, 412–419. [CrossRef]

51. Carafa, I.; Stocco, G.; Franceschi, P.; Summer, A.; Tuohy, K.M.; Bittante, G.; Franciosi, E. Evaluation of autochthonous lactic acid bacteria as starter and non-starter cultures for the production of Traditional Mountain cheese. *Food Res. Int.* **2019**, *115*, 209–218. [CrossRef]

52. Povolo, M.; Pelizzola, V.; Passolungo, L.; Biazzi, E.; Tava, A.; Contarini, G. Characterization of two agrostis-festuca alpine pastures and their influence on cheese composition. *J. Agric. Food Chem.* **2013**, *61*, 447–455. [CrossRef]

53. Ferlay, A.; Bernard, L.; Meynadier, A.; Malpuech-Brugere, C. Production of trans and conjugated fatty acids in dairy ruminants and their putative effects on human health: A review. *Biochimie* **2017**, *141*, 107–120. [CrossRef]

54. Collomb, M.; Schmid, A.; Sieber, R.; Wechsler, D.; Ryhänen, E.L. Conjugated linoleic acids in milk fat: Variation and physiological effects. *Int. Dairy J.* **2006**, *16*, 1347–1361. [CrossRef]

55. Innocente, N.; Praturlon, D.; Corradini, C. Fatty acid profile of cheese produced with milk from cows grazing on mountain pastures. *Ital. J. Food Sci.* **2002**, *14*, 217–224.

56. Lobos-Ortega, I.; Revilla, I.; González-Martín, M.I.; Hernández-Hierro, J.M.; Vivar-Quintana, A.; González-Pérez, G. Conjugated linoleic acid contents in cheeses of different compositions during six months of ripenin. *Czech J. Food Sci.* **2012**, *30*, 220–226. [CrossRef]

57. Coppa, M.; Verdier-Metz, I.; Ferlay, A.; Pradel, P.; Didienne, R.; Farruggia, A.; Monte, M.C.; Martin, B. Effect of different grazing systems on upland pastures compared with hay diet on cheese sensory properties evaluated at different ripening times. *Int. Dairy J.* **2011**, *21*, 815–822. [CrossRef]

58. Chilliard, Y.; Ferlay, A.; Doreau, M. Effect of different types of forages, animal fat or marine oils in cow's diet on milk fat secretion and composition, especially conjugated linoleic acid (CLA) and polyunsaturated fatty acids. *Livest. Prod. Sci.* **2001**, *70*, 31–48. [CrossRef]

59. Collomb, M.; Bisig, W.; Bütikofer, U.; Sieber, R.; Bregy, M.; Etter, L. Fatty acid composition of mountains milk from Switzerland: Comparison of organic and integrated farming systems. *Int. Dairy J.* **2008**, *18*, 976–982. [CrossRef]

60. Bouterfa, A.; Bekada, A.; Homrani, A.; Benguendouz, A.; Homrani, A.; Amrane, A.; Zribi, A.; Benakriche, B.; Dahloum, E.H.; Benabdelmoumene, D.; et al. Influence of lactation stage on lipids and fatty acids profile of artisanal Algerian Camembert-type cheese manufactured with cow's milk. *South Asian J. Exp. Biol.* **2019**, *9*, 17–22.

61. Di Grigoli, A.; Bonanno, A.; Cifuni, G.; Tornambé, G.; Alicata, M. Effetto del regime alimentare delle bovine sulla composizione in acidi grassi del formaggio Caciocavallo Palermitano. *Sci. E Tec. Latt. Casearia* **2008**, *59*, 457–461.

62. Prandini, A.; Sigolo, S.; Cerioli, C.; Piva, G. Survey on conjugated linoleic acid (CLA) content and fatty acid composition of Grana Padano cheese produced in different seasons and areas. *Ital. J. Anim. Sci.* **2009**, *8*, 531–540. [CrossRef]

63. Leiber, F.; Kreuzer, M.; Nigg, D.; Wettstein, H.R.; Scheeder, M.R.L. A study on the causes for the elevated n-3 fatty acids in cows' milk of alpine origin. *Lipids* **2005**, *40*, 191–202. [CrossRef]

64. Schwendel, B.H.; Morel, P.C.H.; Wester, T.J.; Tavendale, M.H.; Deadman, C.; Fong, B.; Shadbolt, N.M.; Thatcher, A.; Otter, D.E. Fatty acid profile differs between organic and conventionally produced cow milk independent of season or milking time. *J. Dairy Sci.* **2015**, *98*, 1411–1425. [CrossRef]

65. Nagpal, R.; Yadav, H.; Puniya, A.K.; Singh, K.; Jain, S.; Marotta, F. Conjugated linoleic acid: Sources, synthesis and potential health benefits—An overview. *Curr. Top. Nutraceut. Res.* **2007**, *5*, 55–65.

66. Lee, Y.J.; Jenkins, T.C. Biohydrogenation of linolenic acid to stearic acid by the rumen microbial population yields multiple intermediate conjugated diene isomers. *J. Nutr.* **2011**, *141*, 1445–1450. [CrossRef]

67. Nudda, A.; McGuire, M.A.; Battacone, G.; Pulina, G. Seasonal variation in conjugated linoleic acid and vaccenic acid in milk fat of sheep and its transfer to cheese and ricotta. *J. Dairy Sci.* **2005**, *88*, 1311–1319. [CrossRef]

68. Meluchová, B.; Blasko, J.; Kubinec, R.; Górová, R.; Dubravská, J.; Margetín, M.; Soják, L. Seasonal variations in fatty acid composition of pasture forage plants and cla content in ewe milk fat. *Small Rumin. Res.* **2008**, *78*, 56–65. [CrossRef]

69. Cozzi, G.; Ferlito, J.; Pasini, G.; Contiero, B.; Gottardo, F. Application of near-infrared spectroscopy as an alternative to chemical and color analysis to discriminate the production chains of Asiago d'Allevo cheese. *J. Agric. Food Chem.* **2009**, *57*, 11449–11454. [CrossRef]

70. Butler, G.; Nielsen, J.H.; Slots, T.; Seal, C.; Eyre, M.D.; Sanderson, R.; Leifert, C. Fatty acid and fat-soluble antioxidant concentrations in milk from high- and low-input conventional and organic system: Seasonal variation. *J. Sci. Food Agric.* **2008**, *88*, 1431–1441. [CrossRef]

71. Zeppa, G.; Giordano, M.; Gerbi, V.; Arlorio, M. Fatty acid composition of Piedmont "Ossolano" cheese. *Lait* **2003**, *83*, 167–173. [CrossRef]

72. Jahangiri, A.; Leifert, W.R.; McMurchie, E.J. Omega-3 polyunsaturated fatty acids: Recent aspects in relation to health benefits. *Food Aust.* **2002**, *54*, 74–77.

73. Segato, S.; Galaverna, G.; Contiero, B.; Berzaghi, P.; Caligiani, A.; Marseglia, A.; Cozzi, G. Identification of lipid biomarkers to discriminate between the different production systems for Asiago PDO cheese. *J. Agric. Food Chem.* **2017**, *65*, 9887–9892. [CrossRef]

74. Renes, E.; Gómez-Cortés, P.; de la Fuente, M.A.; Fernández, D.; Tornadijo, M.E.; Fresno, J.M. Effect of forage type in the ovine diet on the nutritional profile of sheep milk cheese fat. *J. Dairy Sci.* **2020**, *103*, 63–71. [CrossRef]

75. Schingoethe, D.J.; Voelker, H.H.; Beardsley, G.L.; Parsons, J.G. Rumen volatile fatty acids and milk composition from cows fed hay, haylage, or urea-treated corn silage. *J. Dairy Sci.* **1976**, *59*, 894–901. [CrossRef]

76. Abbey, M.; Nestel, P.J. Plasma cholesteryl ester transfer protein is increased when trans-elaidic acid is substituted for cis-oleic acid in the diet. *Atherosclerosis* **1994**, *106*, 99–107. [CrossRef]
77. Vlaeminck, B.; Fievez, V.; Cabrita, A.R.J.; Fonseca, A.J.M.; Dewhurst, R.J. Factors affecting odd and branched-chain fatty acids in milk: A review. *Anim. Feed Sci. Technol.* **2006**, *131*, 389–417. [CrossRef]
78. Bills, D.D.; Morgan, M.E.; Libbey, M.N.; Day, E.A. Identification of compounds responsible for fruity flavor defect of experimental Cheddar cheeses. *J. Dairy Sci.* **1965**, *48*, 1168–1173. [CrossRef]
79. Mc Gugan, W.A.; Blais, J.A.; Boulet, M.; Giroux, R.N.; Elliott, J.A.; Emmons, D.B. Ethanol, ethyl esters, and volatile fatty acids in fruity Cheddar cheese. *Can. Inst. Food Sci. Technol. J.* **1975**, *8*, 196–198. [CrossRef]
80. Lethuaut, L.; Brossard, C.; Meynier, A.; Rousseau, F.; Llamas, G.; Bousseau, B.; Genot, C. Sweetness and aroma perceptions in dairy desserts varying in sucrose and aroma levels and in textural agent. *Int. Dairy J.* **2005**, *15*, 485–493. [CrossRef]
81. Garel, J.P.; Coulon, J.-B. Effect of feeding and breed of dairy cows on Saint-Nectaire cheesemaking. *INRA Prod. Anim.* **1990**, *3*, 127–136.
82. Biede, S.L.; Hammond, G. Swiss cheese favor. II. Organoleptic analysis. *J. Dairy Sci.* **1979**, *62*, 238–248. [CrossRef]

Article

Nutraceutical and Technological Properties of Buffalo and Sheep Cheese Produced by the Addition of Kiwi Juice as a Coagulant

Andrea Serra [1,2,3,*], Giuseppe Conte [1,2,3], Leonor Corrales-Retana [1], Laura Casarosa [1], Francesca Ciucci [1] and Marcello Mele [1,2,3]

[1] Department of Agriculture, Food and Environment, University of Pisa, via del Borghetto 80, 56124 Pisa, Italy; giuseppe.conte@unipi.it (G.C.); leocorretana@gmail.com (L.C.-R.); laura.casarosa@unipi.it (L.C.); francesca.ciucci88@gmail.com (F.C.); marcello.mele@unipi.it (M.M.)

[2] Center of Agricultural and Environmental Studies "E. Avanzi", University of Pisa, via Vecchia di Marina, San Piero a Grado, 6-56122 Pisa, Italy

[3] Research Center of Nutraceuticals and Food for Health, University of Pisa, via del Borghetto 80, 56124 Pisa, Italy

* Correspondence: andrea.serra@unipi.it; Tel.: +39-050-22188949

Received: 1 April 2020; Accepted: 12 May 2020; Published: 15 May 2020

Abstract: Kiwifruit is an interesting alternative to chymosin for milk coagulation. Although the clotting properties of actinidin (the proteolytic agent present in kiwi) have been widely investigated, little is known about the nutraceutical and organoleptic effects of kiwifruit on the characteristics of cheese. We investigated kiwifruit pulp, compared to calf rennet, in cheesemaking using sheep and buffalo milk. Although the kiwifruit extract showed a longer coagulation and syneresis time than calf rennet, it could nevertheless be exploited as a plant coagulant due to its positive effect on the nutraceutical properties. In fact, the sheep and buffalo cheese were higher in polyphenols and phytosterols than the cheese obtained using calf rennet. In addition, the nutraceutical properties were enhanced, with just a slight effect on the aroma of the cheese.

Keywords: milk clotting; cheese; kiwifruit; actinidin; nutraceutical properties

---

1. Introduction

Enzymatic milk coagulation is a key step in cheese manufacturing and involves the addition of chymosin (rennet), an aspartate proteinase that is active in the stomach of non-weaned calves [1], and which hydrolyses the link between amino acids 105 (methionine) and 106 (phenylalanine) of the k-casein. Given various social (i.e., veganism) and religious (Islam, Judaism) issues, which entail limiting or reducing the use of chymosin, new sources of coagulants are needed.

Proteolytic enzymes extracted from plants may be an interesting alternative to animal rennet in dairy technology. In fact, milk-clotting enzymes have been identified in various plant species, such as *Lactuca sativa* [2], *Albizia lebbeck*, *Helianthus annuus* [3] and *Cynara cardunculus* [4].

Actinidin (EC 3.4.22.14) is a cysteine protease from kiwifruit (*Actinidia deliciosa*) with a wide pH activity range (4–10) [5]. Lo Piero et al. [5] demonstrated that actinidin forms milk clots with the typical conditions used in cheese manufacturing (optimum activity at 40–42 °C, mildly acidic pH values). The preferred substrate for actinidin is β-casein, followed by k-casein, and the result of this hydrolysis is the production of a small number of larger peptides [5]. Saha and Hayashi [6] revealed that dairy products that use kiwifruit juice actinidin have a lower perceived off-flavour.

Exploiting kiwifruit in milk cheesemaking could meet the goals of circular agriculture through the use of undersized and/or damaged kiwifruits in a simple and economically sustainable procedure for the production of a clotting mixture. In addition, it could improve the nutritional/nutraceutical characteristics of the cheese, given that kiwifruit contains hydro-soluble components, such as vitamin C and polyphenols [7–9].

Several works have dealt with the clotting properties of kiwifruit. Katsaros et al. [10] reported that 1 mL juice rich in actinidin obtained from peeled, pulped and centrifuged kiwifruit corresponded to 0.42 ± 0.02 units, while 1 g of freeze-dried powder corresponded to 520 ± 20 units. Mazorra-Monzano et al. [11] compared kiwi, ginger and melon with chymosin and found that the kiwi extract produced the most similar curd in terms of texture to curd from chymosin. Lo Piero et al. [5] characterized the role of purified actinidin in casein proteolysis and reported that the activity for total casein was 129.6 U/mg and that the specific milk clotting activity (AC) is higher than the general one (AP); the AC/AP ratio (1.1) was similar to calf rennet (1.0) and much higher than microbiological rennet (0.22). Puglisi et al. [12] reported that kiwi juice shows twice the specific activity for total casein than purified actinidin of 299 U/mg, and speculated that, in kiwi, other proteases assist actinidin in the proteolysis process.

To the best of our knowledge, no data are available on the relationship between using kiwifruit extract in cheesemaking and the nutraceutical substances and flavour of cheese obtained with sheep and buffalo milk. We assessed the technological, nutraceutical and organoleptic properties of sheep and buffalo cheese made from kiwifruit extract compared with animal rennet.

2. Materials and Methods

2.1. Experimental Design

The study was carried out on 12 cheeses obtained from fresh and pasteurized milk, according to the following procedure: three buffalo milk cheeses were made with calf rennet (BM-C); three buffalo milk cheeses were made with kiwifruit extract (BM-K); three sheep milk cheeses were made with calf rennet (SM-C); and three sheep milk cheeses were made with kiwifruit extract (SM-K). The experiment was replicated twice in two independent batches. Milk was obtained from two neighbouring farms located in the province of Grosseto (southern Tuscany, Italy).

The kiwifruit extract was obtained by a modification of the method described by Katsaros et al. [10]. Briefly; 1 kg of kiwi pulp (*A. deliciosa*, cv. Hayward) were filtered through a cotton gauze. The filtrate was centrifuged (10,000 rpm for 15 min at 4 °C), and the supernatant was filtered using a 0.45 μm filter. The purified solution was lyophilized to obtain a powder rich in actinidin; the protein content of the powder was quantified by Bradford's method [13].

The calf rennet used for the coagulation of the BM-C and SM-C cheeses was NATUREN ® PLUS 215 (activity = 215 IMCU/mL; chymosin 63%; pepsin 37%) (CHR HANSEN, Hoersholm, Denmark).

Milk (200 mL) was clotted after the adjustment of the pH value by the addition of *Streptococcus thermophilus* CRV00 LYO 100 L starter (Santamaria srl, Burago di Molgara, Italy) (0.216 g per 100 mL) at 40 °C. Coagulants were added when the milk pH was 6.35–6.38. Calf rennet was diluted in distilled water to have a strength of 40 IMCU/mL. In order to have the same enzymatic activity (measured as Katsaros et al. [10]), for 200 mL of milk we used 80 mg of freeze-dried powder rich in actinidin.

After the milk had clotted, the curd was manually cut with a knife and stirred gently. Then, after about 10 min the cheeses were drained and put into moulds, which were turned and pressed manually. Finally, in order to eliminate the excess water, the miniature cheeses [11] were incubated at 40 ° C for 30 min. The cheeses were not salted. The weight of each cheese was recorded at the end of the cheesemaking and after 24 h. The syneresis of the cheeses was used as one of the indices of the level of coagulation, which was expressed as the volume of the whey (mL) per total mass (100 g) of the coagulated milk product. Cheese yields (initial and after 24 h) were calculated as curd weight (initial or after 24 h)/milk weight × 100.

2.2. Analysis

2.2.1. Milk Clotting

Milk-clotting was evaluated visually when clotting began, in accordance with Uchikoba et al. [14]. During the visual evaluation, the time was measured between the addition of the coagulant solution and the first appearance of solid material against the background (CT, clotting time).

Milk-clotting properties were evaluated by a Formagraph ® (Foss Electric, Hillerød, Denmark). Following the Formagraph ® instructions, milk samples (10 mL) were heated to 35 °C, and 200 μL of a solution of rennet (NATUREN ® PLUS 215-215 IMCU/mL; chymosin 63%; pepsin 37%) (CHR HANSEN, Hoersholm, Denmark) with a strength of 40 IMCU/mL was added; 10 replicates per run and for each sample milk were performed. The same conditions were applied to evaluate the kiwifruit extract; however, 200 μL of a 40mg/mL solution of freeze-dried powder in distilled water was used. Measurements were stopped thirty minutes after the addition of the enzyme.

The principle of lacto-dynamography is based on the control of the oscillation that is driven by an electromagnetic field created by a swinging pendulum. During milk clotting, a pendulum is immersed into the milk container. The greater the extent of the coagulation, the smaller the pendulum swing. This analysis provided measurements of the clotting time (r) in min, curd firming time (k20) in min, and curd firmness (A30) in mm [15].

2.2.2. Physical and Chemical Analysis

Cheese samples were analysed in terms of moisture, protein, fat, and ash following official AOAC methods (AOAC, 2000). For the colour measurements, samples were placed on a standard white tile. Colour readings were taken at four randomly selected locations on the cranial surface of each piece to obtain a representative mean value. The cheese colour was measured in the CIE L*a*b* space (CIE, 19876) with an area diameter of 8 mm, including the specular component, and 0% UV, D65 standard illuminant, observer angle 10°, and a zero and white calibration using a Minolta CM 2006d spectrophotometer (Konica Minolta Holdings, Inc., Osaka, Japan). Lightness (L*), greenness (a*) and yellowness (b*) were recorded [16,17]. The colour parameters were used to calculate the total colour differences between cheeses obtained with calf rennet and cheeses obtained with kiwifruit, using the following formula: $\Delta E^* = ((L^*)^2 + (a^*)^2 + (b^*)^2)^{1/2}$. Values were expressed as the mean ± standard deviation. Following Sanz [18], the colour differences for the human eye are not obvious if $\Delta E^* < 1$; not appreciable if $1 < E^* < 3$; and obvious if $E^* > 3$.

Calcium, iron, sodium, magnesium and potassium were determined by flame atomic-absorption spectroscopy on an iCE 3000 series AA spectrophotometer (Thermo-Scientific, Waltham, MA, USA) equipped with a deuterium lamp as a background-correction system. An acetylene-air flame was used, while the gas flow rates, and the burner height were adjusted in order to obtain the maximum absorbance signal for each element. The organic matter of the samples (0.5 g) was put in a muffle-furnace at 450 °C for 24 h to obtain ash. When cool, the residue was dissolved in 1 mL nitric acid and the volume was diluted to 10 mL with water. The wavelength of the spectrometer was set at 422.7 nm for Ca, 589.6 nm for Na, 248.3 nm for Fe, 766.5 nm for K and 285.2 nm for Mg. The slit width was 0.7 nm for all elements, except Fe with 0.2 nm. The volumes and corresponding concentrations of the samples were selected within the linear range of the instrument used (at least five concentrations).

2.2.3. Lipid Composition of Kiwifruit and Cheeses

Total lipids (TL) of kiwifruit and cheeses were extracted with a chloroform/methanol solution (2:1, v/v), following Rodriguez-Estrada et al. [19].

Unsaponifiable matter was obtained following Sanders et al. [20]. Briefly, 300 mg of TL were cold-saponified by adding 4.5 mL of ethanolic KOH (4.8% w/v) solution and incubated at room temperature for 12 h. The unsaponifiable matter was isolated by two washes with 4.5 mL of water and 9 mL of hexane. The non-polar phase (upper phase) was transferred into a fresh tube and dried by nitrogen gas. Finally, the samples were dissolved once again with 1 mL of methanol. Before saponification, 100 µL of a solution of dihydrocholesterol in chloroform (2 mg/mL) as internal standard for sterols were added to TL.

Sterols were then silylated adding a hexamethyldisilazane/chlorotrimethysilane/pyridine 2/1/5 v/v/v mixture, dried under a nitrogen stream and dissolved in 300 µL of n-hexane. The sterols were identified and quantified using a GC–FID (GC 2000 plus, Shimadzu, Columbia, MD, USA) equipped with a VF 1-ms apolar capillary column (25 m × 0.25 mm i.d., 0.25 µm film thickness; Varian, Palo Alto, CA, USA). A total of 2 µL of the sample in hexane were injected into the column with the carrier gas (hydrogen) flux at 1 mL/min and the split ratio was 1:10. The run was carried out in constant pressure mode. The oven temperature was held at 250 °C for 1 min, and increased to 260 °C over 20 min at the rate of 0.5 °C/min, and then increased to 325 °C over 13 min at the rate of 5 °C/min, and kept at 325 °C for 15 min. The injector and the detector temperatures were set at 325 °C. Chromatograms were recorded with LabSolution (Shimadzu, Columbia, MD, USA). Sterols were calculated by comparing the area of the samples and internal standards and expressed as mg/100 g of cheese.

2.2.4. Phenol Extraction, Quantitation and Characterization

A liquid–liquid extraction was used to isolate the phenolic fraction from the cheeses and kiwifruit, following Suarez et al. [21] with some modifications. Briefly, 10 mL of methanol/water (80/20, *v/v*) were added to 5 g of sample and homogenized for 2 min with an ULTRATURRAX (IKA®-Werke GmbH & Co. KG, Staufen, Germany). After this, two phases were separated by centrifugation at 637× *g* for 10 min and the supernatant (hydroalcoholic phase) was transferred to a balloon. This step was repeated with 5 mL of methanol and the extracts were combined in the balloon. The hydroalcoholic extracts were then rotary evaporated to a syrupy consistency at 31 °C and dissolved in 5 mL of acetonitrile. Subsequently, the extract was washed with 10 mL of n-hexane and the rejected n-hexane was treated with 5 mL of acetonitrile. The acetonitrile solution was finally rotary evaporated to dryness. It was then re-dissolved in 5 mL of methanol and maintained at −20 °C before the chromatographic analysis.

Total phenol concentration extracts were determined spectrophotometrically by the Folin–Ciocalteu assay [22] using gallic acid as a standard. An aliquot of 1 mL of each extract was mixed with 5 mL of H_2O and 1 mL of Folin–Ciocalteu phenol reagent 1N. The reaction had a duration of 7 min. A total of 10 mL of saturated Na_2CO_3 solution (7.5%) and 5 mL of H_2O were then added and allowed to stand for 90 min before the absorbance of the reaction mixture was measured in triplicate at 750 nm. The total phenol content was expressed as mg of polyphenols per 100 g of cheese.

Individual polyphenol profiles by HPLC analysis were determined according to Kim et al. [23] with slight modifications. Briefly, 20 µL of each sample were analysed using a Prostar HPLC (Varian) with UV-DAD and a C18 reverse phase column (ChromSep HPLC Columns SS 250 mm × 4.6 mm including Holder with ChromSep guard column Omnispher 5 C18). The PDA acquisition wavelength was set in the range of 200–400 nm, with an analogue output channel at wavelength 280 nm width 10 nm. The gradient elution was performed by varying the proportion of solvent A (water–acetic acid, 97:3 v/v) to solvent B (methanol), with a flow rate of 1 mL min−1. The initial mobile phase composition was 100% solvent A for 1 min, followed by a linear increase in solvent B to 63% in 27 min. The mobile phase composition was then brought back to the initial conditions in 2 min for the next run. All the solutions prepared were filtered through 0.45 µm membranes.

2.2.5. Volatile Organic Compounds Analysis

The volatile organic compounds were determined by solid phase microextraction–gas chromatography–mass spectrometry (SPME-GC/MS), according to Serra et al. [24]. Briefly, volatile organic compounds (VOCs) were extracted from 5 g of a finely-ground sample in a 20-mL glass vial closed with an aluminium cap equipped with a PTFE-septum. Samples were incubated for 15 min and then VOCs were collected using a divinylbenzene/carboxen/polydimethylsiloxane (DVB/Carboxen/PDMS) Stable Flex SPME fibre (50/30 µm; 2-cm long) (Supelco, Bellefonte, PA, USA).

The SPME fibre was exposed to headspace for 30 min. The conditioning and exposure were carried out at 60 °C [25]. The fibre was inserted into the injector of a single quadrupole GC/MS (TRACE GC/MS, Thermo-Finnigan, Waltham, MA, USA) set at 250 °C, 3 min in splitless mode, keeping the fibre in the injector for 30 min to obtain complete fibre desorption.

The GC programme conditions were the same as those described by Serra et al. [24]. The GC was coupled with a Varian CP-WAX-52 capillary column (60 m × 0.32 mm; coating thickness 0.5 μm). The transfer-line and the ion source were both set at 250 °C. The filament emission current was 70 eV. A mass range from 35 to 270 *m/z* was scanned at a rate of 1.6 amu/s. The acquisition was carried out by electron impact, using the full scan (TIC) mode. Three replicates ($n = 3$) were run per sample. The VOCs were identified in three different ways: (i) comparison with the mass spectra of the Wiley library (version 2.0-11/2008); (ii) injection of authentic standards; and (iii) calculation of the linear retention index (LRI) and matching with reported indexes [26–28]. Data were expressed as the peak percentages of the total VOCs.

2.3. Statistical analysis

JMP software (SAS Institute Inc., Cary, NC, USA) was used for the statistical analysis. Data were analysed with the following mixed linear model:

$$y_{ij} = \mu + C_i + B_j\,(C_i) + e_{ij} \tag{1}$$

where y_{ij} = the dependent variables (physico-chemical component, lactodimography data, fatty acids, sterols, polyphenols) relative to the ith coagulant and to the jth batch; μ = the mean; C_i = the fixed effect of the ith coagulant (BM-C vs. BM-K or SM-C vs. SM-K); $R_j\,(C_i)$ = the random effect of the jth batch (1 or 2) nested within C_i; and e_{ij} = the random residual.

3. Results

3.1. Technological Parameters

The milk-clotting activity could only be assessed visually, as the r value could not be detected (higher than 30 min) either for BM-K or SM-K (Table 1).

Table 1. Technological parameters of the cheeses made with kiwi extract or calf rennet.

	Buffalo				Sheep			
	BM$_C$	BM$_K$	SEM	S	SM$_C$	SM$_K$	SEM	S
Yield (%)	51.25	33.18	2.82	**	24.90	21.27	1.10	*
Whey volume (mL)	84.67	120.67	7.12	**	130.83	143.83	1.77	***
Yield after 24 h (%)	38.95	27.85	1.71	**	23.15	20.52	0.87	*
Weight reduction (%)	23.78	15.18	1.83	**	6.73	3.52	0.87	*
pH whey 24 h	5.23	5.16	0.14	ns	4.62	4.92	0.02	ns
Clotting time (min)	13.50	21.00	0.23	***	11.50	16.50	0.22	***
r (min)	10.96	n.r.	3.7	ne	7.38	n.r.	6.03	ne
k20 (min)	1.95	n.r.	0.41	ne	1.33	n.r.	1.66	ne
a30 (mm)	42.47	n.r.	6.20	ne	41.67	n.r.	2.86	ne

BMC, buffalo–calf rennet cheese; BMK, buffalo–kiwifruit cheese; SMC, sheep–calf rennet cheese; SMK, sheep–kiwifruit cheese; SEM, standard error medium; S, significance. *, $0.01 < P < 0.05$; **, $0.01 < P < 0.001$; ***, $P < 0.001$; ns: not significant; n.r., not reactive; ne: not estimable.

The first appearance of solid material after adding the kiwifruit extract was observed after 20–22 min and 15–17 min in buffalo and sheep milk, respectively. These times were significantly higher than the milk with rennet calf (Table 1).

The kind of coagulant was a significant variation factor with respect to whey volume, and to initial cheese yield (Table 1). These differences became smaller after 24 h (rennet cheese weight decreased more than the kiwi extract) but the cheese yield was still significant.

3.2. Physico-Chemical Composition

The chemical compositions and colour characteristics of the cheeses are reported in Table 2. Dry matter was on average 40% similar to a typical fresh cheese. Buffalo cheeses showed a lower protein, fat and ash content than sheep cheeses.

Table 2. Physico-chemical composition of the cheeses made with kiwi extract or calf rennet.

	Buffalo				Sheep			
	BM_C	BM_K	SEM	S	SM_C	SM_K	SEM	S
Total solid (g/100 g)	35.89	44.44	1.83	+++	43.68	50.69	0.46	+++
Proteins (g/100 g)	10.76	11.74	0.26	+	15.83	15.74	0.23	ns
Lipids (g/100 g)	20.60	27.40	1.24	++	21.72	27.55	0.36	+++
Ashes (g/100 g)	0.99	1.27	0.03	+++	1.33	1.69	0.05	+++
Carbohydrates	4.05	4.53	0.75	ns	5.16	5.52	0.47	ns
Fe (μg/g)	0.71	0.81	0.09	ns	1.57	1.93	0.09	ns
Mg (μg/g)	266.01	254.73	14.05	ns	268.27	271.63	4.58	ns
K (μg/g)	1721.61	2136.11	132.05	ns	1727.81	1696.26	56.39	ns
Ca (μg/g)	7985.18	7938.71	259.73	ns	8131.28	8689.25	285.27	ns
Na (μg/g)	522.83	346.34	42.47	+	1173.25	1004.39	23.37	+
Colour								
L*	93.73	92.15	0.37	+++	93.12	92.09	0.24	+
a*	−1.75	−1.83	0.04	ns	−1.88	−1.73	0.05	ns
b*	8.68	10.62	0.45	+	12.17	12.33	0.17	ns
E*	1.84		0.68		1.05			0.36

BM_C, buffalo–calf rennet cheese; BM_K, buffalo–kiwifruit cheese; SM_C, sheep–calf rennet cheese; SM_K, sheep–kiwifruit cheese; ΔE between cheese from calf rennet—kiwifruit, $\Delta E^* = ((\Delta L^*)2 + (\Delta a^*)2 + (\Delta b^*)2)1/2$. Values are expressed as the mean ± standard deviation; SEM, standard error medium; S, significance. +, $0.01 < P < 0.05$; ++, $0.01 < P < 0.001$; +++, $P < 0.001$; ns: not significant.

The kiwi extract produced a cheese with a higher dry matter, both in buffalo and sheep milk (Table 2), except for proteins, which were not affected by the different holding whey capacity.

Iron, magnesium, potassium and calcium were not affected by cheesemaking. The kind of coagulant affected the sodium content of the cheese. Cheeses produced with the kiwifruit extract had the lowest amount of sodium (-34% and -14% in buffalo and sheep cheese, respectively).

The kiwi extract significantly affected the colour of the cheese (Table 2), reducing the lightness both in sheep and buffalo cheese. The use of kiwi also increased the b* value in buffalo cheese, but not in sheep cheese. Total colour differences (E) were less than 2.

3.3. Polyphenols

Although polyphenols are water-soluble, most were found in the curd but very few in the whey (Figure 1).

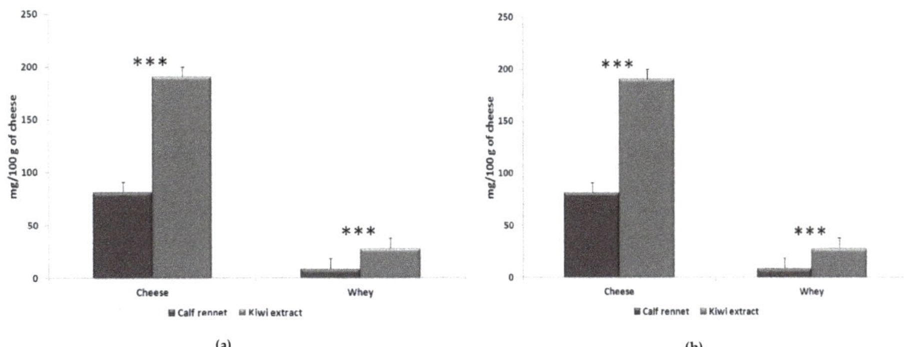

Figure 1. Total polyphenol content of cheese and whey produced with kiwifruit extract and calf rennet. (**a**) Buffalo milk; (**b**) sheep milk. *** $P < 0.001$

Cheeses obtained with the kiwi extract showed a significantly higher content of polyphenols, both in buffalo (+466%) and sheep cheeses (+278%) (Table 3).

Table 3. Polyphenol and sterol content of cheeses made with kiwi extract or calf rennet (mg/100 g of cheese).

	Buffalo				Sheep			
	BM_C	BM_K	SEM	S	SM_C	SM_K	SEM	S
Total polyphenols	32.13	149.94	7.65	***	68.02	189.28	13.40	***
Gallic acid	15.39	15.76	1.99	ns	34.24	15.18	4.97	*
Caffeic acid	4.16	35.97	2.72	*	6.32	11.52	1.50	*
Coumaric acid	-	21.60	1.93	ne	-	33.35	6.32	ne
Cinnamic acid	-	10.78	1.31	ne	-	27.33	6.08	ne
Quercetin	2.93	9.19	0.91	*	10.83	10.03	2.25	ns
Catechin	5.49	24.03	4.44	*	5.17	43.49	3.28	*
Rutin	4.15	43.85	2.91	***	-	58.61	9.85	ne
Sterols								
Cholesterol	5.71	5.51	0.14	ns	5.96	5.76	0.18	ns
Stigmasterol	-	0.17	0.00	ne	-	0.83	0.04	ne
Campesterol	-	0.20	0.00	ne	-	0.69	0.02	ne
β-sitosterol	-	0.63	0.02	ne	-	2.29	0.15	ne

BMC, buffalo–calf rennet cheese; BMK, buffalo–kiwifruit cheese; SMC, sheep–calf rennet cheese; SMK, sheep–kiwifruit cheese; SEM, standard error medium; S, significance. *: $0.01 < P < 0.05$; **: $0.01 < P < 0.001$; ***: $P < 0.001$; ns: not significant; ne: not estimable.

The polyphenol content of SM-C was high, because the sheep milk used for our experiments was obtained from grazing animals. Our results are in line with Hilario et al. [29] in cheese produced with milk from pasture fed goats.

Cheeses obtained with the kiwifruit extract were higher both in flavonoids (quercetin, rutin and catechin) and in phenolic acids (caffeic acid, coumaric acid and cinnamic acid). They also contained coumaric and cinnamic acids, which were not present in the calf rennet cheese. The only polyphenol not affected by the cheesemaking was gallic acid.

3.4. Sterols

The use of kiwi extract promoted the accumulation of a small (but not negligible) quantity of phytosterols: 1.00 and 3.81 mg/100 g of total lipids in buffalo and sheep cheeses, respectively. On the other hand, the sterol profile of the rennet cheeses was characterized by cholesterol alone (Table 3).

The phytosterols we found in cheeses were stigmasterol, campesterol and β-sitosterol, which represent the most abundant sterols in kiwifruit [30].

3.5. Volatile Organic Compounds

As expected, kiwifruit and cheeses were different in terms of the quantity and quality of the odorant compounds (Table 4). The total amount of VOCs from the kiwifruits was about seven times higher than from the cheeses. The most represented category of odorants in kiwi were aldehydes (about 85% of total VOCs), and carboxylic acids and ketones in the cheeses (65% of total VOCs).

More specifically, (E)-2-hexenal and hexanal were the most abundant odorants in the kiwi extracts used in this experiment. These volatile aldehydes are produced during fatty acid oxidation by the lipo-oxygenase enzyme [31]. Other substances affecting odour were the ethyl esters of butyric and caproic acids, two alcohols (2-hexen 1-ol and 1-hexanol) and one ketone, 1-penten-3-one (Table 4).

In terms of cheese, the most represented odorant was 2-butanone (a ketone), which was above 25% of the total VOCs in all the cheeses, followed by acetic acid, 2,3-butenedione (another ketone), caproic, caprylic and nonanoic acids (carboxylic acids). 2-butanone and 2,3-butenedione were not detected in the kiwi pulp, while acetic, caproic, caprylic and nonanoic acids were identified both in the kiwi pulp and in the buffalo and sheep cheeses produced, and both using calf rennet and kiwifruit extract. On the other hand, caproic, caprylic and nonanoic acids are typical components of both buffalo [32] and sheep milk [33].

Terpenes are typical components of kiwi pulp but not of cheese (except for β-phellandrene).

Of the odorants that most characterize kiwifruit, in both buffalo and sheep cheeses we found only (E)-2-hexenal (representing about 80% of VOCs in pulp kiwifruit), ethyl caproate and 2-hexen-1-ol. Interestingly, in the kiwi-cheeses we found 3-methyl eicosane and dibutyl formaldehyde, which were found only in low quantities in the kiwi extract. Finally, cheesemaking was a significant variation factor for the total esters only in buffalo cheese, 3-pentanone 2-hydroxy and β-phellandrene, which was the only terpene we detected in the cheeses.

Table 4. Volatile organic compounds of cheeses made with kiwifruit extract or calf rennet (μg/kg of raw matter) (Part 1).

	Kiwi	Buffalo				Sheep			
		BM$_C$	BM$_K$	SEM	S	SM$_C$	SM$_K$	SEM	S
Acids									
Acetic acid	4.00	9.06	9.06	2.58	ns	2.58	3.03	0.43	ns
Butyric acid	1.56	2.26	1.58	0.48	ns	2.72	3.27	0.74	ns
Caproic acid	3.22	6.28	3.63	1.34	ns	7.65	9.65	2.38	ns
heptanoic acid	1.31	0.95	1.11	0.16	ns	0.79	0.57	0.07	*
caprylic acid	2.31	6.17	7.23	2.88	ns	10.53	13.95	2.83	ns
n-nonanoic acid	3.55	4.97	3.34	1.04	ns	3.83	4.83	0.83	ns
n-decanoic acid	1.69	1.78	2.61	0.76	ns	5.36	8.21	1.60	ns
Total	*17.64*	*31.46*	*28.56*	*6.77*	*ns*	*33.46*	*43.50*	*8.21*	ns
Alcohols									
2-methyl-3-pentanol	-	1.00	2.03	0.28	*	2.15	0.58	0.14	***
2-methyl-1-undecanol	-	1.39	1.98	0.53	ns	1.83	0.90	0.13	***
2-buthyl-1-octanol	-	0.77	0.81	0.19	ns	1.65	0.88	0.35	ns
1-pentanol	2.92	0.51	0.49	0.06	ns	0.87	1.29	0.16	*
1-hexanol	11.12	1.31	0.91	0.16	ns	0.67	0.91	-	ns
3,4-hexane diol	-	3.31	6.11	1.26	ns	3.67	9.52	0.98	**
3-hexen-1-ol	1.67	-	-	-	-	-	-	-	-
2-hexen 1-ol	12.96	-	0.34	0.05	ne	-	0.46	0.07	ne
1-octen-3-ol	0.99	0.30	0.49	0.03	**	0.11	0.15	0.01	*
2,4,7,9-tetramethyl-5-decyne-4,7-diol	1.49	2.84	0.82	0.23	***	1.61	2.73	0.77	ns
Total	*31.11*	*11.42*	*13.97*	*1.64*	*ns*	*12.56*	*17.41*	*1.79*	ns

Table 4. *Cont.*

	Kiwi	Buffalo				Sheep			
		BM$_C$	BM$_K$	SEM	S	SM$_C$	SM$_K$	SEM	S
Aldehydes									
Acetaldehyde	1.65	-	-	-	-	0.16	0.21	0.09	ns
Hexanal	24.48	3.82	3.22	1.01	ns	3.21	2.67	0.70	ns
(*E*)-2-hexenal	710.37	-	1.06	0.09	ne	0.09	1.82	0.26	***
2-heptenal	0.52	-	-	-	-	-	-	-	-
Nonanal	6.66	0.49	0.55	0.13	ns	0.57	0.63	0.07	ns
2-octenal	1.79	-	-	-	-	-	-	-	-
Decanal	0.89	-	-	-	-	-	-	-	-
trans-2-decenal	2.11	-	-	-	ns	-	-	-	-
Undec-2-enal	1.50	-	-	-		-	-	-	-
Dibutyl formaldehyde	7.12	-	0.48	-	ne	-	0.21	0.02	ne
Total	*757.09*	*4.31*	*5.31*	*1.17*	*ns*	*4.03*	*5.54*	*0.88*	*ns*
Alkanes and Alkenes									
2,2 dimethyl decane	-	1.07	1.21	0.37	ns	1.53	1.01	0.34	ns
2,5,6-trimethyldecane	-	2.90	3.79	0.61	ns	4.86	2.79	0.66	*
2,5-dimethylnonane	-	1.52	-	0.01	ne	1.77	-	0.16	ne
2,6,11-trimethyldodecane	-	1.45	-	0.04	ne	0.85	-	0.12	ne
2,5-dimethylundecane	-	1.40	1.14	0.37	ns	1.29	1.07	0.33	ns
2,3 dimethyl nonane	-	1.13	1.32	0.28	ns	1.53	1.30	0.26	ns
3 methyl decane	1.16	1.46	0.67	0.18	*	1.27	0.98	0.24	ns
3-methyl eicosane	5.19	-	0.80	0.01	ne	-	0.75	0.03	ne
5-methyl-undecane	1.65	0.60	0.54	0.08	ns	0.38	0.82	0.07	**
4,5 dipropyloctane	-	1.19	2.11	0.31	*	1.71	1.06	0.30	ns
5-ethyl decane	-	2.30	1.99	0.47	ns	2.27	2.24	0.39	ns
3,5-dimethyl undecane	1.55	0.46	0.47	0.10	ns	0.27	0.34	0.04	ns
2,4-dimethyl-1-heptene	-	1.11	1.64	0.35	ns	3.11	1.68	0.80	ns
5-methyl-1-undecene	-	1.69	-	0.53	ne	1.73	-	0.11	ne
Total	*9.55*	*15.76*	*13.20*	*1.47*	*ns*	*19.74*	*11.66*	*2.26*	*
Aromatic hydrocarbons									
p-xylene	-	1.28	2.16	0.97	ns	2.34	1.40	0.57	ns
o-xylene	-	0.29	1.79	0.61	ns	1.65	1.09	0.47	ns
Total	*-*	*1.57*	*3.95*	*1.54*	*ns*	*3.99*	*2.48*	*1.03*	*ns*
Esters									
Ethyl acetate	7.86	0.50	0.54	0.06	ns	0.88	0.55	0.14	ns
Ethyl butyrate	11.98	-	-	-	-	-	-	-	-
Ethyl caproate	11.43	-	0.42	0.01	ne	-	0.15	-	-
Ethyl caprilate	1.39	-	-	-	-	-	-	-	-
n-heptyl formate	1.84	0.42	0.37	0.06	ns	-	-	-	-
Total	*34.50*	*0.92*	*1.33*	*0.07*	***	*0.88*	*0.70*	*0.13*	*ns*
Ketones									
2,3-butenedione	-	7.66	6.53	1.11	ns	4.94	5.17	0.72	ns
1-penten-3-one	2.05	-	-	-	-	-	-	-	-
2,3 pentanedione	-	4.70	5.20	0.77	ns	2.82	2.23	0.41	ns
2-butanone	-	30.77	38.30	4.69	ns	38.93	40.90	5.36	ns
3-pentanone 2-hydroxy	-	2.14	3.90	0.64	*	2.32	5.70	0.54	**
2-nonanone	-	0.66	0.43	0.07	ns	0.66	0.77	0.12	ns
Total	*2.05*	*46.42*	*52.66*	*6.62*	*ns*	*49.54*	*55.72*	*6.29*	*ns*
Terpenes									
β-phellandrene	2.95	0.12	0.25	0.01	***	0.45	0.66	0.04	**
m-cymene	1.79	-	-	-	-	-	-	-	-
p-mentha-1,4(8)-diene	2.26	-	-	-	-	-	-	-	-
Pinocanphone	2.61	-	-	-	-	-	-	-	-
3-pinanone	6.21	-	-	-	-	-	-	-	-
2-pinen-4-one	4.64	-	-	-	-	-	-	-	-
Total	*20.46*	*0.12*	*0.25*	*0.01*	***	*0.45*	*0.66*	*0.04*	***

Table 4. *Cont.*

	Kiwi	Buffalo				Sheep			
		BM$_C$	BM$_K$	SEM	S	SM$_C$	SM$_K$	SEM	S
Others									
2-ethyl-hexyl tert-butyl ether	-	1.44	1.92	0.28	ns	1.28	0.96	0.21	ns
Dimethyl disulfide	-	0.57	0.95	0.24	ns	-	-	-	-
2 ethyl hexyl chloroformate	-	2.95	6.15	2.80	ns	2.81	3.61	0.83	ns
a-ethyl-furan	8.12	-	-	-	-	0.42	0.33	0.09	ns
m-d-tert-butyl-benzene	-	0.26	0.45	0.09	ns	0.71	0.60	0.15	ns
Ethylhexanol	4.83	-	-	-	-	-	-	-	-
1-cyclopropyl pentane	2.33	0.23	0.22	0.04	ns	0.39	0.33	0.05	ns
1-hexyl-2-methylcyclopropane	1.78	0.42	0.44	0.10	ns	0.40	0.23	0.03	ns
Ethyl benzene carboxylate	3.88	-	-	-	-	-	-	-	-
Total	*20.94*	*5.85*	*10.11*	*2.48*	*ns*	*6.01*	*6.06*	*1.13*	*ns*
Total VOCs	893.34	117.85	129.34	15.61	ns	130.88	143.55	11.91	ns

BMC, buffalo–calf rennet cheese; BMK, buffalo–kiwifruit cheese; SMC, sheep–calf rennet cheese; SMK, sheep–kiwifruit cheese; SEM, standard error medium; S, significance; ns, not significant; ne, not estimable. *: $0.01 < P < 0.05$; **: $0.01 < P < 0.001$; ***: $P < 0.001$;

4. Discussion

4.1. Tecnological Parameters

Table 1 shows that the technological parameters were affected by the coagulant. As demonstrated by the mL of whey released (Table 1) and by the amount of total solid (Table 2), the higher initial cheese yield from calf rennet was due to a higher whey holding capacity and, consequently, a lower and slower curd syneresis. This result is due to the different proteinase action between actinidine and chymosin. Chymosin is an aspartate proteinase that hydrolyses the Met105–Phe106 linkage of K-casein, cleaving out glycol-macropeptide. This leads to a decreasing polarity of the casein micelle and thus promotes the coagulation of casein. On the other hand, actinidin is a cysteine proteinase that cleaves bonds with basic amino acids in position P1, especially in β-casein, with a consequent reduction in milk clotting [5]. Thus, chymosin hydrolysed k-casein only, generating a more elastic and structured curd. Actinidin, instead, cleaves bonds in different sites, producing a curd with small peptones and with a more rapid syneresis.

4.2. Physic-Chemical Composition

The higher syneresis of both buffalo and sheep curd from kiwi extract was responsible for the higher dry matter of these cheeses compared to those obtained by calf rennet (Table 2). Thus, the higher content of lipids and ashes in BM-K and in SM-K are due to a "concentration" effect. However, the cheese protein content was not affected by the type of coagulant used. This could be due to the higher proteolysis of cheeses obtained with kiwi extract, which produces small peptide fragments (5b), which, in turn, are soluble in whey, thus offsetting the wider syneresis of these kinds of cheeses.

The level of carbohydrates was expected to be lower in cheeses from kiwifruit, as these lose whey to a higher extent. In reality the level was higher than expected, perhaps because kiwi extract provided some sugars.

The mineral composition was in line with Cichoscki et al. [34] (Table 2). In spite of the wider syneresis of cheeses from kiwi and their solubility in water, except for sodium, none of the coagulants affected the mineral content of the cheese. Again, these minerals came to some extent from the kiwifruit extract. The fact that calcium was affected by the cheesemaking procedure is difficult to explain, as it is present in milk in three different forms: ionic, soluble as calcium phosphate and colloidal in apatite bridges within casein micelle.

The kind of coagulant affected the sodium content of the cheeses. In fact, cheeses produced by kiwifruit extract showed the lowest sodium amount. Calf rennet is rich in sodium chloride and sodium benzoate, which are added as a preservative, thus partially explaining the different amount of sodium in the cheeses.

The different proteolytic activity of the two kinds of coagulant may explain the differences in colour of the cheese. In fact, both the lower lightness and the higher yellow index, might be related to the occurrence of proteolysis, which, in turn, is related to cheese browning [35]. Is worth noting that the colour differences between two coagulants were detectable only using instruments, as the total colour differences (ΔE) were much lower than 3, which is the discrimination threshold for the human eye [19].

4.3. Polyphenols

The polyphenol content of BM-K and SM-K was very high, as 100 g of these kinds of cheeses have a similar quantity of polyphenols as edible fruits and vegetables such as oranges (217 mg/100 g) and broccoli (290 mg/100 g) [36].

Compared to calf rennet, the kiwifruit extract produced cheeses that were more than 4.5 and 2.7 times richer in polyphenols in buffalo and sheep cheese, respectively. This difference is due not to the total amount of polyphenols (which was almost the same in BM-K and SM-K), but to the kind of milk used for cheesemaking (the sheep milk comes from grazing animals). This is particularly true for gallic acid, which was not negatively affected by cheesemaking. Gallic acid is one of the most abundant polyphenolic substances in plants used for grazing [37] and is not present in the kiwi fruit extract [38].

Coumaric and cinnamic acid were not found in the calf rennet cheese. They can thus be used as good proxies to assess the benefits of using kiwifruit in cheese manufacturing and to conclude that using kiwifruit in cheese coagulation helps to improve the functional features of cheese. In fact, the positive effects on human health of polyphenols are well known as they are able to fight cancer, diabetes, aging, hypertension, asthma and cardiovascular diseases [39]. They also protect against the oxidation of LDL cholesterol and other lipids in the blood [40].

Although polyphenols are water-soluble, most were found in the curd but very few in the whey (Figure 1). It is well known that polyphenols bind caseins using hydrogen bonds [41], thus becoming insoluble in water [42]. The interaction between the polyphenols and proteins is affected by the pH, temperature, phenolic structure and amino acid profile [43–47], and represents a very interesting means to enrich cheese with polyphenols, and, consequently, to increase the nutritional and functional characteristics of cheese [42].

4.4. Sterols

The cheese produced using the kiwi extract contained not only cholesterol (the typical sterol of animal fat) but also some phytosterols, such as stigmasterol, campesterol and β-sitosterol. These substances have several benefits for human health [48–53]. The level of phytosterols observed in BM-K and SM-K cheeses was insufficiently high to obtain a significant reduction in cholesterol absorption (2–3 g/die), or a corresponding reduction in the blood level of LDL-cholesterol (about 6%–15%) [53]. However, they are another positive feature of cheese obtained by kiwifruit coagulant, which helps to improve the overall nutritional and nutraceutical properties. Again, the presence of phytosterols in cheese may be a proxy to trace kiwi extract as a coagulant.

4.5. Volatile Organic Compounds

It was worth studying the effect of a kiwifruit extract on the volatile organic compounds (VOCs) of cheese, since these substances determine the taste and flavour of cheese, and thus influence the consumer's choice. The key odorants of *Actinidia deliciosa* are (E)-2-hexenal, hexanal, ethyl butyrate and 1-penten-3-one, which give a herbal, sweet, marzipan odour [31]; in addition, ethyl butyrate, ethyl caproate, 2-hexen 1-ol and 1-hexanol and 1-penten-3 are contained in non-negligible quantities in kiwifruit (Table 4). They are fat soluble substances and thus we expected them to have been transferred to the cheese during cheesemaking, negatively affecting its organoleptic properties. However, we found that only (E)-2-hexenal (which represents about 80% of VOCs in kiwifruit extract), ethyl caproate and 2-hexen-1-ol, as well as two others volatile substances, 3-methyl eicosane and dibutyl formaldehyde, with a lower content in the kiwi extract, were transferred from kiwifruit into the buffalo and sheep cheese. These substances were found in the "kiwi cheeses" in quantities never higher than 2 µg/kg of cheese and in total accounted for less than 3.4 µg/kg, making up less than 2.5% of the total VOCs of both cheeses.

On the other hand, 2-butanone 2,3-butenedione made up over 30% of the total VOCs in all the cheeses and thus were not impacted by the cheesemaking. Given that the substances that most affected the aroma of the cheeses were the same, irrespectively of the cheesemaking procedure, and that the typical odorants of the kiwi aroma were transferred to the cheese to a very low extent, this would seem to indicate that cheesemaking with the kiwi extract led to a transfer of some volatile substances into the cheese. Further and specific organoleptic tests to assess whether this affects the aroma and taste of cheese should to be done. Finally, VOCs can be effective in tracing of the cheesemaking process.

5. Conclusions

Curd from kiwifruit showed a higher syneresis, giving a lower cheese yield (both initially and after 24 h) than calf-rennet.

The kiwifruit extract improved the nutraceutical properties of cheese by increasing the amount of polyphenols, which were 4.5 times (buffalo) and 3 times (sheep) higher than in cheese made with calf-rennet and phytosterols, which were only detected in cheese obtained with kiwifruit extract. These characteristics represent an important opportunity to produce cheese with a better nutraceutical quality. Finally, cheesemaking with the kiwi extract led to a transfer of some volatile substances into the cheese; specific tests to assess whether this affects the organoleptic properties of cheese are needed.

The results of this research highlight the possibility of using kiwifruit extract as an alternative to rennet in the coagulation of milk.

Author Contributions: Conceptualization, A.S.; methodology, A.S., G.C., L.C.-R., L.C. and F.C.; writing—original draft preparation, A.S. and G.C.; supervision, A.S. and M.M. All authors have read and agreed to the published version of the manuscript.

Funding: This research received no external funding.

Acknowledgments: In this section you can acknowledge any support given which is not covered by the author contribution or funding sections. This may include administrative and technical support, or donations in kind (e.g., materials used for experiments).

Conflicts of Interest: The authors declare no conflict of interest.

References

1. Lopes, A.; Teixeira, G.; Liberato, M.C.; Pais, M.S.; Clemente, A. New vegetal sources for milk clotting enzymes. *J. Mol. Catal. B Enzym.* **1988**, *5*, 63–68. [CrossRef]
2. Lo Piero, A.R.; Puglisi, I.; Petrone, G. Characterization of "Lettucine", a serine-like protease from Lactuca sativa leaves, as a novel enzyme for milk clotting. *J. Agric. Food Chem.* **2002**, *50*, 2439–2443. [CrossRef]
3. Egito, A.S.; Girardet, J.M.; Laguna, L.E.; Poirson, C.; Mollé, D.; Miclo, L.; Humbert, G.; Gaillard, J.L. Milk-clotting activity of enzyme extracts from sunflower and albizia seeds and specific hydrolysis of bovine-casein. *Int. Dairy J.* **2007**, *17*, 816–825. [CrossRef]
4. Heimgartner, U.; Pietrzak, M.; Geertsen, R.; Brodelius, P.; Da Silva Figueiredo, A.C.; Pais, M.S. Purification and partial Characterization of milk clotting proteases from towers of Cynara cardunculus L. (*Cardoon*). *Phytochemistry* **1990**, *29*, 1405–1410. [CrossRef]
5. Lo Piero, A.R.; Puglisi, I.; Petrone, G. Characterization of the purified actinidin as a plant coagulant of bovine milk. *Eur. Food Res. Techol.* **2011**, *233*, 517–524. [CrossRef]
6. Saha, B.C.; Hayashi, K. Debittering of protein hydrolysates. *Biotechol. Adv.* **2001**, *19*, 355–370. [CrossRef]

7. Jung, K.A.; Song, T.C.; Han, D.; Kim, I.H.; Kim, Y.E.; Lee, C.H. Cardiovascular protective properties of kiwifruit extract in vitro. *Biol. Pharm. Bull.* **2005**, *28*, 1782–1785. [CrossRef] [PubMed]

8. Rush, E.C.; Patel, M.; Plank, L.D.; Ferguson, L.R. Kiwifruit promotes laxation in the elderly. *Asia Pac. J. Clin. Nutr.* **2002**, *11*, 164–168. [CrossRef] [PubMed]

9. Park, Y.-S.; Namiesnik, J.; Vearasilp, K.; Leontowicz, H.; Leontowicz, M.; Barasch, D.; Nemirovski, A.; Trakhtenberg, S.; Gorinstein, S. Bioactive compounds and the antioxidant capacity in new kiwi fruit cultivar. *Food Chem.* **2014**, *165*, 354–361. [CrossRef]

10. Katsaros, G.I.; Tavantzis, G.; Taoukis, P.S. Production of novel dairy products using actinidin and high pressure as enzyme activity regulator. *Innov. Food Sci. Emerg. Technol.* **2010**, *11*, 47–51. [CrossRef]

11. Mazorra-Manzano, M.A.; Perea-Gutiérrez, T.C.; Lugo-Sánchez, M.E.; Ramirez-Suarez, J.C.; Torres-Llanez, M.J.; González-Córdova, A.F.; Vallejo-Cordoba, B. Comparison of the milk-clotting properties of three plant extracts. *Food Chem.* **2013**, *141*, 1902–1907. [CrossRef] [PubMed]

12. Puglisi, I.; Petrone, G.; Lo Piero, A.R. A kiwi juice aqueous solution as coagulant of bovine milk and its potential in Mozzarella cheese manufacture. *Food Bioprod. Process.* **2014**, *92*, 67–72. [CrossRef]

13. Bradford, M.M. A rapid and sensitive method for quantitation of microgram quantities of protein utilizing the principle of protein-dye binding. *Anal. Biochem.* **1976**, *72*, 248–254. [CrossRef]

14. Uchikoba, T.; Kaneda, M. Milk-clotting activity of cucumisin, a plant serine protease from melon fruit. *Appl. Biochem. Biotechnol.* **1996**, *56*, 325–330. [CrossRef]

15. Bittante, G.; Penasa, M.; Cecchinato, A. Genetics and modeling of milk coagulation properties. *J. Dairy Sci.* **2012**, *95*, 6843–6870. [CrossRef]

16. MacDougall, D.B. The chemistry of colour and appearance. *Food Chem.* **1986**, *21*, 283–299. [CrossRef]

17. Wyszeck, G.; Stiles, W.S. *Color Science: Concepts and Methods, Quantitative Data and Formulae*, 2nd ed.; John Wiley & Sons: New York, NY, USA, 1982.

18. Sanz, T.; Salvador, A.; Baixauli, R.; Fiszman, S.M. Evaluation of four types of resistant starch in muffins. II. Effects in texture, colour and consumer response. *Eur. Food Res. Technol.* **2009**, *229*, 197–204. [CrossRef]

19. Rodriguez-Estrada, M.T.; Penazzi, G.; Caboni, M.F.; Bertacco, G.; Lercker, G. Effect of different cooking methods on some lipid and protein components of hamburgers. *Meat Sci.* **1997**, *45*, 365–375. [CrossRef]

20. Sander, B.D.; Addis, P.B.; Park, S.W.; Smith, D.E. Quantification of cholesterol oxidation products in a variety of foods. *J. Food Prot.* **1989**, *52*, 109–114. [CrossRef]

21. Suárez, M.; Macià, A.; Romero, M.P.; Motilva, M.J. Improved liquid chromatography tandem mass spectrometry method for the determination of phenolic compounds in virgin olive oil. *J. Chromatogr. A* **2008**, *1214*, 90–99. [CrossRef]

22. Kim, D.O.; Jeong, S.W.; Lee, C.Y. Antioxidant capacity of phenolic phy-tochemicals from various cultivars of plums. *Food Chem.* **2003**, *81*, 321–326. [CrossRef]

23. Kim, D.O.; Chun, O.K.; Kim, Y.J.; Moon, H.-Y.; Lee, C.Y. Quantification of polyphenolic and their antioxidant capacity in fresh plums. *J. Agric. Food Chem.* **2003**, *51*, 6509–6515. [CrossRef] [PubMed]

24. Serra, A.; Buccioni, A.; Rodriguez-Estrada, M.T.; Conte, G.; Cappucci, A.; Mele, M. Fatty acid composition oxidation status and volatile organic com-pounds in "Colonnata" lard from Large White or Cinta Senese pigs as affected by curing time. *Meat Sci.* **2014**, *97*, 504–512. [CrossRef] [PubMed]

25. Yu, A.N.; Sun, B.G.; Tian, D.T.; Qu, W.Y. Analysis of volatile compounds in traditional smoke-cured bacon (CSCB) with different fiber coatings using SPME. *Food Chem.* **2008**, *110*, 233–238. [CrossRef]

26. Lee, S.J.; Umano, K.; Shibamoto, T.; Lee, K.G. Identification of volatile components in basil (*Ocimum basilicum* L.) and thyme leaves (*Thymus vulgaris L.*) and their antioxidant properties (2005). *Food Chem.* **2005**, *91*, 131–137. [CrossRef]

27. Mahattanatawee, K.; Perez-Cacho, P.R.; Davenport, T.; Rouseff, R. Comparison of three lychee cultivar odor profiles using gas chromatography-olfactometry and gas chromatography-sulfur detection. *J. Agric. Food Chem.* **2007**, *55*, 1939–1944. [CrossRef]

28. Povolo, M.; Contarini, G.; Mele, M.; Secchiari, P. Study of the influence of pasture on volatile fraction of sheeps dairy products by solid-phase microextraction and gas chromatography-mass spectrometry. *J. Dairy Sci.* **2007**, *90*, 556–559. [CrossRef]

29. Hilario, M.C.; Delgadillo Puga, C.; Ocaña, A.N.; Pèrez-Gil Romo, F. Antioxidant activity, bioactive polyphenols in Mexican goats' milk cheeses on summer grazing. *J. Dairy Res.* **2010**, *77*, 20–26. [CrossRef]

30. Piironen, V.; Toivo, J.; Puupponen-Pimiä, R.; Lampi, A.M. Plant sterols in vegetables, fruits and berries. *J. Sci. Food Agric.* **2003**, *83*, 330–337. [CrossRef]

31. Garcia, C.V.; Quek, S.Y.; Stevenson, R.J.; Winz, R.A. Kiwifruit flavour: A review. *Trends Food Sci. Technol.* **2012**, *24*, 82–91. [CrossRef]

32. Uzun, P.; Masucci, F.; Serrapica, F.; Napolitano, F.; Braghieri, A.; Romano, R.; Manzo, N.; Esposito, G.; Di Francia, A. The inclusion of fresh forage in the lactating buffalo diet affects fatty acid and sensory profile of mozzarella cheese. *J. Dairy Sci.* **2018**, *101*, 6752–6761. [CrossRef] [PubMed]

33. Serra, A.; Conte, G.; Ciucci, F.; Bulleri, E.; Corrales-Retana, L.; Cappucci, A.; Buccioni, A.; Mele, M. Dietary linseed supplementation affects the fatty acid composition of the sn-2 position of triglycerides in sheep milk. *J. Dairy Sci.* **2018**, *101*, 6742–6751. [CrossRef] [PubMed]

34. Cichoscki, A.J.; Valduga, E.; Valduga, A.T.; Tornadijo, M.E.; Fresno, J.M. Characterization of Prato cheese, a Brazilian semi-hard cow variety: Evolution of physico-chemical parameters and mineral composition during ripening. *Food Control* **2002**, *13*, 329–336. [CrossRef]

35. Mukherjee, K.K.; Hutkins, R.W. Isolation of galactose fermenting thermophilic cultures and their use in the manufacture of low browning Mozzarella cheese. *J. Dairy Sci.* **1994**, *77*, 2839–2849. [CrossRef]

36. Cieslik, E.; Greda, A.; Adamus, W. Contents of polyphenols in fruit and vegetables. *Food Chem.* **2006**, *94*, 135–142. [CrossRef]

37. Cuchillo, H.M.; Puga, D.C.; Wrage-Mönning, N.; Espinosa, M.J.G.; Montaño, B.S.; Navarro-Ocaña, A.; Ledesma, J.A.; Díaz, M.M.; Pérez-Gil, R.F. Chemical composition, antioxidant activity and bioactive compounds of vegetation species ingested by goats on semiarid rangelands. *J. Anim. Feed Sci.* **2013**, *22*, 106–115. [CrossRef]

38. Pinelli, P.; Romani, A.; Fierini, E.; Remorini, D.; Agati, G. Characterisation of the Polyphenol Content in the Kiwifruit (Actinidia deliciosa) Exocarp for the Calibration of a Fruit-sorting Optical Sensor. *Phytochem. Anal.* **2013**, *24*, 460–466. [CrossRef]

39. Pandev, K.B.; Rizvi, S.I. Plant polyphenols as dietary antioxidants in human health and disease. *Oxidative Med. Cell. Longev.* **2009**, *2*, 270–278. [CrossRef]

40. Heinonen, I.M.; Lehtonen, P.J.; Hopia, A.I. Antioxidant activity of berry and fruit wines and liquors. *J. Agric. Food Chem.* **1998**, *46*, 25–31. [CrossRef]

41. Hagerman, A.E. *Phenolic Compounds in Food and Their Effects on Health. I—Analysis, Occurrence and Chemistry*; ACS symposium series 506 237–247; American Chemical Society: Washington, DC, USA, 1992.

42. Han, J.; Britten, M.; St-Gelais, D.; Champagne, C.P.; Fustier, P.; Salmieri, S.; Lacroix, M. Polyphenolic compounds as functional ingredients in cheese. *Food Chem.* **2011**, *124*, 1589–1594. [CrossRef]

43. Bartolomé, B.; Estrella, I.; Hernández, M.T. Interaction of low molecular weight phenolics with proteins (BSA). *J. Food Sci.* **2000**, *65*, 617–621. [CrossRef]

44. Naczk, M.; Oickle, D.; Pink, D.; Shahidi, F. Protein precipitating ca-pacity of crude canola tannins: Effect of pH, tannin and protein concentrations. *J. Agric. Food Chem.* **1996**, *44*, 2144–2148. [CrossRef]

45. Ricardo-da-Silva, J.M.; Cheynier, V.; Souquet, J.M.; Moutounet, M. Interaction of grape seed procyanidins with various proteins in relation to wine fining. *J. Sci. Food Agric.* **1991**, *57*, 111–125. [CrossRef]

46. Serafini, M.; Maiani, G.; Ferro-Luzzi, A. Effect of ethanol on red wine tannin-protein (BSA) interactions. *J. Agric. Food Chem.* **1997**, *45*, 3148–3151. [CrossRef]

47. Spencer, C.M.; Cai, Y.; Martin, R.; Gaffney, S.H.; Goulding, P.N.; Magnolato, D.; Lilley, T.H.; Haslam, E. Polyphenol complexation—Some thoughts and observations. *Phytochemistry* **1988**, *27*, 2397–2409. [CrossRef]

48. Pollack, O.J. Reduction of blood cholesterol in man. *Circulation* **1953**, *7*, 702–706. [CrossRef]

49. Grundy, S.M.; Mok, H.Y. Determination of cholesterol absorption in man by intestinal perfusion. *J. Lipid Res.* **1977**, *18*, 263–271.

50. Mattson, F.H.; Volpenhein, R.A.; Erickson, B.A. Effect of plant sterol esters on the absorption of dietary cholesterol. *J. Nutr.* **1977**, *107*, 1139–1146. [CrossRef]

51. Thompson, G.R. Additive effects of plant sterol and stanol esters to statin therapy. *Am. J. Cardiol.* **2005**, *96*, 37–39. [CrossRef]

52. Kritchevsky, D.; Chen, S.C. Phytosterols—Health benefits and potential concerns: A review. *Nutr. Res.* **2005**, *25*, 413–428. [CrossRef]

53. Casas, R.; Castro-Barquero, S.; Estruch, R.; Sacanella, E. Nutrition and Cardiovascular Health. *Int. J. Mol. Sci.* **2018**, *19*, 3988. [CrossRef] [PubMed]

Article

Applicability of Confocal Raman Microscopy to Observe Microstructural Modifications of Cream Cheeses as Influenced by Freezing

Marcello Alinovi [1,*], Germano Mucchetti [1], Ulf Andersen [2], Tijs A. M. Rovers [2], Betina Mikkelsen [2], Lars Wiking [3] and Milena Corredig [3]

[1]   Food and Drug, University of Parma, Parco Area delle Scienze, 47/A 43124 Parma, Italy; germano.mucchetti@unipr.it

[2]   Arla Innovation Centre, Arla Foods, Agro Food Park 19, 8200 Aarhus, Denmark; ulf.andersen@arlafoods.com (U.A.); tijs.albert.maria.rovers@arlafoods.com (T.A.M.R.); bm@arlafoods.com (B.M.)

[3]   Department of Food Science and iFOOD Center for Innovative Food, Aarhus University, Agro Food Park 48, 8200 Aarhus, Denmark; lars.wiking@food.au.dk (L.W.); mc@food.au.dk (M.C.)

*   Correspondence: marcello.alinovi@studenti.unipr.it; Tel.: +39-334-328-3206

Received: 29 April 2020; Accepted: 22 May 2020; Published: 25 May 2020

Abstract: Confocal Raman microscopy is a promising technique to derive information about microstructure, with minimal sample disruption. Raman emission bands are highly specific to molecular structure and with Raman spectroscopy it is thus possible to observe different classes of molecules in situ, in complex food matrices, without employing fluorescent dyes. In this work confocal Raman microscopy was employed to observe microstructural changes occurring after freezing and thawing in high-moisture cheeses, and the observations were compared to those obtained with confocal laser scanning microscopy. Two commercially available cream cheese products were imaged with both microscopy techniques. The lower resolution (1 µm/pixel) of confocal Raman microscopy prevented the observation of particles smaller than 1 µm that may be part of the structure (e.g., sugars). With confocal Raman microscopy it was possible to identify and map the large water domains formed during freezing and thawing in high-moisture cream cheese. The results were supported also by low resolution NMR analysis. NMR and Raman microscopy are complementary techniques that can be employed to distinguish between the two different commercial formulations, and different destabilization levels.

Keywords: microstructure; Raman spectroscopy; confocal laser scanning microscopy; cheese freezing; cream cheese; NMR spectroscopy; cryoprotectants

---

1. Introduction

When a sample is subjected to a monochromatic light source, a small proportion of the radiation is scattered depending on the physical and chemical properties of its components. The interactions between the laser light and the molecular vibrations cause a shift in energy between the incident and the scattered light. Vibrational spectroscopy methods such as infrared, mid- and near-infrared, and Raman are commonly used to assess structures and identify molecular species. Raman scattering spectra are specific to the chemical functional groups and their bonds' vibrational frequency, and can be used to distinguish molecular structures [1]. In Raman microscopy, the emission spectra are collected at high spatial resolution, and the band pattern is specific to the sample's composition. Furthermore the intensity is related to the concentration of specific component, if the signal of such component is sufficient to overcome the signal-to-noise ratio. Raman microscopy has gained interest as a tool to characterize food products in situ, with minimal sample preparation. This technique has been reported to identify components (fat, proteins, water) in polyphasic systems such as mayonnaise, semi-hard Swiss cheese, and soymilk [2]. Raman microscopy has been employed to investigate differences in composition of milk fat globules [3], to classify cheeses on the basis of their different microstructural organization, to process spectra with the aid of multivariate tools [4], to study the formation and the cause of structural modifications (i.e., crystallization) during ripening in hard and extra-hard cheeses [5]. Furthermore, the microstructural features of a cheese matrix have been analyzed by Raman microscopy, by observing the spatial distribution of ingredients (macromolecules, water, paprika, trisodium citrate, phospholipids) in Cheddar and imitation cheeses [6,7]. Both qualitative and quantitative analysis can be carried out by Raman microscopy. Because of the distinct signal of different chemical species, the use of fluorescence dyes in sample preparation is not necessary. It has been recently shown that it is possible to resolve spatially the NaCl concentration in butter samples [8] and to predict the solid fat content of anhydrous milk fat [1]. In sum, confocal Raman microscopy may be applied not only to observe food microstructure but also to gain information about geometries, distances, angles, and polarizability of the chemical bonds present in the structure for the different components of the matrix [1,9].

These characteristics may represent an important advantage of confocal Raman microscopy if compared to the more conventional non-disruptive technique used to study the microstructure of complex food matrices [10]: confocal laser scanning microscopy. Confocal laser scanning microscopy necessitates specific dyes to observe different compounds and therefore it requires some sample preparation. Furthermore, the observations would be limited by the dies and their interactions with the molecules, and the number of wavelengths (i.e., laser sources) available for the observation [11].

In this work, confocal Raman microscopy was applied together with laser scanning microscopy to investigate the microstructure of high moisture cheese, and to identify changes in the distribution of the molecular species caused by processing and storage; the comparison of the obtained results can be useful to benchmark and to compare the benefits and disadvantages related to both microscopy techniques. Despite the high potential of Raman microscopy, this technique has some drawbacks: it can be time-consuming, and the laser beam applied in Raman spectroscopy can heat, dry or damage the sample [12]. Long analysis times and the increase in temperature of the sample can result in analytical challenges (e.g., moisture losses, free surface flows, thickness variations), especially in high moisture samples. Furthermore, in spite of the chemical specificity, some of the components may not be detected in mixed systems, because of low signal thresholds [13,14].

Cream cheese is a soft, fresh cheese with a slightly acidic taste [15,16]. This cheese variety is characterized by a moisture content usually higher than 65%. Cream cheeses can also be categorized according to their fat content [16]. To obtain the right structure, the fat to protein ratio with the high moisture content are often balanced by the application of heat treatments of the cheese, in order to increase the interactions between whey proteins, caseins and water [16]. As for yogurt products, to encounter the preference of a larger number of consumers the acidic taste of cream cheese may be modified by addition of sugars.

Depending on commercial standards, this product may contain various stabilizers and emulsifiers that modify its structure and functionality, properties that are especially important when cream cheese is used for further processing. Freezing is an important process applied to cream cheese in cases where robust supply chains are needed, for example, to extend market reach, to provide food service customers with a consistent product, or to decrease waste and improve shelf life by reducing the rate of degradation during refrigerated storage [17–19]. Hence, freezing is becoming a convenient tool for export or when the product is used as an ingredient for further manufacturing [20]. However, from a physical point of view, freezing can have a negative impact on the cheese matrix: it can cause the rupture of the casein matrix as a consequence of ice crystals formation, with the creation of voids and large serum channels in the structure [21,22], the modification of the water status and distribution (e.g., protein dehydration phenomena) [23], and some rheological and sensory changes [17].

In cream cheese products, non-dairy ingredients are often added to improve technological functionalities. To preserve the structure of the protein network, which partially gives the cheese its characteristics, and to limit freezing-destabilization, different additives are used. Sorbitol, polyphosphate, gelatin, polyols, sucrose and other sugars or polysaccharides are often incorporated as cryoprotectants [24] and act via different mechanisms. Sugars and polyols stabilize proteins through their effect on the water fraction and modification of solvent quality, critical to keep the protein network intact. Carbohydrates or polyols can also cause the formation of hydrogen bonds with protein side chains and increase protein hydration while decreasing protein-protein interactions. Polysaccharides can bind water and form large complexes, and increase the viscosity of the continuous phase.

This project focuses on evaluating the potential application of confocal Raman microstructure tools to observe changes in the structure and in the distribution of the components. In particular, to evaluate the ability to observe changes in the quality of cream cheese after freezing and thawing with confocal Raman microscopy, compared to confocal laser scanning microscopy. The observations are supported by measurements of water mobility using low resolution NMR, for products subjected to freezing and thawing. Two commercial cream cheeses with a different protein to fat ratio and containing different amounts and types of stabilizers, and expected to have different stabilities during freezing and thawing, were used as model systems in this study.

2. Materials and Methods

2.1. Cream Cheese Treatments

Two different commercial full fat cream cheese products (A and B) were selected as having a different composition in terms of macro components and stabilizers, their composition is reported in Table 1. In particular, cream cheese B had a higher protein and carbohydrates content, and a higher number of stabilizers in its formulation, than cream cheese A. The cheese samples were collected and processed in portions of 180 g and stored in polypropylene plastic boxes (2.5 cm × 10 cm × 7 cm). Samples were frozen at −20 °C in still air conditions; after 7 days of frozen storage at −20 °C, cheeses were thawed at 4 °C overnight. Analyses were conducted on non-frozen cheese before freezing (control sample), and after thawing. For each treatment, at least 5 replicates were performed, or 5 images were obtained for each measured sample.

Table 1. Compositional information of full fat cream cheese samples.

Composition (g/100 g)	Cream Cheese A	Cream Cheese B
Fat	23.0	24.0
Carbohydrates	3.4	15.0
Fibre	1.1	1.5
Protein	5.4	8.7
Salt	0.72	0.58
Stabilizers	Carrageenan, locust bean gum	Carrageenan, locust bean gum, gelatin, citrus fibre

2.2. Cheese Preparation for Microscopy

Cream cheese samples stored at 4 °C were sectioned with a knife in a thin layer of approximately 5 mm and were then positioned on a microscope slide. Another microscope slide was placed on the upper layer and sample's outer layer was coated with paraffin oil to avoid moisture evaporation during analysis. Samples were left to equilibrate at 22 °C for 15 min in order to reduce possible changes of thickness caused to by temperature changes during the analysis. For laser scanning confocal measurements, Nile red and fluorescein isothiocyanate (FITC) were used as fluorescent staining agents for fats and proteins, respectively. Both staining agents were dissolved in acetone (0.01%). One or two drops of staining solution were placed on a microscope slide and acetone was allowed to evaporate before sample addition, as previously described [25]. Samples were left in contact with the dyes for approximately 5 min before imaging.

2.3. Microstructural Analyses

Microstructural analyses of the cheeses were carried out using a Nikon Eclipse Ti2 confocal laser scanning microscope (Minato, Tokyo, Japan) equipped with 10×, 40× and 100× objectives. The Ar/Kr and HeNe laser beam were set at 488 nm and 561 nm, respectively. The micrographs were acquired using the 40× objective (with the exception of Figures 6 and 7 that were acquired with the 10× objective) in 1024 × 1024 pixels sections, with each image covering an area of 375 × 375 μm; each image was obtained as the average of four frames. As the laser penetration depth was lower than 10–15 μm according to Auty [26], samples were imaged close to the surface of each specimen.

Raman images obtained using an Alpha 300 R instrument (WITec, Ulm, Germany) were produced by taking 50 × 50 Raman spectra in a uniform 50 × 50 μm grid, with a final resolution of 1 μm/px. A laser wavelength of 532 nm with a laser power of 20 mW was used. Laser intensity value was chosen as it was the highest one that did not cause an excessive temperature increase of the sample, that could possibly alter the cheese characteristics during the analysis. Despite the maximum depth penetration of the laser source was higher than 40–50 μm, the measurements were performed at a penetration depth of ~10 μm, in order to maximize the signal intensity [27,28]. A 50× objective was used to observe microstructure of samples; an integration time of 0.20–0.10 s was used, depending on sample characteristics and stability. Analysis time was about 5–10 min depending on the integration time used. Data were computed into microstructure images using WITec Suite FIVE software package. Cosmic ray removal, Savitzky Golay smoothing and baseline offset correction were used as pre-processing methods to improve spectra quality and to reduce noise, without the loss of any important information [6]. Microstructural images were created from the hyperspectral datasets by performing a multivariate clustering analysis and the identification of the components was made by comparing spectral data of the samples with a reference database of cheese ingredients. Reference spectra were collected for all the ingredients present in the cheese formulations, as indicated in Table S1 (Supplementary Material). Raman spectra were collected using a 10× objective, a laser power of 30 mW, an integration time of 5 s and performing 30 accumulations. Reference spectra were finally obtained by performing three single acquisitions in different sampling points of the surface of the reference sample and by averaging the signals.

2.4. NMR Analysis

Proton self-diffusion coefficient (D) measurements were carried out using a low-field NMR spectrometer (Bruker MiniSpec, Bruker BioSpin GmbH, Rheinstetten, Germany) operating at 20 MHz and at a temperature of 20 °C with a Pulsed Field Gradient Spin-Echo (PFG-SE) method. The probe was calibrated with a solution containing 1.25 g/L $CuSO_4$ $5H_2O$ and pure water (D = 1.98×10^{-9} m^2 s^{-1} at 20 °C) at 10, 30, 50% pulsed gradient amplitude. For each measurement, a total of 16 echoes were acquired as a function of the gradient pulse duration using a recycle delay of 2 s. The gradient pulse width was set at 0.5 ms, and the gradient pulse separation (Δ) was set to 7.5 ms. Approximately 4.5 g of samples were placed into NMR tubes (10 mm external diameter). The tubes were filled with sample to about 15 mm height. A gradient pulse of 40% was used for the analyses.

2.5. Statistical Analysis

^1H self-diffusion coefficient (D) data were statistically analyzed for both the cream cheese formulations to observe possible effects related to the freezing and thawing processes. One-way Analysis of Variance (ANOVA) was carried out using SPSS v.25 (IBM, Armonk, NY, USA), considering a significance level $\alpha = 0.05$.

3. Results

3.1. Raman Spectra Database Collection

Reference spectra were collected for the ingredients present in the cheese samples. The whole spectra dataset is reported as Supplemental Information (Figures S1–S14, Supplementary Material). Figures 1 and 2 show the spectra collected for the main components present in the cream cheese samples: water, casein proteins (measured as micellar casein powder), and the lipid fraction, that was constituted by both butter and rapeseed oil. Carrageenan, citrus fibre, gelatin and locust bean gum did not show Raman emission peaks when measured in reference samples. For these components, it was just possible to observe a variation of the baseline at different wavenumbers, as already reported by Yang et al. [13]. On the other hand, it was possible to identify peaks of sorbic acid molecules (Figure S2, Supplementary Material), a strong peak was observable at 1635 cm^{-1} which was attributed to C=O or C-C stretches; identification was made according to Kai et al. [29].

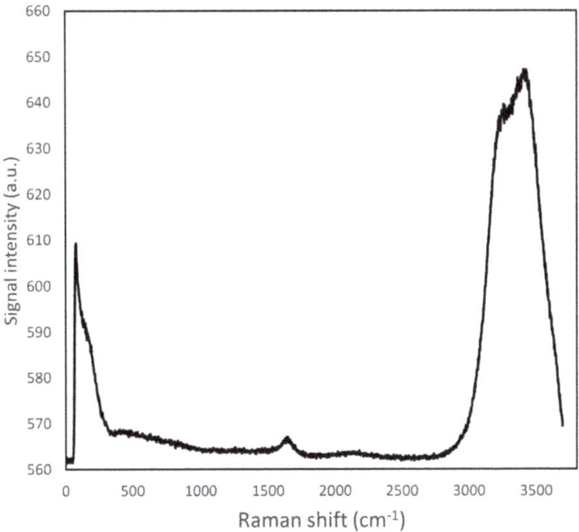

Figure 1. Raman spectrum of ultra-pure water; a large emission band can be viewed in the region around 2900–3800 cm^{-1}.

Sugars (lactose as a residue of fermentation and sucrose, mainly in cheese B) that were present in both formulations, showed several emission peaks in the range between 300 and 1700 cm^{-1} (Figures S8 and S9, Supplementary Material).

The Raman emission spectrum of water showed a strong, broad peak around 2900–3800 cm^{-1} (Figure 1), far away from the fingerprint region of many molecules, including fat and protein [6]. Fat and protein components showed a large peak around 2800–3000 cm^{-1} that can be attributed to CH$_2$, CH$_3$ stretching; however, because of this overlapping, this signal cannot be used to identify and to discriminate the single components. The most important region of the spectra is located in the range between 500–1800 cm^{-1}. This spectral region contains fingerprint peaks that are typical of a class of components (i.e., fat or protein) or of single molecules [2] and that can be useful to identify and map the presence and the distribution of the different components in a complex matrix. In particular, proteins are characterized by 7 important peaks: tryptophan indole ring (750–760 cm^{-1}), phenylalanine ring breathing (1000–1005 cm^{-1}), β-sheets and α-helices structures related to amide III (1270 cm^{-1} and 1320–1340 cm^{-1}, respectively), CH$_2$ bending peak (1450 cm^{-1}), tyrosine (1614 cm^{-1}), and Amide I

(1650–1670 cm^{-1}), and, as it can be observed in details in Figures S13 and S14 (Supplementary Material) and in Figures 2A and 3 [6,30–32].

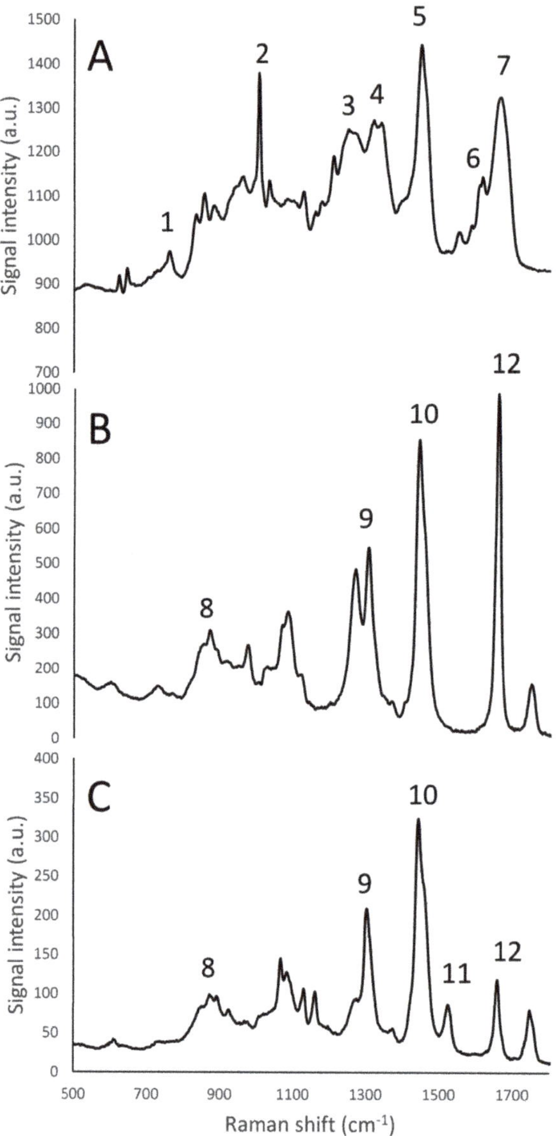

Figure 2. Raman fingerprint region (500–1800 cm^{-1}) of micellar casein isolate (**A**), rapeseed oil (**B**), and butter (**C**). Panel A: (1) tryptophan indole ring (750–760 cm^{-1}); (2) phenylalanine benzene ring breathing (1000–1005 cm^{-1}); (3) amide III β-sheet structure (1270 cm^{-1}); (4) amide III α-helices structure (1320–1340 cm^{-1}) (5) CH$_2$ bending peak (1450 cm^{-1}); (6) tyrosine (1614 cm^{-1}); (7) amide I (1650–1670 cm^{-1}). Panels B, C: (8) phospholipids headgroup (870 cm^{-1}); (9) CH$_2$ twisting of phospholipids (1300–1310 cm^{-1}); (10) CH$_2$ scissoring of sterols (1442 cm^{-1}); (11) Carotenoids stretching band (1525 cm^{-1}); (12) C=C *cis* stretching of phospholipids (1655 cm^{-1}).

Of these Raman emission peaks, the phenylalanine peak has been proved to be the most useful to visualize the protein distribution in cheeses, because of the absence of other interfering emission peaks in its spectral range [6].

The whey proteins α-lactalbumin and β-lactoglobulin showed different relative amide III peak intensities that are related to differences in terms of secondary protein structure, as the α-lactalbumin structure is mainly composed by α-helices, and β-lactoglobulin is mainly characterized by β-sheets [33]; in particular, the band located at 1240 cm^{-1}, corresponding to β-sheet secondary structures of amide III [32] had an higher intensity in β-lactoglobulin than in α -lactalbumin spectrum, while conversely the band located at 1320–1340 cm^{-1}, corresponding to α-helix secondary structures of amide III [32], showed a higher intensity in the case of α -lactalbumin than β-lactoglobulin (Figure 3).

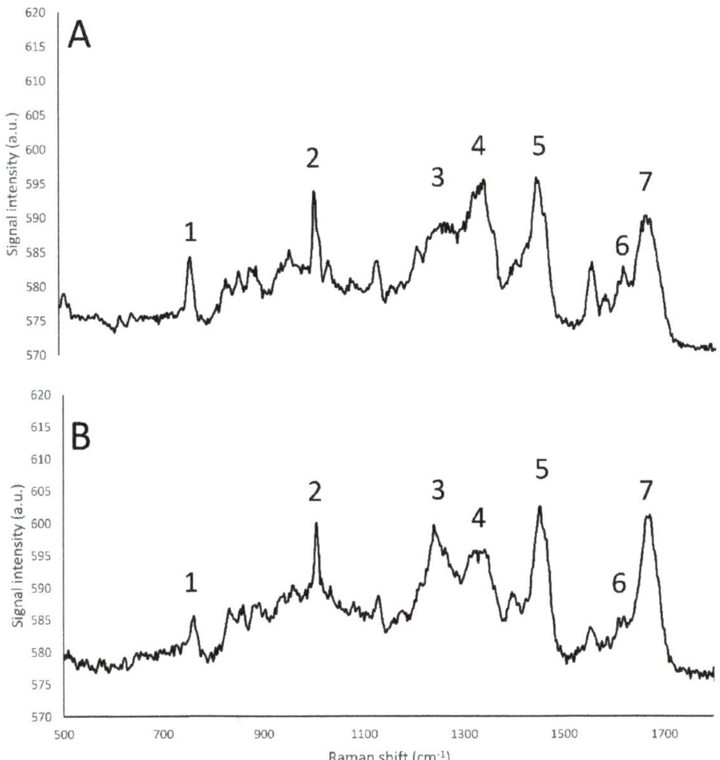

Figure 3. Raman fingerprint region (500–1800 cm^{-1}) of α-lactalbumin (**A**), and β-lactoglobulin (**B**). (1) tryptophan indole ring (750–760 cm^{-1}); (2) Phenylalanine benzene ring breathing (1000–1005 cm^{-1}); (3) amide III β-sheet structure (1270 cm^{-1}); (4) amide α -helices structure (1320–1340 cm^{-1}); (5) CH$_2$ bend peak (1450 cm^{-1}); (6) tyrosine (1614 cm^{-1}); (7) amide I (1650–1670 cm^{-1}).

The lipid fraction showed an important peak, representing the CH$_2$ scissoring of sterols at 1442 cm^{-1} (Figure 2B,C). Other peaks that could be identified [6] corresponded to: the CH$_2$ twisting of phospholipids (1300–1310 cm^{-1}), phospholipids headgroup (870 cm^{-1}) and C=C cis stretching of phospholipids (1655 cm^{-1}). Furthermore, rapeseed oil and butter, containing different quantities of saturated and unsaturated fatty acids, show different intensities of the peak at 1655 cm^{-1}. Butter is also characterized by the presence of a carotenoid stretching band at 1525 cm^{-1}. These peaks could also be used to quantify or identify and discriminate the presence of different lipid components in products with unknown composition, by the development of proper calibrations [34]. From Figure 2B,C it was possible to note one of the most intense fat emission peak was the CH_2 scissoring of sterols, in accordance with Gallier et al. [3]. This peak has already been used to map with Raman microscopy the fat distribution in cheese matrices [6].

In the following Raman microscopy observations of cream cheese, the proteins were identified using the phenylalanine's benzene ring breathing peak (1000–1005 cm^{-1}) or the amide III peak related to the β-sheets structure (1270 cm^{-1}), while the lipid molecules were visualized by the CH_2 scissoring band of sterols (1442 cm^{-1}).

3.2. Microstructural Observations

The microstructural observations of the cheese samples obtained by confocal Raman microscopy (Figure 4) were in agreement with those obtained with conventional laser scanning microscopy (Figure 5) in terms of microstructural changes caused by freezing and thawing for the protein and fat domain. Moreover, these observations were representative of the cheese samples, as observable from microstructure observations made on different sampling aliquots (Figures S15 and S16). In the case of confocal Raman microscopy, it was possible to obtain additional information regarding the distribution of water in the samples, thanks to the peak outside the fingerprint region (3000–3800 cm^{-1}). This information was not obtainable with confocal laser scanning microscopy: in this case the appearance of voids in between the protein and fat domains may correspond to the serum domain and other water-soluble compounds present in the cheeses.

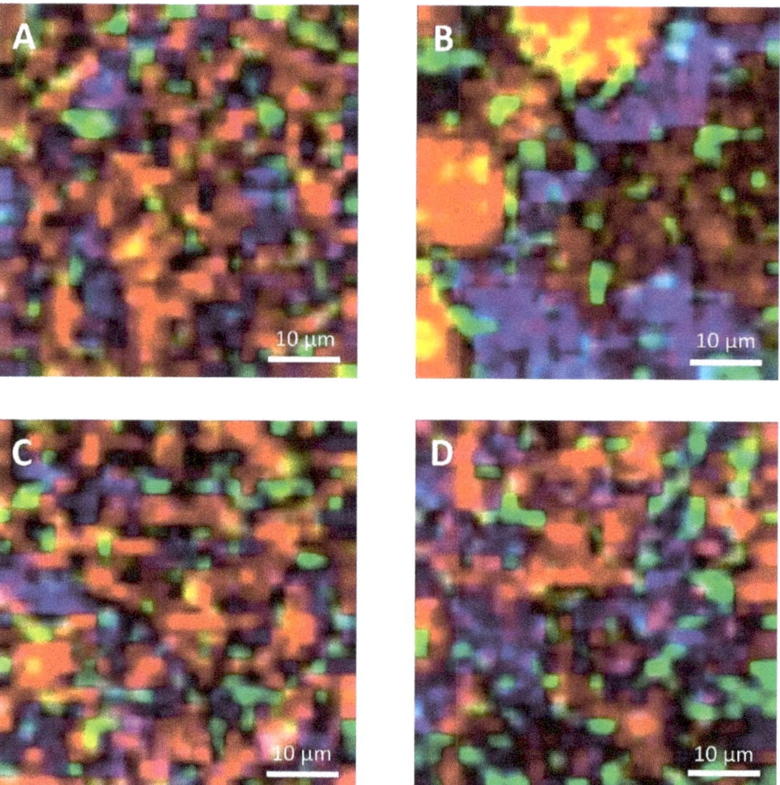

Figure 4. Confocal Raman micrographs of cream cheeses. Cream cheese A before freezing (**A**), after freezing (**B**). Cream cheese B before freezing (**C**), after freezing (**D**). Red: fat. Green: protein. Blue: water.

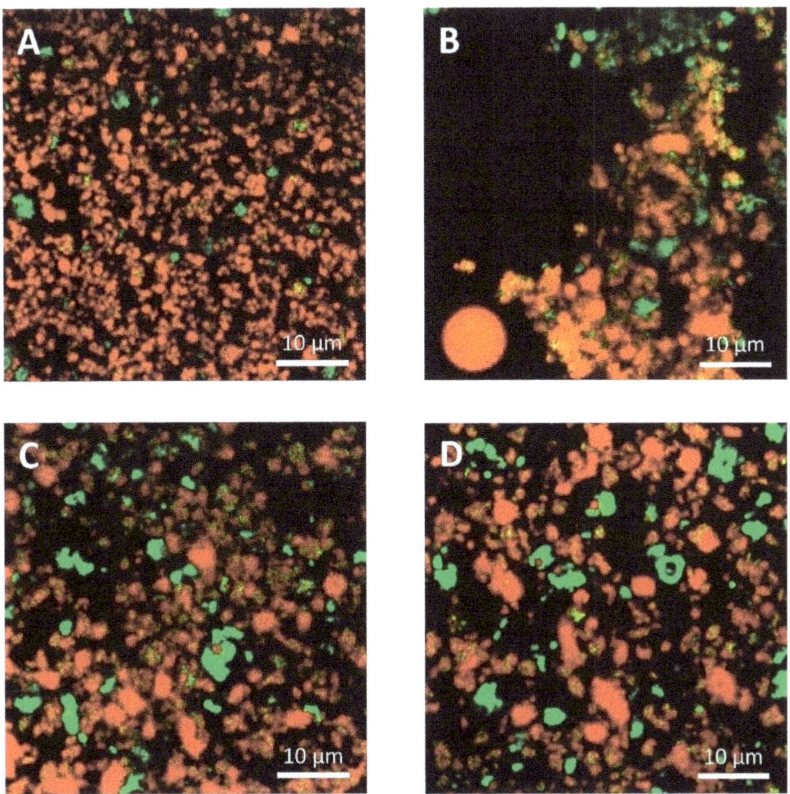

Figure 5. Confocal laser scanning micrographs of cream cheeses. Cream cheese A before freezing (**A**), after freezing (**B**). Cream cheese B before freezing (**C**), after freezing (**D**). Red: fat. Green: protein.

On the contrary, because of the limitations imposed to acquisition parameters that were selected to guarantee the samples' stability during analyses and also to limit the time required for each measurement, Raman micrographs were characterized by a limited and lower image resolution than conventional laser scanning microscopy. Because of this limitation, with Raman microscopy it was not possible observe fat globules, protein aggregates or other components having dimensions lower than 1 µm, that were observable with confocal laser scanning microscopy (Figure 5). Moreover, according to this limitation it was also not possible to detect carbohydrates (mainly sugars) that were present in high concentrations in both formulations, and some of the additives present in low concentrations. For example, it was not possible to detect sucrose, lactose, and sorbic acid, with the last one that showed a strong intensity of its signal with a peak at 1629 cm^{-1} in the reference sample (Figure S2, Supplementary Material).

With confocal laser scanning microscopy, fresh cream cheeses A and B showed a slightly different microstructural organization: cheese B compared to cheese A showed the presence of slightly bigger fat globules and fat globule aggregates that form clusters (Figure 5A,C); moreover, the lipid fraction of cheese B was surrounded by a denser protein matrix, compared to cheese A, in accordance with their higher protein content (Table 1). These differences were not visible in Raman observations (Figure 4A,C), as a consequence of the limited resolution of the acquired images.

With both microscopy techniques, it was possible to observe that in both formulations, a proteins fraction was adsorbed on the surface of fat globules. Cheese B showed a higher amount of adsorbed protein probably because the higher protein/fat ratio (0.36, if compared to 0.23 of cheese A) [35].

In Figure 3A,C and Figure 4A,C, it was also possible to visualize voids around fat globules and the protein matrix. These voids probably contained other ingredients (stabilizers, sugars, etc.) dispersed in the interstitial solution. These components were not directly detected neither by the confocal Raman or the conventional laser scanning microscopy procedures. The differences in the protein distribution at the interface, and the organization of the fat globules clusters, may have a role in modulating the freeze-thawing stability of the different cheese formulations; for example, a higher amount of adsorbed proteins on the surface of fat globules/clusters may result in a reduction of fat globules clustering and coalescence, and an improved stability.

Cheese A showed significant destabilization after thawing, fat aggregation and the formation of clumps/granules of proteins/fat (Figure 6) with large channels of free water (Figure 4B). Thus, it was clear that in this sample, a strong separation of water from the solid components of the cheese occurred after freezing. On the other hand, cream cheese B appeared less modified by freezing (Figure 7) and its microstructure was substantially different from that of frozen thawed cheese A (Figure 6). There were smaller changes in this sample: it was still possible to observe slightly larger voids between the protein/fat structure in the frozen-thawed sample (Figures 4 and 5D) compared to the unfrozen control (Figures 4 and 5C), and this microstructure difference was attributed to a slight separation of water and the formation of larger serum channels (Figure 4D).

The two model systems were chosen as differences in formulations would result in differences in freezing stability: cheese B, contained more stabilizers, proteins and a higher carbohydrates and sugar content (Table 1) than cheese A. As expected, Cheese B was more stable, and showed only slight structural modifications, because of the effect of these ingredients acting as cryoprotectants and in general as stabilizers during freezing and thawing.

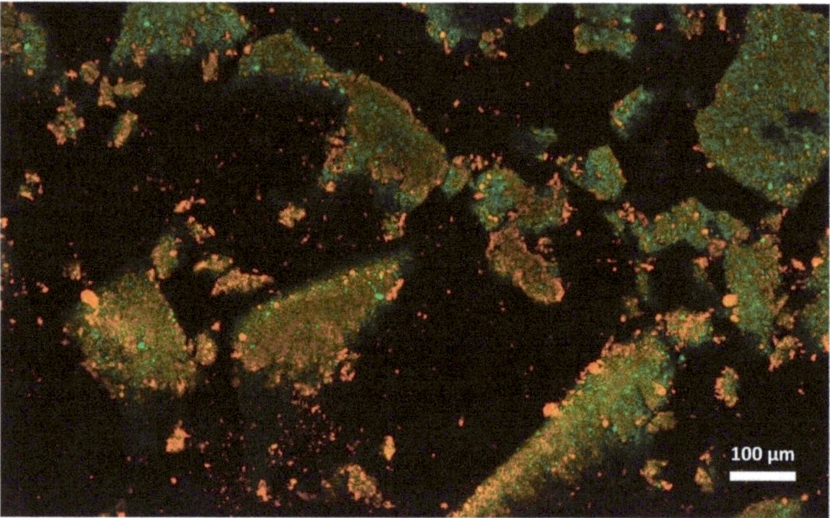

Figure 6. Microstructure of cream cheese A observed with confocal laser scanning microscope after freezing and thawing. Red: fat. Green: protein.

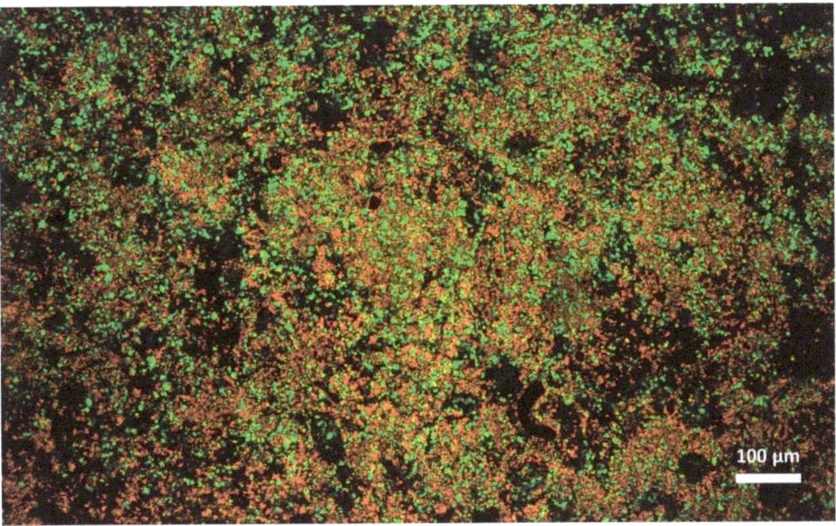

Figure 7. Microstructure of cream cheese B observed with confocal laser scanning microscope after freezing and thawing. Red: fat. Green: protein.

3.3. Water Mobility Changes in Cream Cheese after Thawing

The microstructure observations described in the previous section (3.2) were supported by low resolution NMR measurements on the two cheese samples. As reported in Table 2, an increase in the translational mobility of the water protons, was reflected by the increased D values after freezing. The ^{1}H self-diffusion coefficient (D) measured by low resolution NMR showed a significant increase ($p < 0.05$) for cheese B, from 0.624×10^{-9} m^2 s^{-1} of the control, unfrozen cheese, to 0.651×10^{-9} m^2 s^{-1} after freezing and thawing. However, in the case of cheese A, the diffusion coefficient did not show a significant change ($p > 0.05$). This was due to the large variability of the matrix after freezing and thawing, resulting in a high standard deviation of the ^{1}H self-diffusion coefficient (30% coefficient of variation). It is well known that freezing process can damage the structure of cheese [36–38]. In particular, it has been reported that casein gels may rearrange due to the growth of ice crystals; as water migrates to form ice crystals, protein-protein interactions occur [39]. After thawing, the protein structures may not be able to rebind the amount of water depleted, resulting in a higher value of D and a larger amount of unbound water. Hori [20] reported that the amount of unbound water in frozen-thawed cream cheeses measured by NMR is inversely proportional to the rates of freezing-thawing processes, and directly proportional to the extent of freeze and thaw-induced damages.

In cream cheese A, the freezing-thawing treatments caused a high degree of destabilization and non-homogeneity in the physical structure of the cheese, as already pointed out by the high standard deviation of D coefficient; this phenomenon was also highlighted by Raman and laser scanning microscopical observations that showed the presence of solid clumps of proteins-fat surrounded by channels of free water; conversely, this was not observed in the case of cheese B (Figures 4–7). While the lack of homogeneity was clearly shown in microstructure measurements, NMR analyses underestimated the D values in the solid portion of the cheese, and a higher D value in the areas rich in free serum, in non-homogenous samples.

Table 2. Averaged ^{1}H self-diffusion coefficient (D) ($\times 10^{-9}$ m^2 s^{-1}) of cream cheese formulations A and B before and after freezing and thawing.

Treatment	Cheese A (*n* = 5)		Cheese B (*n* = 5)	
	Mean D ($\times10^{-9}$ m^2 s^{-1})	Standard Deviation	Mean D ($\times10^{-9}$ m^2 s^{-1})	Standard Deviation
Control cheese	0.927	0.005	0.624 *	0.007
Frozen-thawed cheese	1.05	0.31	0.651 *	0.019

* Asterisks within the same column indicate significant difference ($p < 0.05$).

4. Conclusions

Confocal Raman microscopy showed the potential to be applied to study process-related microstructural modifications in high moisture spreadable dairy products, such as cream cheese. With Raman microscopy it was possible to observe structural differences, especially in terms of phase separation and water pockets re-organization. On the other hand, under the conditions reported in this work, with confocal laser scanning microscopy better results were obtained, in terms of fat and protein structure visualization. The use of confocal Raman microscopy allows one to better distinguish the effect of freezing-thawing processes of the two different cream cheeses. This was coherent with the low resolution of Raman microscopy obtained with the measurement parameters applied in this study; the parameters were limited in order to avoid any possible sample damage or modification that could be expected by a higher extent of heat generated by the laser source in the case of a slower but more resolute scan. Therefore, it was not possible to visualize the presence and distribution of sugars and different stabilizers, which probably contributed to the different freezing stability of the cheeses. It is also important to note that the Raman microscopy data complemented the information derived from low resolution NMR.

These first results suggest that confocal Raman microscopy may become in the future an additional key analytical tool because it can provide relatively rapid, non-destructive and online food quality evaluation, that could be complementary to confocal laser scanning microscopy.

Further studies need to be carried out on the quantitative assessment of components and ingredients, by image analysis techniques and/or by further elaborating spectral information.

Supplementary Materials: The following are available online at http://www.mdpi.com/2304-8158/9/5/679/s1, Figure S1–S12: Raman spectra of reference samples: (1) Carrageenan, (2) Sorbic acid, (3) Citrus fibre, (4) Gelatine, (5) α-lactalbumin, (6) β-lactoglobulin (7) Whey protein concentrate, (8) Sucrose, (9) Lactose, (10) Skim milk powder, (11) Micellar casein isolate, (12) Calcium caseinate, Figure S13–S14: Raman fingerprint region (500-1800 cm^{-1}) of protein components: (13) Calcium caseinate, (14) Whey protein concentrate. (A) Phenylalanine's benzene ring breathing (1000-5 cm^{-1}); (B) Amide III (1270 cm^{-1}); (C) Tyrosine (1614 cm^{-1}); (D) Amide I (1650-70 cm^{-1}), Figure S15: Supplementary Confocal Raman micrographs of cream cheeses. Cream cheese A before freezing (A), after freezing (B). Cream cheese B before freezing (C), after freezing (D). Red: fat. Green: protein. Blue: Water, Figure S16. Supplementary Confocal laser scanning micrographs of cream cheeses. Cream cheese A before freezing (A), after freezing (B). Cream cheese B before freezing (C), after freezing (D). Red: fat. Green: protein. Table S1: Reference samples used to acquire Raman spectroscopy peaks.

Author Contributions: Conceptualization, M.A., U.A., L.W. and M.C.; methodology, M.A.; software, M.A., T.A.M.R.; formal analysis, M.A.; investigation, M.A., T.A.M.R., B.M.; resources, G.M., U.A. and M.C.; data curation, M.A.; writing—original draft preparation, M.A.; writing—review and editing, G.M., U.A., T.A.M.R., L.W. and M.C.; visualization, M.A.; supervision, G.M. and M.C.; project administration, U.A., L.W. and M.C.; funding acquisition, G.M. and M.C. All authors have read and agreed to the published version of the manuscript.

Funding: This research received no external funding.

Acknowledgments: This research did not receive any specific grant from funding agencies, in the public, commercial or non-for-profit sectors. Authors would like to acknowledge the University of Parma and iFOOD for the support of M. A. mobility grant.

Conflicts of Interest: The authors declare no conflict of interest.

References

1. McGoverin, C.M.; Clark, A.S.S.; Holroyd, S.E.; Gordon, K.C. Raman spectroscopic prediction of the solid fat content of New Zealand anhydrous milk fat. *Anal. Method.* **2009**, *1*, 29–38. [CrossRef]
2. Roeffaers, M.B.J.; Zhang, X.; Freudiger, C.W.; Saar, B.G.; van Ruijven, M.; van Dalen, G.; Xiao, C.; Xie, X.S. Label-free imaging of biomolecules in food products using stimulated Raman microscopy. *J. Biomed. Opt.* **2011**, *16*, 021118. [CrossRef] [PubMed]
3. Gallier, S.; Gordon, K.C.; Jiménez-Flores, R.; Everett, D.W. Composition of bovine milk fat globules by confocal Raman microscopy. *Int. Dairy J.* **2011**, *21*, 402–412. [CrossRef]
4. Burdikova, Z.; Svindrych, Z.; Hickey, C.; Wilkinson, M.G.; Auty, M.A.E.; Samek, O.; Bernatova, S.; Krzyzanek, V.; Periasamy, A.; Sheehan, J.J. Application of advanced light microscopic techniques to gain deeper insights into cheese matrix physico-chemistry. *Dairy Sci. Technol.* **2015**, *95*, 687–700. [CrossRef]
5. D'Incecco, P.; Limbo, S.; Faoro, F.; Hogenboom, J.; Rosi, V.; Morandi, S.; Pellegrino, L. New insight on crystal and spot development in hard and extra-hard cheeses: Association of spots with incomplete aggregation of curd granules. *J. Dairy Sci.* **2016**, *99*, 6144–6156. [CrossRef]
6. Smith, G.P.S.; Holroyd, S.E.; Reid, D.C.W.; Gordon, K.C. Raman imaging processed cheese and its components. *J. Raman Spectrosc.* **2017**, *48*, 374–383. [CrossRef]
7. Hickey, C.D.; Diehl, B.W.K.; Nuzzo, M.; Millqvist-Feurby, A.; Wilkinson, M.G.; Sheehan, J.J. Influence of buttermilk powder or buttermilk addition on phospholipid content, chemical and bio-chemical composition and bacterial viability in Cheddar style-cheese. *Food Res. Int.* **2017**, *102*, 748–758. [CrossRef]
8. Jensen, B.B.; Glover, Z.J.; Pedersen, S.M.M.; Andersen, U.; Duelund, L.; Brewer, J.R. Label free noninvasive spatially resolved NaCl concentration measurements using Coherent Anti-Stokes Raman Scattering microscopy applied to butter. *Food Chem.* **2019**, *297*, 124881. [CrossRef]
9. Falcone, P.M.; Baiano, A.; Conte, A.; Mancini, L.; Tromba, G.; Zanini, F.; Del Nobile, M.A. Imaging Techniques for the Study of Food Microstructure: A Review. *Adv. Food Nutr. Res.* **2006**, *51*, 205–263. [CrossRef]
10. El-Bakry, M.; Sheehan, J. Analysing cheese microstructure: A review of recent developments. *J. Food Eng.* **2014**, *125*, 84–96. [CrossRef]
11. Betz, T.; Teipel, J.; Koch, D.; Härtig, W.; Guck, J.; Käs, J.; Giessen, H. Excitation beyond the monochromatic laser limit: Simultaneous 3-D confocal and multiphoton microscopy with a tapered fiber as white-light laser source. *J. Biomed. Opt.* **2005**, *10*, 054009. [CrossRef] [PubMed]
12. Thygesen, L.G.; Løkke, M.M.; Micklander, E.; Engelsen, S.B. Vibrational microspectroscopy of food. Raman vs. FT-IR. *Trends Food Sci. Technol.* **2003**, *14*, 50–57. [CrossRef]
13. Yang, M.; Liu, X.; Qi, Y.; Sun, W.; Men, Y. Preparation of κ-carrageenan/graphene oxide gel beads and their efficient adsorption for methylene blue. *J. Colloid Interface Sci.* **2017**, *506*, 669–677. [CrossRef] [PubMed]
14. Pudney, P.D.; Hancewicz, T.M.; Cunningham, D.G.; Brown, M.C. Quantifying the microstructures of soft solid materials by confocal Raman spectroscopy. *Vib. Spectrosc.* **2004**, *34*, 123–135. [CrossRef]
15. Phadungath, C. Cream cheese products: A review. *Songklanakarin J. Sci.* **2005**, *27*, 191–199.
16. Schulz-Collins, D.; Senge, B. Acid- and acid/rennet-curd cheeses part A: Quark, cream cheese and related varieties. In *Cheese: Chemistry, Physics and Microbiology*; Fox, P.F., McSweeney, P.L.H., Cogan, T.M., Guinee, T.P., Eds.; Elsevier Academic Press: London, UK, 2004; Volume 2, pp. 301–328. ISBN 0122636538.
17. Alinovi, M.; Mucchetti, G. Effect of freezing and thawing processes on high-moisture Mozzarella cheese rheological and physical properties. *LWT-Food Sci. Technol.* **2020**, *124*, 109137. [CrossRef]
18. Alinovi, M.; Mucchetti, G. A coupled photogrammetric-finite element method approach to model irregular shape product freezing: Mozzarella cheese case. *Food Bioprod. Process.* **2020**. [CrossRef]
19. Alvarenga, N.; Ferro, S.; Almodôvar, A.; Canada, J.; Sousa, I. Shelf-life extension of cheese: Frozen storage. In *Handbook of Cheese in Heatlh: Production, Nutrition and Medical Sciences*; Preedy, V.R., Watson, R.R., Patel, V.B., Eds.; Wageningen Academic Publishers: Wageningen, The Netherlands, 2013; Volume 6, pp. 73–86. ISBN 978-90-8686-211-5.
20. Hori, T. Effects of Freezing and Thawing Green Curds Before Processing on the Rheological Properties of Cream Cheese. *J. Food Sci.* **1982**, *47*, 1811–1817. [CrossRef]
21. Graiver, N.G.; Zaritzky, N.E.; Califano, A.N. Viscoelastic Behavior of Refrigerated Frozen Low-moisture Mozzarella Cheese. *J. Food Sci.* **2004**, *69*, 123–128. [CrossRef]

22. Kuo, M.-I.; Anderson, M.E.; Gunasekaran, S. Determining Effects of Freezing on Pasta Filata and Non-Pasta Filata Mozzarella Cheeses by Nuclear Magnetic Resonance Imaging. *J. Dairy Sci.* **2003**, *86*, 2525–2536. [CrossRef]
23. Diefes, H.A.; Rizvi, S.S.H.; Bartsch, J.A. Rheological Behavior of Frozen and Thawed Low-Moisture, Part-Skim Mozzarella Cheese. *J. Food Sci.* **1993**, *58*, 764–769. [CrossRef]
24. Xiong, Y.L. Protein Denaturation and Functionality Losses. In *Quality in Frozen Foods*; Erickson, M.C., Hung, Y.C., Eds.; Springer: Boston, MA, USA, 1997; pp. 111–140.
25. Karadağ, A.; Hermund, D.B.; Jensen, L.H.S.; Andersen, U.; Jónsdóttir, R.; Kristinsson, H.G.; Alasalvar, C.; Jacobsen, C. Oxidative stability and microstructure of 5% fish-oil-enriched granola bars added natural antioxidants derived from brown alga Fucus vesiculosus. *Eur. J. Lipid Sci. Technol.* **2017**, *119*, 1500578. [CrossRef]
26. Auty, M.A.E. Confocal microscopy: Principles and applications to food microstructures. In *Food Microstructures: Microscopy, Measurement and Modelling*; Woodhead Publishing Limited: Cambridge, UK, 2013; pp. 96–139. ISBN 9780857095251.
27. Gallardo, A.; Spells, S.; Navarro, R.; Reinecke, H. Confocal Raman microscopy: How to correct depth profiles considering diffraction and refraction effects. *J. Raman Spectrosc.* **2007**, *38*, 1538–1553. [CrossRef]
28. Brillante, A.; Bilotti, I.; Della Valle, R.G.; Venuti, E.; Masino, M.; Girlando, A. Characterization of phase purity in organic semiconductors by lattice-phonon confocal Raman mapping: Application to pentacene. *Adv. Mater.* **2005**, *17*, 2549–2553. [CrossRef]
29. Kai, S.; Chaozhi, W.; Guangzhi, X. Multiple adsorbed states and surface enhanced Raman spectra of crotonic and sorbic acids on silver hydrosols. *J. Raman Spectrosc.* **1989**, *20*, 267–271. [CrossRef]
30. Kizil, R.; Irudayaraj, J. Spectroscopic Technique: Fourier Transform Raman (FT-Raman) Spectroscopy. In *Modern Techniques for Food Authentication*; Sun, D.W., Ed.; Elsevier Inc.: New York, NY, USA; Academic Press: Cambridge, MA, USA, 2018; pp. 193–217. ISBN 9780128142646.
31. Takeuchi, H. Raman structural markers of tryptophan and histidine side chains in proteins. *Biopolym.-Biospectroscopy Sect.* **2003**, *72*, 305–317. [CrossRef]
32. Ngarize, S.; Herman, H.; Adams, A.; Howell, N. Comparison of changes in the secondary structure of unheated, heated, and high-pressure-treated β-lactoglobulin and ovalbumin proteins using Fourier transform Raman spectroscopy and self-deconvolution. *J. Agric. Food Chem.* **2004**, *52*, 6470–6477. [CrossRef]
33. Gulzar, M.; Bouhallab, S.; Jardin, J.; Briard-Bion, V.; Croguennec, T. Structural consequences of dry heating on alpha-lactalbumin and beta-lactoglobulin at pH 6.5. *Food Res. Int.* **2013**, *51*, 899–906. [CrossRef]
34. Tao, F.; Ngadi, M. Applications of spectroscopic techniques for fat and fatty acids analysis of dairy foods. *Curr. Opin. Food Sci.* **2017**, *17*, 100–112. [CrossRef]
35. Wendin, K.; Langton, M.; Caous, L.; Hall, G. Dynamic analyses of sensory and microstructural properties of cream cheese. *Food Chem.* **2000**, *71*, 363–378. [CrossRef]
36. Fontecha, J.; Kaláb, M.; Medina, J.; Peláez, C.; Juárez, M. Effects of freezing and frozen storage on the microstructure and texture of ewe's milk cheese. *Z. für Lebensm. Unters. Forsch.* **1996**, *203*, 245–251. [CrossRef]
37. Ribero, G.G.; Rubiolo, A.C.; Zorrilla, S.E. Microstructure of Mozzarella cheese as affected by the immersion freezing in NaCl solutions and by the frozen storage. *J. Food Eng.* **2009**, *91*, 516–520. [CrossRef]
38. Reid, D.S.; Yan, H. Rheological, melting and microstructural properties of cheddar and mozzarella cheeses affected by different freezing methods. *J. Food Qual.* **2004**, *27*, 436–458. [CrossRef]
39. Bertola, N.C.; Califano, A.N.; Bevilacqua, A.E.; Zaritzky, N.E. Textural changes and proteolysis of low-moisture Mozzarella cheese frozen under various conditions. *LWT-Food Sci. Technol.* **1996**, *29*, 470–474. [CrossRef]

 *foods*

Article

Effect of Black Tea Infusion on Physicochemical Properties, Antioxidant Capacity and Microstructure of Acidified Dairy Gel during Cold Storage

Han Chen [1], Haotian Zheng [2,3,*], Margaret Anne Brennan [1], Wenpin Chen [4], Xinbo Guo [5] and Charles Stephen Brennan [1,*]

[1] Department of Wine, Food and Molecular Biosciences, Lincoln University, Lincoln 7647, New Zealand; jackchen.chen@lincolnuni.ac.nz (H.C.); margaret.brennan@lincoln.ac.nz (M.A.B.)
[2] Department of Food, Bioprocessing and Nutrition Sciences, Southeast Dairy Foods Research Center, North Carolina State University, Raleigh, NC 27695, USA
[3] Dairy Innovation Institute, California Polytechnic State University, San Luis Obispo, CA 93407, USA
[4] Tea Science Department, College of Horticulture, South China Agricultural University, Guangzhou 510642, China; michaelcwp@163.com
[5] School of Food Science of Engineering, South China University of Technology, Guangzhou 510641, China; guoxinbo@scut.edu.cn
* Correspondence: haotian.zheng@ncsu.edu (H.Z.); charles.brennan@lincoln.ac.nz (C.S.B.)

Received: 9 April 2020; Accepted: 22 June 2020; Published: 25 June 2020

Abstract: The impacts of black tea infusion on physicochemical properties, antioxidant capacity and microstructure of stirred acidified dairy gel (ADG) system have not been fully explored. These impacts were studied during a 28-day cold storage (4 °C) period to explore the feasibility and technical boundaries of making acidified dairy gels in which black tea infusion (BTI) is incorporated. Reconstituted skim milks containing different proportions of BTI were acidified by GDL (glucono-δ-lactone) at 35 °C for making ADG systems. Both textural properties and structural features were characterized; antioxidant capacity was determined through three assays. They are (1) free radical scavenging ability by DPPH (2,2-diphenyl-1-picrylhydrazyl) assay; (2) ABTS [2,2′-azino-bis-(3-ethylbenzothiazoline-6-sulphonic acid)] assay and (3) ferric reducing antioxidant power (FRAP) assay. The microstructure of the ADGs was observed using SEM (scanning electron microscopy) and CLSM (confocal laser scanning microscopy). Results showed that BTI significantly increased the antioxidant capacity of the gel systems and the gel containing 15% BTI was as stable as the control gel in terms of syneresis rate. However lower phase stability (higher syneresis rate) was observed in the ADG with a higher portion of BTI (30% to 60%). The microstructure of the ADGs observed may explain to the phase stability and textural attributes. The results suggested that tea polyphenols (TPs) improved antioxidant capacity in all samples and the interactions between BTI and dairy components significantly altered the texture of ADGs. Such alterations were more pronounced in the samples with higher proportion of BTI (60%) and/or longer storage time (28 days).

Keywords: black tea; acidified dairy gel; textural property; antioxidant capacity; microstructure

1. Introduction

Free radicals, by-products of metabolism, such as reactive oxygen species (ROS) are constantly being generated in our body such as hydroxyl radical, superoxide radical, hydrogen peroxide and lipid peroxides [1]. Normally, there is an antioxidant defence system in our body, comprising of several enzymes such as iron-dependent catalase, superoxide dismutase (copper/zinc and manganese-dependent) and selenium-dependent glutathione peroxidase to detoxify these free radicals [2]. However, when there is large number of free radicals, there is a disorder between the generation and removal of free radicals in the body, in which case oxidative stress will occur. This may result in oxidative damage to cellular metabolism and biomolecules, and create the onset of many chronic diseases related to aging such as cardiovascular disease, diabetic disease, neurodegenerative diseases or even cancer [3]. As antioxidants play important roles in preventing or inhibiting oxidation of cellular components, adequate intake of these compounds is beneficial to protect cells from oxidative damages. In this regard, extracts of many polyphenol-rich plants or herbs, such as tea, are used more often either as additive in food industry or consumed directly as a natural source of antioxidants [4].

Tea extracts are rich in phenolic compounds [5], these components including flavonoids are considered as antioxidants [6,7]. Yoghurt has been consumed as a healthy food for a long time since 6000 BCE in central Asia [8] due to its nutritional properties, taste and health benefits as results of fermentation of lactic acid bacteria. Acidified dairy protein gels (ADG) induced by glucono-δ-lactone (GDL) as robust model systems have been commonly used in research for studying ingredient functionalities in yogurt like gels and for studying structure and texture features of such gels [9–12]. To explore the dairy components-tea infusion interactions in an acidic environment, GDL induced gels were used in this study to remove the influence of unnecessary impacts of live culture [13,14].

Although different analytical methods for antioxidant capacity may lead to different results for the same antioxidant [6], previous study showed that supplementing 5–15% of green tea in yogurt may result up to 31-fold higher radical scavenging activity compared to the control yogurt [15]. Muniandy and co-authors reported that during refrigeration storage condition, antioxidant activity for tea enriched yogurt is stable and it remained constant as shown in a study [2]. However, this study only showed comparison of antioxidant activity between blank control yogurt and tea infusion enriched yogurts. It is not clear that if there is any change of antioxidant activity of the original tea infusion when it incorporated in the dairy protein gel system. In another study, the authors showed that there is nearly no change of ferric reducing antioxidant power of the green tea extract either in yogurt or in its original state (based on back calculated values) [15]. However, regarding to black tea extract, such information is unknown and the situation may not necessarily as same as of green tea extract because black tea infusion contains more than 20 times of gallic acid than that of green tea [5]. It was found that radical scavenging activity of the black tea was lowered by adding milk [16]. Such phenomenon might be attributed to the interaction between gallic acid and milk proteins [17], because research showed that interactions between flavonoids and proteins affect their antioxidant capacity [18]. Ryan and Petit also concluded that the addition of neutral pH milk components may reduce total antioxidant activity of black tea [19]. In general, casein micelles and whey proteins can bind to tea catechins [20,21] via hydrogen bonds between peptide carbonyl and phenolic hydroxyl [22]. The unique structure of protein–catechin complex may reduce the bioaccessibility of the original tea catechin [23]. The phenolic compound composition is different between green and black tea [5]. The later has relatively lower total phenolic content [5]. Besides gallic acid-protein interaction, the lower content of phenolic compounds may be another major reason explaining why black tea added yogurt showed the lower DPPH scavenging activity than green tea yogurt [2]. However, it is still not clear that if the original antioxidant activity of black tea fusion may be affected after incorporating in acid dairy protein gel.

In a relatively recent review article, the authors consolidated previously published results regarding the functionality of polyphenols interacting with milk proteins and pointed out that some of these findings are conflicted [24]. The authors therefore suggested more research is needed for understanding the true mechanism of interaction between phenolic compounds and dairy matrices. Numerous studies have been done on the mixture systems such as milk–tea infusion systems [16,19,21,22,25,26] and dairy gel–tea infusion systems [2,15,27–31]. Regarding to the topic of tea infusion enriched yogurt, many studies focused on the impact of tea fortification on the antioxidant activity and microbial growth [15,20–22,29,31]. Few efforts have been done to investigate the impact of storage time on key characteristics of tea infusion enriched ADG systems. Rheology and texture analyser techniques may be useful tools for characterizing textural changes of semi-solid foods such as yogurt gels [32,33]. Najgebauer-Lejko and co-workers found that the mechanical properties and syneresis rate of tea infusion enriched yogurts depended on the type of tea. The study concluded that green tea incorporation resulted relatively favourable texture and lower syneresis compared to Pu-erh tea infusion [27]. However, black tea infusion was not included in that study. It is clear that the overall antioxidant activity of tea–yogurt system is positively correlated with the volume fraction of tea infusion incorporated in the yogurt gel [15]. However, from application point of view, it is still unclear about what is the highest supplementation rate of tea infusion that a stirred ADG can be tolerant without compromising phase stability and textural characteristics. To the best of our knowledge and according to the aforementioned referenced research, the tea-enriched stirred ADG has never been studied with higher supplementation rates of tea (>15%, *w/w*).

Based on the mentioned research questions, the current study aimed to investigate the impact of black tea supplementation on phase stability, textual characteristics, and antioxidant capacity of stirred ADG made from reconstituted milk system. The phase stability of ADG was investigated through cold storage (4 °C for up to 28 days). Tea infusion were incorporated in the ADG systems at different volume fractions (15–60%, *w/w*) for testing the formulation boundary. This work provides an in-depth understanding of interaction between ADG network and black tea infusion. Such knowledge may contribute to the development of dairy products with enhanced nutritional benefits and without compromised texture.

2. Materials and Methods

2.1. Materials

Skimmed milk powder (SMP, fat 1.2%, protein 33.0%, carbohydrate 54%, sodium 0.39%, calcium 1.24%, Pams, Auckland, New Zealand) and bagged black tea (premium Ceylon tea, Dilmah Co., Peliyagoda, Sri Lanka) were purchased from local supermarket (Christchurch, New Zealand). DPPH (2,2-Diphenyl-1-picrylhydrazyl), ABTS (2,2-Diphenyl-1-picrylhydrazyl), TPTZ (2,4,6-Tris(2-pyridyl)-s-triazine), gallic acid were purchased form Sigma-Aldrich Co., New York, NY, USA). Trolox (6-hydroxy-2,5,7,8-tetramethylchloromane-2-carboxylic acid) was purchased from Acros Organic (NS, New York, NY, USA). Phosphate buffered saline was purchased from Oxoid Ltd. (Hampshire, England). Methanol was purchased from ECP-Laboratory Reagent (Auckland, New Zealand). Ultrapure water was generated by a Milli-Q water purification system.

2.2. Preparation of Black Tea Infusion (BTI)

Bagged black tea was used to make tea infusion. One tea bag (2 g-equivalent of tea leaves) per 100 mL RO (Reverse osmosis) water was used to extract polyphenol rich infusions, with the mixture being soaked at 85 °C for 15 min in a water bath. After the extraction, the bags were removed and the solution was cooled to 40 °C in ice water for further experiments [34].

2.3. (Stirred) ADG Preparation

In general, tea-enriched ADG samples were made by acidification of tea-enriched reconstituted skimmed milks (T-RSM). Regarding the preparation of a T-RSM, SMP was dissolved in the water-BTI mixture solvent; the volume fraction of BTI in the water-BTI mixture solvent was manipulated for achieving the targeted proportion of BTI in the overall T-RSM system. The targeted BTI proportions in individual T-RSM samples were 15%, 30%, 45% and 60% (*w/w*). Also, each of the yield T-RSM sample should contain 10% (*w/w*) milk dry matter. Then, ADG samples was prepared directly from the T-RSM samples containing different proportions of tea infusion. RSM samples without the supplement of BTI were used to prepare the negative control ADG samples. The ADG samples were prepared following Vega and Grover [35] method with slight modification. Briefly, for instance, for each replicate milk sample, 200 g of SMP was dissolved in water-BTI mixture solvent to yield 2000 g T-RSM under vortex condition for 30 min and then refrigerated at 4 ± 1 °C overnight before further use. As aforementioned, the volume fraction of BTI in T-RSM was controlled for achieving the targeted proportion. The T-RSM was heated in water bath at 85 °C for 20 min and then rapidly cooled to 40 °C in an ice-water bath.

GDL (1.3% *w/w*) was added and dissolved in the T-RSM system with gentle stir. Different T-RSM (including RSM control) samples were then incubated at 35 °C until pH reached pH 4.55. At this point, the set ADG samples were prepared. Twelve replicate set gels (batches) were made for each sample. Subsequently, the individual set gels were homogenized (using IKA T25 Digital Ultra-Turrax, Werke GmbH & Co. KG, Staufen, Germany, at 4500 rpm, for 1 min) [36,37] for making stirred gels. These stirred gels containing 0%, 15%, 30%, 45% and 60% (*w/w*) of black tea infusion were the key samples studied in this research and these samples are named as $ADG_{0\%}$ (negative control), $ADG_{15\%}$, $ADG_{30\%}$, $ADG_{45\%}$ and $ADG_{60\%}$ respectively. The stirred gels were stored at 4 ± 1 °C for further analysis. Related measurements were carried out on day 1, 7, 14, 21 and 28 of cold storage.

2.4. pH and Ca^{2+} Content Measurement

The pH of ADGs was determined at 20 ± 2 °C using a digital pH meter after calibration (SevenEasy pH, Mettler-toledo GmbH, Schwerzenbach, Switzerland).

The free Ca^{2+} content in ADG system was determined by a portable Calcium Ion-Selective Electrode (B-751, LAQUA Twin Calcium Ca++ Ion Meter, Horiba, Japan) according to Kosasih, et al. [38] and the instruction of the Ca^{2+} meter. The Ca^{2+} meter was calibrated by a 2-point calibration using 150 ppm and 2000 ppm Ca^{2+} standards before use and the error was below 20%. Samples stayed for 1 h at ambient temperature and temperature was measured by a thermometer right before pH and free Ca^{2+} content were measured.

pH and free Ca^{2+} content of stirred ADG samples were carried out on the day 1, day 7, day 14, day 21 and day 28 of cold storage. Measurements were carried out in triplicate.

2.5. Texture Characteristics

Textural Attributes of Stirred Gel. The textural properties of stirred gel were characterized by back-extrusion method described by Ciron, et al. [14]. A 5-kg load cell was used and the samples were tested in cylinder pot (50 mm internal diameter) at 15 °C, using an extrusion disc (Φ = 35 mm) operating at a set-speed of 1.0 mm/s to a 30 mm depth. Firmness and cohesiveness values were calculated from the obtained profiles using the software provided by Stable Microsystems. Textural profile of stirred ADG samples were carried out on the day 1, day 7, day 14, day 21 and day 28 of cold storage. Textural profiles were tested in triplicate for stirred gel samples.

2.6. Phase Stability

The water-holding capacity (WHC) and syneresis rate of ADGs were measured according to the method of Ciron et al. [32] with minor modifications using a centrifuge (Heraeus® Multifuge X3R, Heraeus Co., Hanau, Germany). For measuring WHC, the stirred ADG samples (1.0 g) in 1.5-mL microtubes (Axygen®) were centrifuged at 15,000× *g* for 15 min at 25 °C. The WHC was expressed as % (pellet/sample, *w/w*). The extent of syneresis (EOS) was determined by centrifuging stirred ADG samples (1.0 g) in 1.5 mL tubes at 101× *g* for 60 min at 5 °C. The EOS was expressed as % (supernatant/sample, *w/w*). WHC and EOS measurements of the stirred ADGs were conducted in quadruplicate. WHC and EOS measurements of stirred ADG samples were carried out on the day 1, day 7, day 14, day 21 and day 28 of cold storage.

2.7. Antioxidant Capacity Evaluation

Sample Extract. The sample (1 g for both tea infusion and gel samples) was placed in a 50-mL plastic pot with 20 mL 50% methanol solution and stirred overnight (speed 3, RT 15 Power 15-position analogue hotplate stirrer, IKA, Staufen im Breisgau, Germany) at ambient temperature. Extracts were stored at −20 °C until required.

Total Phenolic Content (TPC). The total phenolic content (TPC) of the sample extracts was determined by a spectrophotometer in triplicate using Folin-Ciocalteu (F-C) reagent according to the method described in previous studies [39,40] with slight modifications. 500 µL of sample was added to the test tubes followed by 2.0 mL of sodium carbonate (7.5 g/100 mL) and 2.5 mL of 0.2 mol/L Folin-Ciocalteu reagent. The samples were mixed thoroughly and stored in the dark for 2 h before the absorbance was measured at 760 nm using VWR V-1200 Spectrophotometer (VWR International Co., Pennsylvania, USA). TPC was expressed as mg gallic acid equivalents (GAE) per 100 g of fresh material.

DPPH Assay. The DPPH radical-scavenging activity was assayed by the method reported by Al-Dabbas et al. [41]. DPPH was dissolved in methanol to get a concentration of 0.1 mM. To 500 µL of sample extract, 1 mL 0.1 mM DPPH solution and 1.5 mL of methanol were added and mixed by vortex. The absorbance was measured at 517 nm (VWR International Co., Pennsylvania, USA) after the solution was kept at room temperature for 30 min in the dark, methanol was used as the control. The DPPH radical-scavenging activity was expressed as mg Trolox equivalents (TE) per 100 g of fresh material. Each sample was analysed in triplicate.

Ferric Reducing/Antioxidant Power (FRAP) Assay. FRAP was assessed according to Khanizadeh, Tsao, Rekika, Yang and DeEll [40] with slight modifications. A fresh working solution of FRAP reagent was prepared each time by mixing acetate buffer (300 µM, pH 3.6), a solution of 10 mM TPTZ in 40 mM HCL, and 20 mM $FeCl_3 \bullet 7H_2O$ at 10:1:1 (*v/v/v*). 250 µL of standards of iron (II) sulphate ($FeSO_4 \bullet 7H_2O$) or sample extracts were added to 2.5 mL of the FRAP reagent and the absorbance at 593 nm recorded immediately after the addition of the sample and after 2 h incubation at 37 °C [42] (p.39). The results were expressed as µmol Fe2+/g sample. Each sample was analysed in triplicate.

ABTS radical scavenging capacity. The ABTS radical scavenging assay was based on the method of Elfalleh, et al. [43]. ABTS working solution was prepared by mixing colourless ABTS stock solution (7 mM in water) with 2.45 mM potassium persulfate (1:1) and then maintain the reaction for 16 h in the dark at room temperature. Before analysis, the working solution was diluted to an absorbance of 0.70 (±0.02) at 734 nm with PBS (pH 7.4) and 3 mL of ABTS·+ transferred to a cuvette. After the addition of 300 µL Trolox or sample extract, the mixture was well mixed by blowing with pipette, allow to stand 6 min and absorbance read at 734 nm. All samples were assayed in triplicate. The results were expressed as Trolox equivalents (TE).

TPC, DPPH, FRAP and ABTS measurements of stirred ADG samples were carried out on the day 1, day 7, day 14, day 21 and day 28 of cold storage.

2.8. Microstructure of ADG

Scanning Electron Microscope (SEM). The bulk microstructure characterization method was developed according to Kalab [44]. Approximately 3 mm^3 cubes of chilled stirred gel samples were coated in a thin layer of low melting point agarose (3%) before being placed into primary fixative (3% glutaraldehyde 0.1 M sodium cacodylate buffer, pH 7.2) for at least 8 hours at ambient temperature. The samples were then washed three times (10–15 min each) in sodium cacodylate buffer (0.1 M, pH 7.2) followed by post fixation in 1% osmium tetroxide (in sodium cacodylate buffer) for 1 hour at room temperature and another three buffer washes. Finally, the samples were dehydrated in gradient ethanol series (25%, 50%, 75%, 95% and 100%) for 10–15 min each before a final 100% ethanol wash for 1 h.

Samples were critical point dried using liquid CO_2 as the CP fluid and 100% ethanol as the intermediary (Polaron E3000 series II critical point drying apparatus, Quorum Technologies, Laughton, UK). Samples were torn to expose structure, mounted on to aluminium stubs, sputter coated with approximately 200 nm of gold (Baltec SCD 050 sputter coater, Schalksmühle, Germany) and viewed in the FEI Quanta 200 Environmental Scanning Electron Microscope with the EDS Detector (EDAX Genesis, Sydney, Australia) at an accelerating voltage of 20 kV. Analyses of ADG microstructure were conducted at Massey University, Palmerston North, New Zealand [45–48].

Confocal Laser Scanning Microscopy (CLSM). The planar microstructure of the protein arrangement in stirred ADG was investigated using CLSM. The method was adapted from Ciron, Gee, Kelly and Auty [32] with modifications. A drop of stirred gel sample was placed in a concave slide and 40 µL each of 0.2 g/L Nile red (in methanol) and fast green (in water) were added before being covered with a coverslip.

Imaging was carried out using the Leica DM6000B SP5 confocal laser scanning microscope system with LAS AF software (version 2.7.3.9723; Leica Microsystems Berlin, CMS GmbH, Berlin, Germany). Images were acquired with a HCX PL APO CS 10x (N.A. 0.40), HCX PL FLUOTAR 40x (N.A. 0.75) and HCX PL APO CS 100x oil (N.A. 1.40). Nile red and fast green were sequentially imaged through excitation at 488 nm (argon laser) and 633 nm (HeNe 633 laser) (respectively) and emission collection at 498–569 nm and 643–787 nm (respectively).

2.9. Statistical Analysis

One-way analysis of variance (ANOVA) (with Fisher comparison) and principle component analysis (PCA – as of Figure 1) were carried out using Minitab 17.0 (Minitab, State College, PA, USA) and the significance level was set at $p \leq 0.05$.

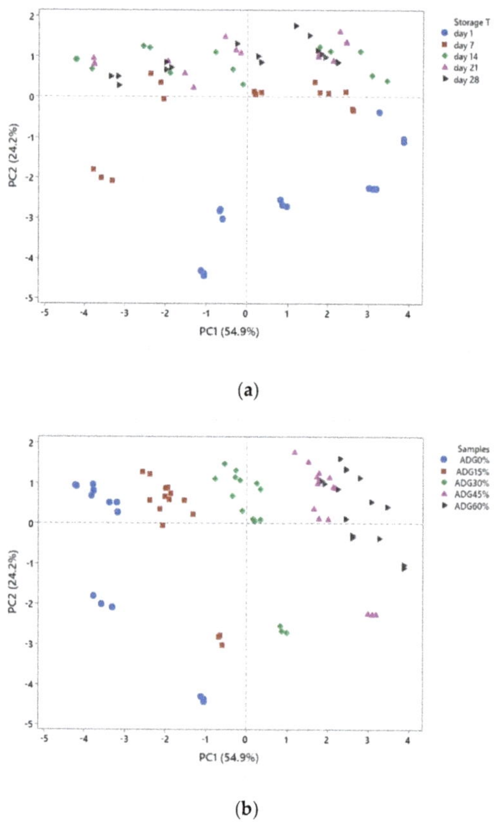

(a)

(b)

Figure 1. Score plots of principle component analysis (PCA). $ADG_{0\%}$: plain acidified dairy gel, $ADG_{15\%}$: acidified dairy gel containing 15% black tea infusion, $ADG_{30\%}$: acidified dairy gel containing 30% black tea infusion, $ADG_{45\%}$: acidified dairy gel containing 45% black tea infusion, $ADG_{60\%}$: acidified dairy gel containing 60% black tea infusion. PC1 strongly associated with TPC, DPPH (2,2-Diphenyl-1-picrylhydrazyl), ABTS (2,2-Diphenyl-1-picrylhydrazyl) and ferric reducing antioxidant power (FRAP); PC2 strongly associated with values of EOS, pH and Ca^{2+}. (**a**) is PCA plot based on different storage time; (**b**) based on different formula.

3. Results and Discussion

3.1. Physicochemical Characteristics

The physicochemical characteristics of ADGs are shown in Table 1. As it is shown in Table 1, plain $ADG_{0\%}$ showed the highest values in EOS in comparison with other ADG samples on day 1. Throughout the cold storage period, in general, decrease trends were observed in pH, Ca^{2+} concentration, and EOS for all ADG samples. Significant decreases of Ca^{2+} concentration were only observed in $ADG_{45\%}$ and $ADG_{60\%}$ on day 1 (in comparison to $ADG_{0\%}$). The decrease of Ca^{2+} concentration may due to the interaction between calcium ion and polyphenols or oxalate in tea. Charrier and coworkers [49] stated that the oxalate in teas may bind with calcium ion. Yamada et al. [50] used infrared spectroscopy (IR) and matrix-assisted laser desorption ionization technique with a time-of-flight mass spectrometer (MALDI-TOF-MS) found out that the calcium ion increased the amount of tea stain greatly due to the combination of phenolic compounds and calcium ions. Such finding provided an evidence that calcium-bridged polyphenols complex may be formed. Another study [51] also evidenced occurrence of $EGCG$-Ca^{2+}-$EGCG$ bridging effect. These Ca^{2+}-phenol interaction mechanisms help explained our results regarding the observed decrease of Ca^{2+} from a molecular point of view. The slight decrease trend of pH upon increasing the incorporation amount of tea infusion (Table 1) is attributed to the relatively lower pH of black tea infusion (pH 5.18) [21].

Table 1. Physicochemical characteristics between different formulations of black tea enriched acidified dairy gel (ADG) during a cold storage of 28 days (4 °C).

Physicochemical Properties	Storage Time (Days)	Gel Samples [1]				
		$ADG_{0\%}$	$ADG_{15\%}$	$ADG_{30\%}$	$ADG_{45\%}$	$ADG_{60\%}$
pH value	1	4.55 ± 0.01 [Aa]	4.53 ± 0.01 [Ab]	4.54 ± 0.01 [Aab]	4.54 ± 0.01 [Aab]	4.53 ± 0.01 [Ab]
	7	4.35 ± 0.01 [Bc]	4.37 ± 0.01 [Bb]	4.36 ± 0.01 [Bc]	4.34 ± 0.01 [Cd]	4.40 ± 0.01 [Ba]
	14	4.29 ± 0.01 [Db]	4.28 ± 0.01 [Dbc]	4.36 ± 0.01 [Ba]	4.36 ± 0.01 [Ba]	4.28 ± 0.01 [Dc]
	21	4.30 ± 0.01 [Db]	4.33 ± 0.01 [Ca]	4.32 ± 0.01 [Cb]	4.28 ± 0.02 [Dc]	4.29 ± 0.01 [Dc]
	28	4.32 ± 0.01 [Ca]	4.28 ± 0.02 [Dc]	4.25 ± 0.01 [Dd]	4.29 ± 0.01 [Dbc]	4.31 ± 0.01 [Cab]
Ca^{2+} concentration	1	450.00 ± 0.00 [Ba]	443.33 ± 5.77 [Aa]	446.67 ± 15.28 [Aa]	353.33 ± 5.77 [Ab]	273.33 ± 15.28 [Bc]
	7	500.00 ± 10.00 [Aa]	356.67 ± 11.55 [Bb]	316.67 ± 5.77 [Cc]	253.33 ± 5.77 [Cd]	326.67 ± 15.28 [Ab]
	14	356.67 ± 5.77 [Ca]	323.33 ± 5.77 [Cb]	326.67 ± 5.77 [Bb]	253.33 ± 5.77 [Cc]	263.33 ± 5.77 [Cc]
	21	370.00 ± 10.00 [Ca]	326.67 ± 15.28 [Bb]	293.33 ± 5.77 [Dc]	266.67 ± 5.78 [Bd]	246.67 ± 11.55 [Ce]
	28	330.00 ± 10.00 [Dab]	336.67 ± 5.77 [Ba]	316.67 ± 5.77 [Cb]	246.67 ± 5.77 [Cd]	290.00 ± 10.00 [Bc]
EOS^2/%	1	39.95 ± 0.89 [Aa]	24.05 ± 0.78 [Ad]	29.48 ± 0.73 [Bc]	35.09 ± 1.15 [Bb]	29.01 ± 1.10 [Ac]
	7	37.74 ± 1.21 [Ba]	22.99 ± 0.52 [Ad]	31.16 ± 0.90 [Ab]	39.02 ± 1.22 [Aa]	27.50 ± 0.80 [Bc]
	14	21.28 ± 0.80 [Dc]	23.06 ± 1.23 [Ab]	24.83 ± 0.20 [Db]	20.34 ± 0.59 [Ec]	28.42 ± 0.52 [Aa]
	21	17.21 ± 0.39 [Ee]	22.10 ± 0.92 [Bd]	24.68 ± 0.68 [Dc]	28.92 ± 0.60 [Da]	26.36 ± 0.78 [Bb]
	28	23.37 ± 0.33 [Cd]	23.30 ± 0.96 [Ad]	28.25 ± 0.33 [Cb]	30.20 ± 0.27 [Ca]	26.87 ± 0.14 [Bc]
WHC^3/%	1	17.13 ± 0.81 [Eb]	19.56 ± 1.04 [Ba]	21.10 ± 0.63 [Ba]	15.63 ± 0.12 [Dc]	15.30 ± 0.48 [Cc]
	7	24.33 ± 0.69 [Ba]	22.69 ± 0.23 [Ab]	23.50 ± 0.78 [Aa]	20.49 ± 0.79 [Ac]	21.01 ± 0.65 [Ac]
	14	27.31 ± 0.72 [Aa]	21.93 ± 0.65 [Ab]	20.69 ± 0.34 [Cc]	16.34 ± 0.15 [Cd]	15.22 ± 0.63 [Ce]
	21	18.82 ± 0.25 [Db]	16.84 ± 1.16 [Cc]	23.89 ± 0.58 [Aa]	18.07 ± 0.51 [Bc]	18.54 ± 0.94 [Bc]
	28	20.15 ± 0.55 [Cb]	22.20 ± 0.32 [Aa]	19.42 ± 0.37 [Cb]	19.82 ± 0.32 [Ab]	19.37 ± 0.84 [Bb]

[A-E] Means $\pm$ SD within a column with different superscripts differ ($p \leq 0.05$). [a-e] Means $\pm$ SD within a row with different superscripts differ ($p \leq 0.05$). [1] Formulation: $ADG_{0\%}$: acidified dairy gel without supplementation of tea infusion; $ADG_{15\%}$: acidified dairy gel containing 15% (w/w) black tea infusion; $ADG_{30\%}$: acidified dairy gel containing 30% (w/w) black tea infusion; $ADG_{45\%}$: acidified dairy gel containing 45% (w/w) black tea infusion; $ADG_{60\%}$: acidified dairy gel containing 60% (w/w) black tea infusion; [2] EOS: Extend of syneresis; [3] WHC: Water holding capacity.

Stirred ADGs are viscoelastic semi-solid material, poor gel structure and poor stability are associated with higher risk of shrinkage and subsequent expulsion of whey solution [32], so a relatively phase stable ADG should be able to retain a certain amount water over a storage time [52]. Interestingly, by including tea infusion in the stirred ADG, the syneresis rate tended to decrease when the stirred gel was just made (day 1, Table 1). A study demonstrated an opposite phenomenon, in which the authors found that polyphenols especially at higher concentrations are able to reduce the elasticity of acidified milk gels and induce extensive shrinkage of gel system, therefore, more syneresis [53]. Such mechanism can explain our EOS results on day 28. Larger polyphenols with more aromatic rings and hydroxyl groups have stronger affinity to casein for forming protein–phenol complex [1]. Such type of complex may result in tighter gel network then cause relatively more release of whey fraction [27]. In the same study [27], the coworkers found two types of protein–phenol complex systems when milk proteins interact with phenolic compounds from either green tea or Pu-erh tea. These two different complex systems resulted in different texture and syneresis rate. Green tea infusion was able to reduce syneresis; however, Pu-erh tea promoted syneresis in acidified dairy gel [27]. The current research showed that black tea had even stronger ability than green tea (as shown in the reference [27]) in terms of reducing the syneresis rate of ADG on day 1 (Table 1). After 28 days of storage time, $ADG_{0\%}$ and $ADG_{15\%}$ were not significantly different for EOS; however, for gels with higher levels of tea infusion inclusion the EOS were significantly increased indicating time is an important factor for developing the impact of protein–phenol complex on gel structure (Table 1). Such explanation is confirmed by (SEM) microstructure images (e.g., Figure 2e vs. Figure 3e) which will be discussed later.

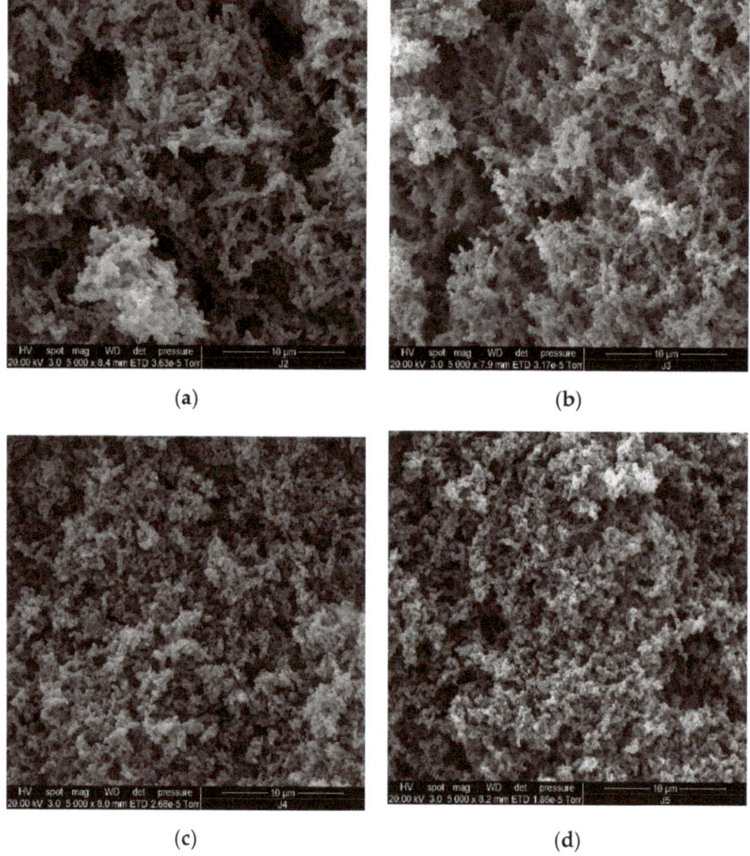

(a)

(b)

(c)

(d)

Figure 2. *Cont.*

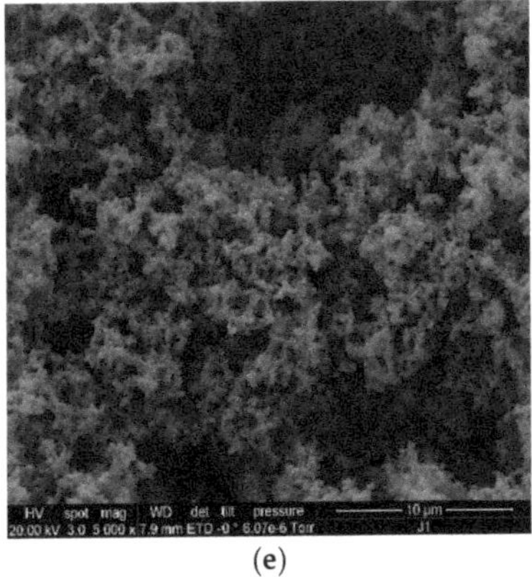

(**e**)

Figure 2. SEM images of plain $ADG_{0\%}$ (**a**), ADG with 15% black tea infusion (**b**), ADG with 30% black tea infusion (**c**), ADG with 45% black tea infusion (**d**), ADG with 60% black tea infusion (**e**) (day 1).

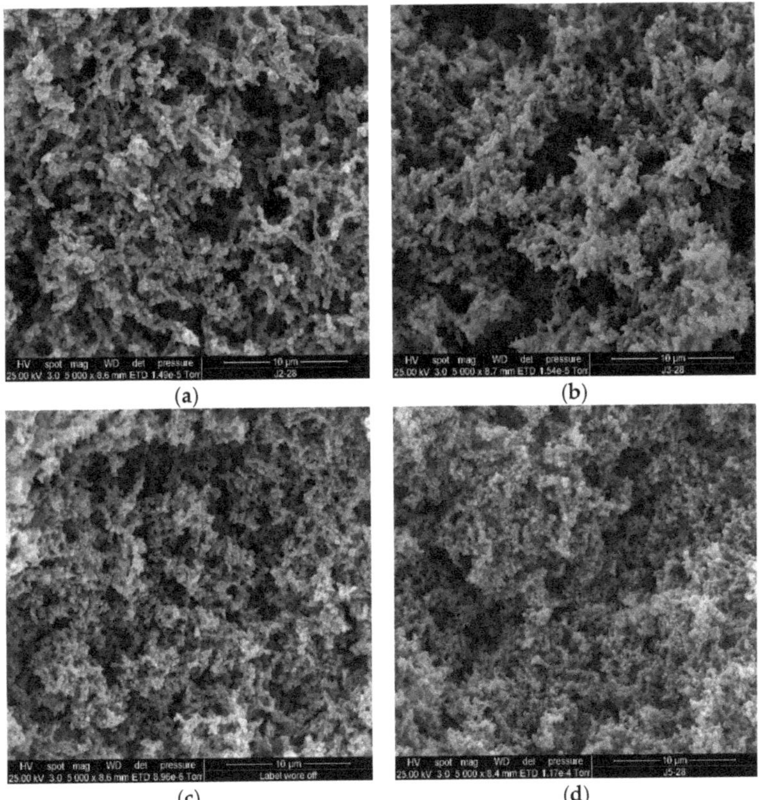

(**a**)   (**b**)

(**c**)   (**d**)

Figure 3. *Cont.*

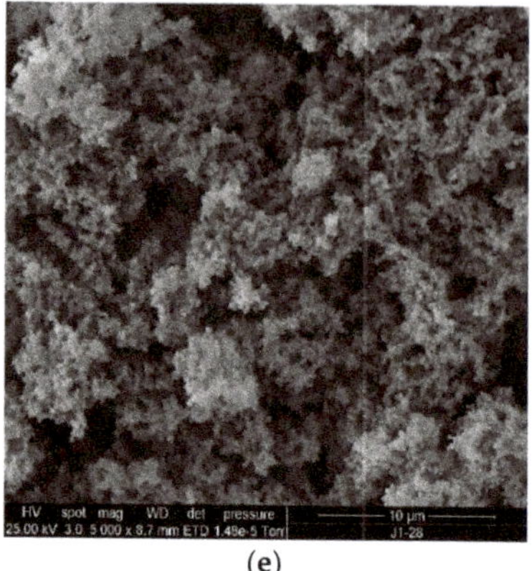

(**e**)

Figure 3. SEM images of plain ADG$_{0\%}$ (**a**), ADG with 15% black tea infusion (**b**), ADG with 30% black tea infusion (**c**), ADG with 45% black tea infusion (**d**), ADG with 60% black tea infusion (**e**) (day 28).

As for WHC, ADG$_{15\%}$ and ADG$_{30\%}$ acquired the highest values for day 1; however, after 28 days ADG$_{15\%}$ had the highest WHC than other gel samples including the control gel without inclusion of tea infusion ($p < 0.05$), thus the relatively lower amount of tea infusion (15%) may help to maintain the gel structure of ADGs. Reports have shown that exopolysaccharides (EPSs) can act as an agent which is capable of thickening, stabilizing, emulsifying and gelling as well as water-binding in the food system and EPSs can be produced by certain strains, such as *Lactobacillus paracasei*, therefore an increase in EOS was observed through cold storage [14,52]. However, in this research, no starter culture was used and the gelation was induced by (slow) direct acidification due to the hydrolysis of GDL. Such results suggest that not only starter culture and its products but also physicochemical interactions between dairy components and tea phenolic compounds can alter gel structure and texture, therefore, the gel physical stability over a period. The same mechanism has been discussed previously elsewhere [27], although, in that research the dairy gel was induced by fermentation of starter culture, the author attributed the texture and stability features to the protein–phenol interactions.

3.2. Texture Characteristics

The textural characteristics of stirred ADGs are shown in Table 2. In general, firmness describes the force needed for causing a certain deformation. Cohesiveness represents for the threshold of extent of a material deformation for triggering an irreversible rupture [54]. The former parameter provides good indication about how hard or how soft a gel material, whereas, the later indicates the tenacity of a gel material (e.g., brittleness). On day 1, incorporation of tea infusion regardless the addition level resulted no change in firmness in comparison to the control (Table 2). Over the storage time (> day 1), all gel samples showed an increase in gel firmness in comparison to day 1, this is due to the reformation of gel structure after shearing the gel on day 0 and ongoing fusion of casein particles such as dissociation of casein micelles at relatively low pH and rearrangement of bonds and strands [55]. The similar phenomenon was observed in a previous work in which the authors observed significant increase of storage modulus of stirred yogurt gel over a period of storage (up to 35 days) [56]. Apparently, the current results showed that the higher addition amount of tea infusion (ADG$_{60\%}$, Table 2) retarded the fusion process of casein particles in the gel system over the 28 days storage time. In summary, including tea infusion at extremely high level (e.g., 60%) tends to make dairy gel relatively softer over storage time. The reason is not clear that why ADG$_{45\%}$ had slightly higher firmness than ADG$_{30\%}$ after 28 days of storage. The impact of tea infusion on dairy gel firmness is not fully clear and the published data is lack of consistency. Opposite observations were reported, for instance, one group found that green tea extracts were able to increase the firmness of yogurt.

And the phenomenon was attributed to flavonoid–protein cross-linking [57]. However, the authors did not mention the details of the gel preparation procedure. It is important to point out that the shearing process of set gel for the preparation of stirred is a critical step, which determines the firmness of the final sample. If the gel shearing process was not consistent, consequently, it is impossible to make relevant discussion about the causality for experimental observations. Moreover, it was found that green tea extracts significantly increased hardness of cheese gel but decreased its cohesiveness and this is due to the net effect of moisture reduction and alter microstructure of cheese gel in which green tea extract was included [30]. Similar to our findings, Najgebauer-Lejko et al. [27] also found that tea infusion may be able to decrease firmness of acidified dairy gel, although green tea (rather than black tea) and lower addition levels were used (5–15%) in that research.

Table 2. Texture characteristics of stirred gel samples.

Texture Properties	Storage Time (Days)	Gel Samples [1]				
		ADG$_{0\%}$	ADG$_{15\%}$	ADG$_{30\%}$	ADG$_{45\%}$	ADG$_{60\%}$
Firmness, g	1	14.77 ± 0.46 ^{Ba}	14.73 ± 0.35 ^{Ba}	14.57 ± 0.31 ^{Ba}	14.07 ± 0.42 ^{Ba}	14.70 ± 1.04 ^{Aa}
	7	20.10 ± 0.90 ^{Aa}	20.43 ± 1.25 ^{Aa}	17.13 ± 0.72 ^{Ab}	16.40 ± 0.72 ^{Ab}	15.17 ± 1.08 ^{Ab}
	14	21.13 ± 0.66 ^{Aa}	20.07 ± 0.86 ^{Aa}	18.60 ± 0.87 ^{Ab}	16.37 ± 0.83 ^{Bb}	15.33 ± 1.60 ^{Ab}
	21	21.93 ± 0.55 ^{Aa}	19.83 ± 1.15 ^{Aa}	17.70 ± 0.27 ^{Ab}	16.30 ± 0.87 ^{Bb}	16.73 ± 0.15 ^{Ab}
	28	20.53 ± 0.32 ^{Aa}	18.80 ± 0.82 ^{Aa}	18.20 ± 0.87^{Ab}	19.03 ± 1.61 ^{Aa}	16.57 ± 0.85 ^{Ab}
Cohesiveness, g	1	−9.50 ± 0.27 ^{Ca}	−9.07 ± 0.06 ^{Bb}	−8.67 ± 0.40 ^{Bb}	−8.53 ± 0.23 ^{Bb}	−9.40 ± 0.61 ^{Bb}
	7	−12.43 ± 0.50 ^{Ba}	−12.63 ± 1.01 ^{Aa}	−10.93 ± 0.35 ^{Ab}	−10.53 ± 0.67 ^{Ac}	−9.37 ± 0.31 ^{Bc}
	14	−13.27 ± 0.42 ^{Ba}	−12.83 ± 1.24 ^{Aa}	−12.47 ± 1.25 ^{Aa}	−10.27 ± 0.40 ^{Ab}	−8.83 ± 0.71 ^{Bb}
	21	−14.13 ± 0.51 ^{Aa}	−12.80 ± 0.27 ^{Aa}	−12.33 ± 1.10 ^{Ab}	−10.07 ± 0.59 ^{Bb}	−10.13 ± 0.06 ^{Bb}
	28	−13.73 ± 0.12 ^{Aa}	−12.57 ± 0.15 ^{Ab}	−11.80 ± 1.21 ^{Ac}	−10.43 ± 0.67 ^{Ac}	−10.43 ± 0.40 ^{Bc}

$^{A–E}$ Means ± SD within a column with different superscripts differ ($p \leq 0.05$). $^{a–e}$ Means ± SD within a row with different superscripts differ ($p \leq 0.05$). [1] Gel samples: ADG$_{0\%}$: acidified dairy gel without supplementation of tea infusion; ADG$_{15\%}$: acidified dairy gel containing 15% (w/w) black tea infusion; ADG$_{30\%}$: acidified dairy gel containing 30% (w/w) black tea infusion; ADG$_{45\%}$: acidified dairy gel containing 45% (w/w) black tea infusion; ADG$_{60\%}$: acidified dairy gel containing 60% (w/w) black tea infusion.

Unlike the situation for firmness, incorporation of tea infusion (15–60%) in ADG significantly reduced the cohesiveness on day 1 (Table 2). Such results indicated that addition of tea infusion may make the acidified dairy gel slightly more brittle. The opposite results were reported for green tea and Pu–erh tea enriched yogurt systems, in which cohesiveness was increased upon addition of tea infusion (up to 15%) [27]. However, it is important to point out that not only were different teas used in the two studies, but also relatively lower tea infusion addition levels were investigated in the mentioned reference. Najgebauer–Lejko et al. believed that the increase of cohesiveness is due to the existence of protein–phenol interactions and such interactions strengthened the food internal bonds [27]. However, such deduced mechanism does not explain the contradictory trends between firmness and cohesiveness as found in the mentioned study. In the current research, the changing trend of cohesiveness over 28 days of storage was similar to the firmness-changing trend. The change of firmness and cohesiveness upon addition of tea infusion may be attributed to the alteration of microstructure during storage period [30]. The mechanism will be elaborated later in the microstructure section.

3.3. Antioxidant Capacity

To determine the antioxidant abilities of the obtained stirred ADGs and tea infusion, we chose three methods which allowed us to measure both the ability to reduce pro-oxidant metal ions (FRAP assay) and radical scavenging activity (DPPH assay and ABTS assay) along with TPC measurement. The results are shown in Table 3.

Table 3. Total phenolic content and antioxidant capacity of stirred ADGs.

Physicochemical Properties	Storage Time (Days)	Gel Samples				
		ADG$_{0\%}$	ADG$_{15\%}$	ADG$_{30\%}$	ADG$_{45\%}$	ADG$_{60\%}$
TPC (GAE µg/g)	1	151.42 ± 1.94 [Ae]	229.75 ± 6.71 [Ad]	257.90 ± 11.66 [Ac]	454.98 ± 17.23 [Ab]	529.28 ± 8.92 [Aa]
	7	117.22 ± 1.71 [Cd]	224.08 ± 9.85 [Ac]	228.66 ± 10.26 [Cc]	305.48 ± 8.09 [Cb]	503.80 ± 7.38 [Ba]
	14	119.30 ± 0.99 [Ce]	159.08 ± 6.19 [Dd]	211.79 ± 8.20 [Dc]	348.37 ± 7.88 [Bb]	441.32 ± 10.71 [Ca]
	21	121.41 ± 6.44 [Ce]	208.55 ± 9.31 [Bd]	236.00±4.24 [Bc]	303.85 ± 6.03 [Cb]	422.99 ± 2.60 [Da]
	28	129.25 ± 7.14 [Be]	194.81 ± 3.34 [Cd]	244.67 ± 5.91 [Bc]	301.10 ± 6.03 [Cb]	388.34 ± 5.99 [Ea]
DPPH (TE µmol/g)	1	0.94 ± 0.06 [Ad]	1.35 ± 0.07 [Ac]	4.58 ± 0.29 [Ab]	4.77 ± 0.07 [Ab]	5.17 ± 0.06 [Ba]
	7	0.86 ± 0.02 [Ae]	1.27 ± 0.03 [Bd]	4.53 ± 0.04 [Ac]	5.02 ± 0.13 [Ab]	5.17 ± 0.01 [Ba]
	14	0.98 ± 0.04 [Ae]	1.26 ± 0.07 [Bd]	4.33 ± 0.14 [Ac]	4.89 ± 0.08 [Ab]	5.14 ± 0.06 [Ba]
	21	0.93 ± 0.08 [Ad]	1.18 ± 0.01 [Bd]	4.48 ± 0.15 [Ac]	4.85 ± 0.26 [Ab]	5.24 ± 0.11 [Aa]
	28	0.97 ± 0.13 [Ae]	1.29 ± 0.09 [Bd]	4.56 ± 0.15 [Ac]	5.01 ± 0.07 [Ab]	5.36 ± 0.08 [Aa]
FRAP (Fe^{2+} equivalent µmol/g)	1	0.63 ± 0.03 [Ce]	2.74 ± 0.12 [Cd]	7.17 ± 0.13 [Dc]	11.56 ± 0.28 [Cb]	13.62 ± 0.23 [Ba]
	7	0.72 ± 0.03 [Be]	4.42 ± 0.18 [Bd]	8.73 ± 0.28 [Ac]	11.73 ± 0.13 [Cb]	12.65 ± 0.05 [Ca]
	14	0.67 ± 0.02 [Be]	4.53 ± 0.06 [Bd]	7.82 ± 0.2 [Cc]	13.01 ± 0.23 [Ab]	13.54 ± 0.11 [Ba]
	21	0.81 ± 0.03 [Ae]	5.17 ± 0.09 [Ad]	8.29 ± 0.12 [Bc]	12.48 ± 0.07 [Bb]	14.43 ± 0.19 [Aa]
	28	0.80 ± 0.03 [Ae]	5.24 ± 0.13 [Ad]	8.66 ± 0.18 [Ac]	12.62 ± 0.08 [Bb]	12.94 ± 0.19 [Ca]
ABTS (TE µmol/g)	1	0.30 ± 0.05 [Ae]	1.78 ± 0.03 [Bd]	3.03 ± 0.10 [Ac]	3.68 ± 0.03 [Bb]	4.34 ± 0.05 [Aa]
	7	0.24 ± 0.03 [Ae]	1.86 ± 0.04 [Ad]	3.09 ± 0.04 [Ac]	3.70 ± 0.16 [Bb]	4.09 ± 0.11 [Ba]
	14	0.31 ± 0.05 [Ae]	1.88 ± 0.02 [Ad]	3.07 ± 0.07 [Ac]	3.84 ± 0.06 [Bb]	4.15 ± 0.07 [Ba]
	21	0.31 ± 0.06 [Ae]	1.90 ± 0.02 [Ad]	3.08 ± 0.05 [Ac]	4.15 ± 0.12 [Aa]	3.99 ± 0.13 [Bb]
	28	0.29 ± 0.03 [Ae]	1.86 ± 0.04 [Ad]	3.11 ± 0.04 [Ac]	4.21 ± 0.05 [Aa]	4.07 ± 0.12 [Bb]

[A–E] Means ± SD within a column with different superscripts differ ($p \leq 0.05$). [a–e] Means ± SD within a row with different superscripts differ ($p \leq 0.05$). [1] Formulation: ADG$_{0\%}$: acidified dairy gel; ADG$_{15\%}$: acidified dairy gel containing 15% (*w/w*) black tea infusion; ADG$_{30\%}$: acidified dairy gel containing 30% (*w/w*) black tea infusion; ADG$_{45\%}$: acidified dairy gel containing 45% (*w/w*) black tea infusion; ADG$_{60\%}$: acidified dairy gel containing 60% (*w/w*) black tea infusion; EOS: Extend of syneresis; WHC: Water holding capacity.

Based on the results from FRAP, DPPH and ABTS radical assays, the strongest antioxidant capacity appeared in tea infusion (TPC: 1632.56 ± 9.99 GAE µg/g; FRAP: 17.20 ± 0.09 Fe^{2+} equivalent µmol/g; DPPH: 9.26 ± 0.08 TE µmol/g and ABTS: 9.30 ± 0.04 TE µmol/g) compared to dairy gel samples. The difference in antioxidant activity is due to the different supplement volume of the original tea infusion. Based on the original antioxidant capacity of black tea infusion and the proportion of tea infusion supplementation, we calculated the theoretical antioxidant capacities for tea-enriched ADGs. By comparing the theoretical and practical values, we found that the acidified dairy gels did not reduce the antioxidant activity from the tea infusion. The higher proportion of tea infusion presented in ADGs the more antioxidant capacity and TPC. Interestingly, the increasing rate in TPC was much lower than the increasing rate of tea infusion supplementation (e.g., comparison between ADG$_{0\%}$, ADG$_{15\%}$, and ADG$_{30\%}$) which indicated that the polyphenols in tea were binding with calcium ion or protein as reported by Tanizawa, et al. [58] and Spiro and Chong [59], respectively. These previous findings are in a good agreement with our results in which we observed that the free Ca^{2+} were decreased upon the increase of the addition level of tea infusion (Table 1). Surprisingly, some antioxidant parameters were remained at higher level at day 28 compared to results from day 7, 14 and 21. Although the mechanism has not yet been clear, similar results were reported by Najgebauer-Lejko, et al. [15]. The authors studied yogurt gels with incorporation of different levels of either green tea infusion or Pu-erh tea infusion, and their results showed that the antioxidant activities decreased until day 21 for both types of tea but such activity increased back to nearly the initial value on day 28.

Overall, the incorporation of black tea infusion in stirred ADG significantly enhanced the antioxidant potential of ADG as a 4 to 12-fold increase in DPPH, an 8 to 40-fold in FRAP and a 6 to 18-fold in ABTS value were observed in comparison to plain ADG$_{0\%}$. The antioxidant activities were stable over the 28 days shelf life. Moreover, such activities may be effectively increased by incorporation of increased proportion of black tea infusion.

3.4. Principle Component Analysis (PCA)

Plots of PCA is shown in Figure 1. PC1 and PC2 including physicochemical properties and antioxidant properties explained 79.1% of the variation with 54.9% and 24.2% for the two PCs respectively. The main difference between PC1 and PC2 is that PC1 strongly associated with TPC (PC1 0.394; PC2 0.037), DPPH (PC1 0.385; PC2 0.156), ABTS (PC1 0.395; PC2 0.191) and FRAP (PC1 0.398; PC2 0.217) but values of EOS (PC1 0.145; PC2 −0.355), pH (PC1 0.100; PC2 −0.559) and Ca^{2+} (PC1 −0.234; PC2 −0.490) were the dominant variables for PC2. Furthermore, Figure 1a. showed the sample differentiation-based storage time (shelf life); Figure 1b. showed that gels with different proportions of tea infusion are clearly separated when taking both physicochemical and antioxidant activity associated factors into account. As it is shown in the biplot figure (in the Supplemental Material, Supplement Figure S1), the antioxidant properties were mainly determined by the incorporation rate of tea infusion in stirred ADGs, but the textural and physicochemical properties were mainly influenced by storage period.

3.5. Microstructure

The microstructures of the stirred ADG samples are shown in Figures 2 and 3 (SEM) and Figures 4 and 5 (CLSM).

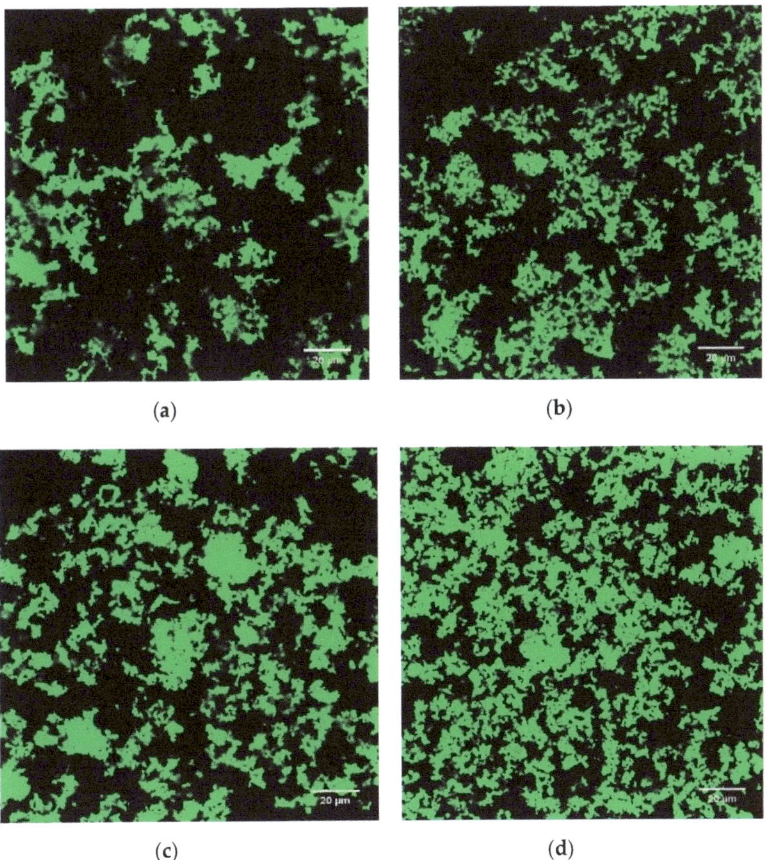

(a)　　　　　　　　　　(b)

(c)　　　　　　　　　　(d)

Figure 4. *Cont.*

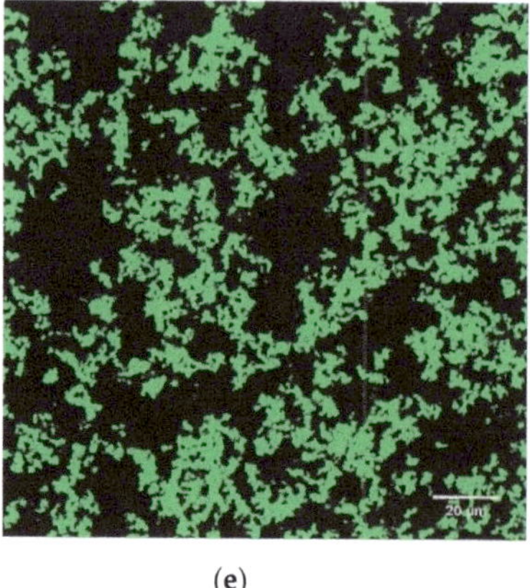

(**e**)

Figure 4. CLSM (confocal laser scanning microscopy) images of plain ADG$_{0\%}$ (**a**), ADG with 15% black tea infusion (**b**), ADG with 30% black tea infusion (**c**), ADG with 45% black tea infusion (**d**), ADG with 60% black tea infusion (**e**) (day 1). Protein stained by Fast Green FCF appears as green and non-fluorescent areas (dark areas) correspond to the serum pores.

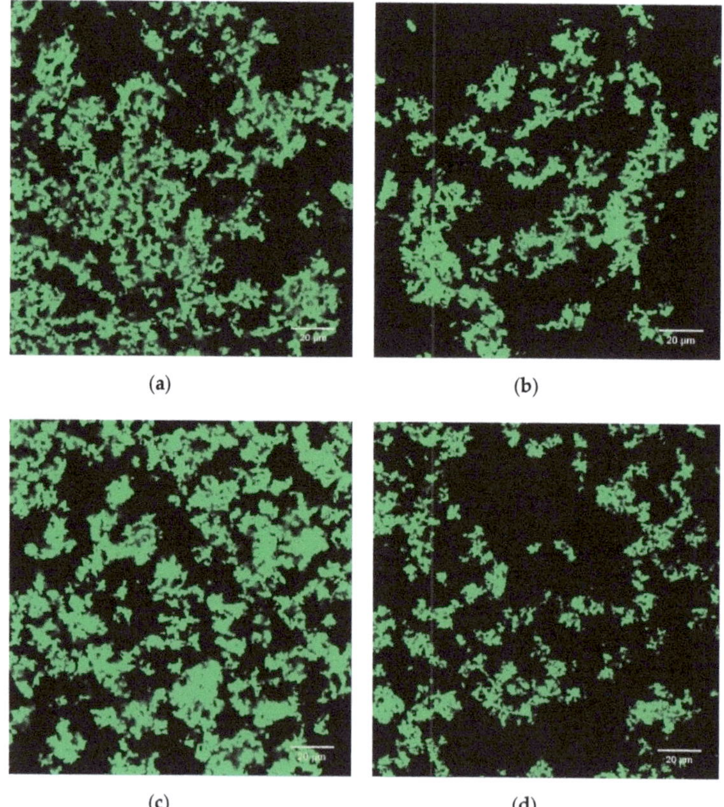

(**a**)　　　　　　　　　　　　(**b**)

(**c**)　　　　　　　　　　　　(**d**)

Figure 5. *Cont.*

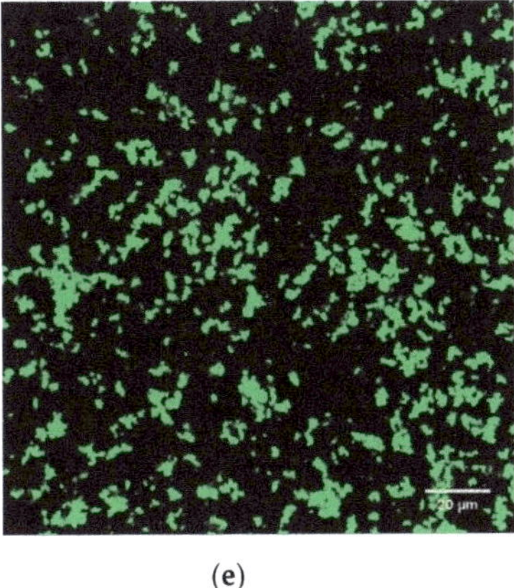

(**e**)

Figure 5. CLSM images of plain ADG$_{0\%}$ (**a**), ADG with 15% black tea infusion (**b**), ADG with 30% black tea infusion (**c**), ADG with 45% black tea infusion (**d**), ADG with 60% black tea infusion (**e**) (day 28). Protein stained by Fast Green FCF appears as green and non-fluorescent areas (dark areas) correspond to the serum pores.

SEM micrographs showed the 3D (including z-depth) organization of protein gels (stirred ADG samples). In Figure 2, comparing the morphology and the segregated structure between ADG$_{0\%}$ and ADG$_{15\%}$, the z-depth structure is more densely packed in the ADG$_{15\%}$ (Figure 2a vs. Figure 2b). Also, the later had slightly finer protein arrangement, as shown in Figure 2, ADG$_{0\%}$ had relatively more large cavities (white arrows in Figure 2a) than those in ADG$_{15\%}$ (Figure 2b). The structure of ADG$_{30\%}$ (Figure 2c) was relatively thicker and denser among other gel samples on day 1. Figure 2c showed that the sample had fine protein arrangement resulting very small pores in the gel structure. Such microstructure explains the better phase stability (WHC, Table 1) in the gel sample containing 30% BTI. Generally, ADG$_{15\%}$, ADG$_{30\%}$ and ADG$_{45\%}$ three samples had similar spongy-like interior with few air cells and highly branched-structure. Fiszman, et al. [60] found that the smooth bridge with double network structures of dairy gel seemed to be located at the inside of network of casein micelles that could maintain the aqueous phase more effectively and reduce EOS. Although the authors did not study the impact of tea infusion on dairy protein gelation, the work clearly demonstrated that smaller pore sizes within the 3D gel network resulted in lower syneresis rate. Such observation about the relation between 3D structure of gel and its phase stability is in a good agreement with our observations. For instance, including even a relatively small volume of BTI may result in relatively smaller pore sizes for acidified milk gel (ADG$_{15\%}$, Figure 2b, white arrow) than the control gel (Figure 2a, white arrow); consequently, such structure may resulted in relatively lower EOS for ADG$_{15\%}$ on day 1 in comparison with the control gel (Table 1). Although the impact of tea addition to yogurt (or ADG) on the gel texture, antioxidant activity, lactic acid bacteria and quality related characteristics have been studied elsewhere [15,22,27,61,62], few research has been done to investigate the impact of tea infusion on microstructure of acidified milk protein gel systems. However, other phenolic compounds rich materials have been incorporated in yogurt gels and the microstructure of the gels was studied. For instance, a recent research showed that 3% apple pomace in stirred yogurt resulted in more compacted protein network and larger cavities in the gel in comparison with negative control gel [63]; moreover, Pan and co-workers found that inclusion of 5% pomegranate juice powder in set yogurt resulted in denser protein gel packing and inhomogeneous size distribution pores [64,65]. These observations in literature are in good agreement with results found in this research. Some aggregated and/or lumpy protein networks (white circles in Figure 3d,e) appeared in gels with high incorporation rate of BIT (45% & 60%) after 28 days of cold storage; these structural features cannot be observed in the same samples on day 1 (white circles in Figure 2d,e). These changes in structural organization of ADGs explain the instability of $ADG_{45\%}$ and $ADG_{60\%}$ (EOS and WHC results in Table 1).

CLSM was used for characterizing laminar microstructure of ADG samples. On day 1, $ADG_{30\%}$ (Figure 4c) and $ADG_{45\%}$ (Figure 4d) had denser protein gel blocks as pointed by white arrows. By comparing between Figure 4d,e and Figure 5d,e, more and bigger pores appeared in samples containing large volume of BTI (45% and 60%) after 28 days period of storage. The results suggested that significant transformations of gel structure took place during storage in which incorporation rate of BTI is high (45% and 60%). The structure characteristics of $ADG_{30\%}$ remained nearly the same on day 1 and day 28. Such result is consistent in both SEM (Figures 2c and 3c) and CLSM (Figures 4c and 5c) images. The poorest gel network was found in $ADG_{60\%}$ after 28 days storage, such structure resulted in worst cohesiveness among all gel samples (Table 2, $p < 0.05$).

In general, the structural changing trend was consistent between SEM images and CLSM images. The interactions between polyphenols and dairy proteins may be responsible to the microstructure changes of acid induced milk protein gel. The impact of tea extract components on the acidified gelation process is rather complicated as they affect both micro- and macro- structures at the same time.

4. Conclusions

The addition of black tea infusion as a nutritional ingredient to the acidified milk gel significantly reduced EOS on day 1 at all addition levels. However, such advantage started disappearing for $ADG_{45\%}$ after 7 days of storage; after 28 days, only $ADG_{15\%}$ showed similar EOS as the negative control gel and other gels samples all had higher EOS. At 30% or higher incorporation rates of BTI, the texture of the stirred gel became relatively softer and more brittle after 28 days of cold storage. Both phase stability results and changing trend of texture can be explained by the micrographs, SEM and CLSM images provided complementary information regarding the structural characteristics of BTI enriched milk gel systems. These micrographs may be used as good references for future research about tea-enriched ADG system (e.g., yogurt), since the microstructure features of such type of gel system have not yet been extensively reported. The inclusion of BTI led to remarkable increases of the antioxidant activity for the ADG samples. Such increased antioxidant activity is attributed to the increased TPC derived from BTI. The antioxidant capacity obtained from BTI was relatively stable during a 4-week cold storage. Overall, we recommend that 15% incorporation rate of BTI in ADG is the optimum. At this rate, after 28 days storage, EOS and gel firmness was not compromised and WHC was even higher in comparison with the negative control gel. Moreover, at the end of shelf life the TPC was increased nearly 50% in $ADG_{15\%}$ compared to $ADG_{0\%}$. Both similar and contradictory results were found in literature resources regarding the impact of supplement of tea infusion on the texture of acidified milk gel. Also, detailed mechanisms of interactions between tea infusion components and dairy proteins remain unclear. Therefore, further research is still needed for revealing these mechanisms.

Supplementary Materials: The following are available online at http://www.mdpi.com/2304-8158/9/6/831/s1, Figure S1: Supplement Biplot of the PCA.

Author Contributions: Conceptualization, C.S.B., M.A.B. and H.C.; methodology, H.Z., H.C., X.G. and C.S.B.; software, H.C. and H.Z.; validation, H.C., H.Z. and M.A.B.; formal analysis, H.C., H.Z. and M.A.B.; investigation, H.C. and X.G.; resources, H.Z., M.A.B., W.C. and X.G.; data curation, H.C; writing—original draft preparation, H.C., H.Z. C.S.B. and M.A.B.; writing—review and editing, H.C., H.Z. C.S.B. and M.A.B.; visualization, H.C., H.Z. C.S.B. and M.A.B.; supervision, C.S.B., M.A.B., H.Z. and G.X; project administration, C.S.B.; funding acquisition, C.S.B. All authors have read and agreed to the published version of the manuscript."

Funding: This research received no external funding.

Conflicts of Interest: The authors declare no conflict of interest.

References

1. Ben Hlel, T.; Borges, T.; Rueda, A.; Smaali, I.; Marzouki, M.N.; Seiquer, I. Polyphenols bioaccessibility and bioavailability assessment in ipecac infusion using a combined assay of simulated in vitro digestion and Caco-2 cell model. *Int. J. Food Sci. Technol.* **2019**, *54*, 1566–1575. [CrossRef]
2. Muniandy, P.; Shori, A.B.; Baba, A.S. Influence of green, white and black tea addition on the antioxidant activity of probiotic yogurt during refrigerated storage. *Food Packag. Shelf Life* **2016**, *8*, 1–8. [CrossRef]
3. Chen, D.; Sun, J.; Dong, W.; Shen, Y.; Xu, Z. Effects of polysaccharides and polyphenolics fractions of Zijuan tea (Camellia sinensis var. kitamura) on α-glucosidase activity and blood glucose level and glucose tolerance of hyperglycaemic mice. *Int. J. Food Sci. Technol.* **2018**, *53*, 2335–2341. [CrossRef]
4. Garcia Santos, F.A.; Freire, S.A.; Vieira, D.P.; Papa, P.d.C.; De Barros, G.F.; Castilho, C.; Guaberto, L.M.; Souza, L.F.A.d.; Laposy, C.B.; Nogueira, R.M.B.; et al. White tea intake interferes with the expression of angiogenic factors in the corpora lutea of superovulated rats. *Int. J. Food Sci. Technol.* **2018**, *53*, 1666–1671. [CrossRef]
5. Del Rio, D.; Stewart, A.J.; Mullen, W.; Burns, J.; Lean, M.E.J.; Brighenti, F.; Crozier, A. HPLC-MSn Analysis of Phenolic Compounds and Purine Alkaloids in Green and Black Tea. *J. Agric. Food Chem.* **2004**, *52*, 2807–2815. [CrossRef]
6. Gadow, A.V.; Joubert, E.; Hansmann, C.F. Comparison of the antioxidant activity of rooibos tea (Aspalathus linearis) with green, oolong and black tea. *Food Chem.* **1997**, *60*, 73–77. [CrossRef]
7. Islam, S.N.; Farooq, S.; Sehgal, A. Effect of consecutive steeping on antioxidant potential of green, oolong and black tea. *Int. J. Food Sci. Technol.* **2018**, *53*, 182–187. [CrossRef]
8. Donovan, S.M.; Shamir, R. Introduction to the Yogurt in Nutrition Initiative and the First Global Summit on the Health Effects of Yogurt. *Am. J. Clin. Nutr.* **2014**, *99*, 1209–1211. [CrossRef]
9. Zhang, S.; Zhang, Y.; Li, M.; Luo, X.; Xiao, M.; Sun, Q.; Xie, F.; Zhang, L. The effect of Lactobacillus delbrueckii subsp bulgaricus proteinase on properties of milk gel acidified with glucono-δ-lactone. *Int. J. Food Sci. Technol.* **2019**, *54*, 2094–2100. [CrossRef]
10. Eshpari, H.; Tong, P.S.; Corredig, M. Changes in the physical properties, solubility, and heat stability of milk protein concentrates prepared from partially acidified milk. *J. Dairy Sci.* **2014**, *97*, 7394–7401. [CrossRef]
11. Le Maux, S.; Nongonierma, A.B.; Lardeux, C.; FitzGerald, R.J. Impact of enzyme inactivation conditions during the generation of whey protein hydrolysates on their physicochemical and bioactive properties. *Int. J. Food Sci. Technol.* **2018**, *53*, 219–227. [CrossRef]
12. Lucey, J.A.; Tamehana, M.; Singh, H.; Munro, P.A. A comparison of the formation, rheological properties and microstructure of acid skim milk gels made with a bacterial culture or glucono-δ-lactone. *Food Res. Int.* **1998**, *31*, 147–155. [CrossRef]
13. Lourens-Hattingh, A.; Viljoen, B.C. Yogurt as probiotic carrier food. *Int. Dairy J.* **2001**, *11*, 1–17. [CrossRef]
14. Ng, E.W.; Yeung, M.; Tong, P.S. Effects of yogurt starter cultures on the survival of Lactobacillus acidophilus. *Int. J. Food Microbiol.* **2011**, *145*, 169–175. [CrossRef]
15. Najgebauer-Lejko, D.; Sady, M.; Grega, T.; Walczycka, M. The impact of tea supplementation on microflora, pH and antioxidant capacity of yoghurt. *Int. Dairy J.* **2011**, *21*, 568–574. [CrossRef]
16. Sharma, V.; Vijay Kumar, H.; Jagan Mohan Rao, L. Influence of milk and sugar on antioxidant potential of black tea. *Food Res. Int.* **2008**, *41*, 124–129. [CrossRef]
17. Cornell, D.G.; De Vilbiss, E.D.; Pallansch, M.J. Binding of Antioxidants by Milk Proteins. *J. Dairy Sci.* **1971**, *54*, 634–637. [CrossRef]
18. Arts, M.; Haenen, G.; Voss, H.-P.; Bast, A. Masking of antioxidant capacity by the interaction of flavonoids with protein. *Food Chem. Toxicol.* **2001**, *39*, 787–791. [CrossRef]
19. Ryan, L.; Petit, S. Addition of whole, semiskimmed, and skimmed bovine milk reduces the total antioxidant capacity of black tea. *Nutr. Res.* **2010**, *30*, 14–20. [CrossRef]
20. Haratifar, S.; Corredig, M. Interactions between tea catechins and casein micelles and their impact on renneting functionality. *Food Chem.* **2014**, *143*, 27–32. [CrossRef]
21. Ye, J.; Fan, F.; Xu, X.; Liang, Y. Interactions of black and green tea polyphenols with whole milk. *Food Res. Int.* **2013**, *53*, 449–455. [CrossRef]
22. Rashidinejad, A.; Birch, E.J.; Sun-Waterhouse, D.; Everett, D.W. Addition of milk to tea infusions: Helpful or harmful? Evidence from in vitro and in vivo studies on antioxidant properties. *Crit. Rev. Food Sci. Nutr.* **2017**, *57*, 3188–3196. [CrossRef]

23. Jöbstl, E.; Howse, J.R.; Fairclough, J.P.A.; Williamson, M.P. Noncovalent cross-linking of casein by epigallocatechin gallate characterized by single molecule force microscopy. *J. Agric. Food Chem.* **2006**, *54*, 4077–4081. [CrossRef]

24. Yildirim-Elikoglu, S.; Erdem, Y.K. Interactions between milk proteins and polyphenols: Binding mechanisms, related changes, and the future trends in the dairy industry. *Food Rev. Int.* **2018**, *34*, 665–697. [CrossRef]

25. Brown, P.J.; Wright, W.B. An Investigation of the interactions between milk proteins and tea polyphenols. *J. Chromatogr. A* **1963**, *11*, 504–514. [CrossRef]

26. Van der Burg-Koorevaar, M.C.D.; Miret, S.; Duchateau, G.S.M.J.E. Effect of Milk and Brewing Method on Black Tea Catechin Bioaccessibility. *J. Agric. Food Chem.* **2011**, *59*, 7752–7758. [CrossRef]

27. Najgebauer-Lejko, D.; Żmudziński, D.; Ptaszek, A.; Socha, R. Textural properties of yogurts with green tea and Pu-erh tea additive. *Int. J. Food Sci. Technol.* **2014**, *49*, 1149–1158. [CrossRef]

28. Lamothe, S.; Langlois, A.; Bazinet, L.; Couillard, C.; Britten, M. Antioxidant activity and nutrient release from polyphenol-enriched cheese in a simulated gastrointestinal environment. *Food Funct.* **2016**, *7*, 1634–1644. [CrossRef]

29. Amirdivani, S.; Baba, A.S.H. Green tea yogurt: Major phenolic compounds and microbial growth. *J. Food Sci. Technol.* **2015**, *52*, 4652–4660. [CrossRef]

30. Giroux, H.J.; De Grandpré, G.; Fustier, P.; Champagne, C.P.; St-Gelais, D.; Lacroix, M.; Britten, M. Production and characterization of Cheddar-type cheese enriched with green tea extract. *Dairy Sci. Technol.* **2013**, *93*, 241–254. [CrossRef]

31. Jaziri, I.; Ben Slama, M.; Mhadhbi, H.; Urdaci, M.C.; Hamdi, M. Effect of green and black teas (*Camellia sinensis* L.) on the characteristic microflora of yogurt during fermentation and refrigerated storage. *Food Chem.* **2009**, *112*, 614–620. [CrossRef]

32. Ciron, C.I.E.; Gee, V.L.; Kelly, A.L.; Auty, M.A.E. Comparison of the effects of high-pressure microfluidization and conventional homogenization of milk on particle size, water retention and texture of non-fat and low-fat yoghurts. *Int. Dairy J.* **2010**, *20*, 314–320. [CrossRef]

33. Zheng, H. Introduction: Measuring Rheological Properties of Foods. In *Rheology of Semisolid Foods*; Joyner, H.S., Ed.; Springer International Publishing: Cham, Switzerland, 2019; pp. 3–30.

34. Yuksel, Z.; Avci, E.; Erdem, Y.K. Characterization of binding interactions between green tea flavanoids and milk proteins. *Food Chem.* **2010**, *121*, 450–456. [CrossRef]

35. Vega, C.; Grover, M.K. Physicochemical Properties of Acidified Skim Milk Gels Containing Cocoa Flavanols. *J. Agric. Food Chem.* **2011**, *59*, 6740–6747. [CrossRef]

36. Weidendorfer, K.; Bienias, A.; Hinrichs, J. Investigation of the effects of mechanical post-processing with a colloid mill on the texture properties of stirred yogurt. *Int. J. Dairy Technol.* **2008**, *61*, 379–384. [CrossRef]

37. Morell, P.; Chen, J.; Fiszman, S. The role of starch and saliva in tribology studies and the sensory perception of protein-added yogurts. *Food Funct.* **2017**, *8*, 545–553. [CrossRef]

38. Kosasih, L.; Bhandari, B.; Prakash, S.; Bansal, N.; Gaiani, C. Physical and functional properties of whole milk powders prepared from concentrate partially acidified with CO2 at two temperatures. *Int. Dairy J.* **2016**, *56*, 4–12. [CrossRef]

39. Lim, Y.Y.; Murtijaya, J. Antioxidant properties of Phyllanthus amarus extracts as affected by different drying methods. *Lwt-Food Sci. Technol.* **2007**, *40*, 1664–1669. [CrossRef]

40. Khanizadeh, S.; Tsao, R.; Rekika, D.; Yang, R.; DeEll, J. Phenolic composition and antioxidant activity of selected apple genotypes. *J. Food Agric. Environ.* **2007**, *5*, 61–66.

41. Al-Dabbas, M.M.; Al-Ismail, K.; Kitahara, K.; Chishaki, N.; Hashinaga, F.; Suganuma, T.; Tadera, K. The effects of different inorganic salts, buffer systems, and desalting of Varthemia crude water extract on DPPH radical scavenging activity. *Food Chem.* **2007**, *104*, 734–739. [CrossRef]

42. Le, H.M. *Antioxidative Effects of Mango Wastes on Shelf Life of Pork Products*; Lincoln University: Lincoln, New Zealand, 2012.

43. Elfalleh, W.; Nasri, N.; Marzougui, N.; Thabti, I.; M'Rabet, A.; Yahya, Y.; Lachiheb, B.; Guasmi, F.; Ferchichi, A. Physico-chemical properties and DPPH-ABTS scavenging activity of some local pomegranate (Punica granatum) ecotypes. *Int. J. Food Sci. Nutr.* **2009**, *60*, 197–210. [CrossRef]

44. Lin, S.Y.; Lo, L.C.; Chen, I.Z.; Chen, P.A. Effect of shaking process on correlations between catechins and volatiles in oolong tea. *J. Food Drug Anal.* **2016**, *24*, 500–507. [CrossRef]

45. Kalab, M. Microstructure of Dairy Foods. 1. Milk Products Based on Protein1. *J. Dairy Sci.* **1979**, *62*, 1352–1364. [CrossRef]
46. Domagala, J. Instrumental Texture, Syneresis and Microstructure of Yoghurts Prepared from Goat, Cow and Sheep Milk. *Int. J. Food Prop.* **2009**, *12*, 605–615. [CrossRef]
47. Wells, H.C.; Sizeland, K.H.; Kirby, N.; Hawley, A.; Mudie, S.; Haverkamp, R.G. Collagen Fibril Structure and Strength in Acellular Dermal Matrix Materials of Bovine, Porcine, and Human Origin. *ACS Biomater. Sci. Eng.* **2015**, *1*, 1026–1038. [CrossRef]
48. Nowak, J.S. *Identification and Understanding the Roles of Biofilm Formation-Related Genes in Listeria Monocytogenes Isolated from Seafood*; Palmerston North, Messy University: Palmerston North, New Zealand, 2017.
49. Wells, H.C.; Sizeland, K.H.; Kirby, N.; Hawley, A.; Mudie, S.; Haverkamp, R.G. Acellular dermal matrix collagen responds to strain by intermolecular spacing contraction with fibril extension and rearrangement. *J. Mech. Behav. Biomed. Mater.* **2018**, *79*, 1–8. [CrossRef]
50. Charrier, M.J.S.; Savage, G.P.; Vanhanen, L. Oxalate content and calcium binding capacity of tea and herbal teas. *Asia Pac. J. Clin. Nutr.* **2002**, *11*, 298–301. [CrossRef]
51. Yamada, K.; Abe, T.; Tanizawa, Y. Black tea stain formed on the surface of teacups and pots. Part 2—Study of the structure change caused by aging and calcium addition. *Food Chem.* **2007**, *103*, 8–14. [CrossRef]
52. Carnovale, V.; Labaeye, C.; Britten, M.; Couillard, C.; Bazinet, L. Effect of various calcium concentrations on the interactions between β-lactoglobulin and epigallocatechin-3-gallate. *Int. Dairy J.* **2016**, *59*, 85–90. [CrossRef]
53. Pimentel, T.C.; Garcia, S.; Prudencio, S.H. Effect of long-chain inulin on the texture profile and survival of Lactobacillus paracasei ssp paracasei in set yoghurts during refrigerated storage. *Int. J. Dairy Technol.* **2012**, *65*, 104–110. [CrossRef]
54. Hasni, I.; Bourassa, P.; Hamdani, S.; Samson, G.; Carpentier, R.; Tajmir-Riahi, H.-A. Interaction of milk α-and β-caseins with tea polyphenols. *Food Chem.* **2011**, *126*, 630–639. [CrossRef]
55. Rawson, H.L.; Marshall, V.M. Effect of 'ropy' strains of Lactobacillus delbrueckii ssp. bulgaricus and Streptococcus thermophilus on rheology of stirred yogurt. *Int. J. Food Sci. Technol.* **1997**, *32*, 213–220. [CrossRef]
56. Lucey, J.A.; Singh, H. Formation and physical properties of acid milk gels: A review. *Food Res. Int.* **1997**, *30*, 529–542. [CrossRef]
57. Damin, M.R.; Minowa, E.; Alcântara, M.R.; Oliveira, M.N. Effect of cold storage on culture viability and some rheological properties of fermented milk prepared with yogurt and probiotic bacteria. *J. Texture Stud.* **2008**, *39*, 40–55. [CrossRef]
58. Avci, E.; Yuksel, Z.; Erdem, Y. Green yoghurt revolution. Identification of interactions between green tea polyhenols and milk proteins and resultant functional modifications in yoghurt gel. In Proceedings of the 5th Central European Congress on Food, Bratislava, Slovak Republik, 20 May 2020; pp. 19–22.
59. Tanizawa, Y.; Abe, T.; Yamada, K. Black tea stain formed on the surface of teacups and pots. Part 1— Study on the chemical composition and structure. *Food Chem.* **2007**, *103*, 1–7. [CrossRef]
60. Spiro, M.; Chong, Y.Y. Kinetics and equilibria of tea infusion .14. Surface films formed in hard water by black tea brews containing milk. *Food Chem.* **1997**, *59*, 247–252. [CrossRef]
61. Fiszman, S.M.; Lluch, M.A.; Salvador, A. Effect of addition of gelatin on microstructure of acidic milk gels and yoghurt and on their rheological properties. *Int. Dairy J.* **1999**, *9*, 895–901. [CrossRef]
62. Dönmez, Ö.; Mogol, B.A.; Gökmen, V. Syneresis and rheological behaviors of set yogurt containing green tea and green coffee powders. *J. Dairy Sci.* **2017**, *100*, 901–907. [CrossRef]
63. Liu, D. Effect of Fuzhuan brick-tea addition on the quality and antioxidant activity of skimmed set-type yoghurt. *Int. J. Dairy Technol.* **2018**, *71*, 22–33. [CrossRef]
64. Wang, X.; Kristo, E.; LaPointe, G. Adding apple pomace as a functional ingredient in stirred-type yogurt and yogurt drinks. *Food Hydrocoll.* **2020**, *100*, 105453. [CrossRef]
65. Pan, L.-H.; Liu, F.; Luo, S.-Z.; Luo, J.-p. Pomegranate juice powder as sugar replacer enhanced quality and function of set yogurts: Structure, rheological property, antioxidant activity and in vitro bioaccessibility. *LWT* **2019**, *115*, 108479. [CrossRef]

 *foods*

Article

Antioxidant, Antibacterial Activities and Mineral Content of Buffalo Yoghurt Fortified with Fenugreek and *Moringa oleifera* Seed Flours

Faten Dhawi [1], Hossam S. El-Beltagi [1,2,*], Esmat Aly [3] and Ahmed M. Hamed [4,*]

[1] Agricultural Biotechnology Department, College of Agriculture and Food Sciences, King Faisal University, P.O. Box 420, Al-Ahsa 31982, Saudi Arabia; falmuhanna@kfu.edu.sa

[2] Biochemistry Department, Faculty of Agriculture, Cairo University, Giza 12613, Egypt

[3] Dairy Technology Research Department, Food Technology Research Institute, Agricultural Research Center, Giza 12613, Egypt; esmat_rayan_2010@yahoo.com

[4] Dairy Science Department, Faculty of Agriculture, Cairo University, Giza 12613, Egypt

[*] Correspondence: helbeltagi@kfu.edu.sa (H.S.E.-B.); ahmed.hamed@agr.cu.edu.eg (A.M.H.); Tel.: +966-541775875 (H.S.E.-B.); +201-153763241 (A.M.H.)

Received: 21 July 2020; Accepted: 19 August 2020; Published: 21 August 2020

Abstract: Recently, there is an increasing demand for functional yoghurts by consumer, especially those produced through the incorporation of food of plant origin or its bioactive components. The current research was devoted to formulating functional buffalo yoghurt through the addition of 0.1 and 0.2% of fenugreek (*Trigonella foenum-graecum*) seed flour (F1 and F2) and *Moringa oleifera* seed flour (M1 and M2). The effects of fortification were evaluated on physicochemical, total phenolic content (TPC), antioxidant activity (AOA), the viability of yoghurt starter, and sensory acceptability of yoghurts during cold storage. *Moringa oleifera* seed flour had higher contents of TPC (140.12 mg GAE/g) and AOA (31.30%) as compared to fenugreek seed flour (47.4 mg GAE/g and 19.1%, respectively). Values of TPC and AOA significantly increased in fortified yoghurts, and M2 treatment had the highest values of TPC (31.61, 27.29, and 25.69 mg GAE/g) and AOA (89.32, 83.5, and 80.35%) at 1, 7, and 14 days of storage, respectively. M2 showed significantly higher antibacterial activity against *E. coli*, *S. aureus*, *L. monocytogenes*, and *Salmonella* spp. and the zones of inhibition were 12.65, 13.14, 17.23 and 14.49 mm, respectively. On the other hand, control yoghurt showed the lowest antibacterial activity and the zones of inhibition were (4.12, 5.21, 8.55, and 8.39 mm against *E. coli*, *S. aureus*, *L. monocytogenes*, and *Salmonella* spp., respectively). Incorporation of 0.1% and 0.2% of moringa seed flour (M1 and M2) led to a higher content of Ca, P, K, and Fe and lower content of Mg and Zn as compared to F1 and F2, respectively. Thus, it could be concluded that fenugreek and *Moringa oleifera* seed flour can be exploited in the preparation of functional novel yoghurt.

Keywords: functional yogurt; fenugreek and *Moringa oleifera* seed flours; total phenolic content; antioxidant activity; antibacterial activity; mineral content

---

1. Introduction

Worldwide, yoghurt is considered one of the most popular fermented dairy products due to not only for its nutritional value but also for its health benefits [1]. Buffalo milk is much preferred by consumers for its rich nutrition and is drunk or transformed into valuable products such as cheese, curd, yogurt, and ice cream. Buffalo milk contains about twice as much butterfat as cow milk and higher amounts of total solids and casein, making it highly suitable for processing various types of yogurt and resulting in creamy textures and rich flavor profiles. Although its many healthy and nutritious impacts are well-established, milk and its products are generally not regarded as a rich source for particular bioactive ingredients such as polyphenols and antioxidants [2]. Thus, the formulation of novel dairy products using medicinal herbs or their extracts has gotten more attention to meet the demand of health-conscious consumers [3]. In this context, several new fermented dairy products enhanced with plant-derived foods (fruit, vegetables, or even their by-products) have been created and assessed [4]. Fenugreek (*Trigonella foenum-graecum*) is an annual plant indigenous to India and North Africa which has a lengthy background of using a range of circumstances, including diabetes and hyperlipidemia, as traditional herbal medicine. Fenugreek seeds and leaves are used in food as well as in medicinal applications, which is an old practice of human history [5]. It is widely known for its high fiber, gum, and other phytochemical components. Fenugreek seed dietary fiber, forms about 25%, has beneficial effects on digestion and can also actually change the texture of the food. In addition, polyphenol compounds such as rhaponticin and isovitexin [6], flavonoids, alkaloids, amino acids, coumarins, vitamins, saponins, and other antioxidants are thought to be the main bioactive elements in fenugreek seeds [7]. Fenugreek seeds are also a rich source for vitamins, minerals, and antioxidants. Other fenugreek components include carbohydrates, principally mucilaginous fiber (galactomannans), fixed oils (lipids), volatile oils, free amino acids, calcium, and iron, ..., etc. [8]. Antidiabetic, antioxidant, anticarcinogenic, hypoglycemic activity, hypocholesterolemic activity are the major medicinal properties of the fenugreek demonstrated in various studies. Based on these several healthful benefits, fenugreek can be recommended and be a part of our daily diet and incorporated into foods to produce functional foods [9]. Thus, methi, ground fenugreek seed drink, was used in ancient Egypt to ease birth and increase milk flow, and is still used by modern Egyptian women today to ease menstrual cramps and in making hilba tea out of it to ease other types of abdominal pain. *Moringa* (*Moringa oleifera*), a Moringaceae species drumstick plant, has several medicinal advantages including injury healing, antitumor, hypotensive, anti-hepatotoxic, anti-inflammatory, antiulcer, hypocholesterolaemic, antibacterial and anti-diabetes activities. *Moringa oleifera* has large amounts of vitamin A, proteins, carbohydrates, minerals. Therefore; it is commonly used to improve nutritional status. *Moringa oleifera* has multifunctional use and has vital nutritional, industrial and medicinal applications [10]. Moreover, *Moringa oleifera* Lam. is a fast-growing tree with interesting benefits for human health [11] Nutritionally, it is possible to combine all parts of *Moringa oleifera* (leaves, seeds, fruits, immature pods, and flowers) with traditional food for human consumption [12]. In this sense, *Moringa oleifera* or its extracts have been used to improve the nutritional value of yoghurt and cottage cheese [13], with special reference to protein, fiber, and minerals [14]. *Moringa oleifera* is an abundant source of polyphenols, flavonoids, minerals, alkaloids and proteins. Substances such as bioactive carotenoids, tocopherols and vitamin C showed health-promoting potential in maintaining a balanced diet and protecting against free-radical damage that might initiate many diseases [15].

Unlike moringa, which is deeply studied and incorporated in several forms to yoghurt formulations, fenugreek has not yet been incorporated in yoghurt formulations. Consequently, the main objective of this study was to develop a functional yoghurt fortified with fenugreek seed flour (0.1 and 0.2%) and, for comparative purposes, yoghurt formulations fortified with *Moringa oleifera* seed flour (0.1 and 0.2%). Then, the effect of fortification was assessed in both formulated yoghurts by exploring the physicochemical characteristics, the viability of starter culture, as well as the mineral content and antioxidant activity, during cold storage. Moreover, the antibacterial effect of yoghurt supernatant was conducted against some pathogenic bacteria including *E. coli*, *S. aureus*, *L. monocytogenes and Salmonella* spp.

2. Materials and Methods

2.1. Chemicals, Reagents, and Culture

Buffalo milk was obtained from the herd of the Faculty of Agriculture, Cairo University, Cairo, Egypt. DVS ThermophilicYoFlex®starter culture consisting of *Streptococcus thermophiles* and *Lactobacillus delbrueckii subsp. bulgaricus* (Chr. Hansen, Horsholm, Denmark) was used in yoghurt manufacturing. MRS agar and M17 agar, the media of agar plates used for pathogenic bacteria, Salmonella Shigella agar for *S. Typhimurium*, mannitol salt agar for *S. aureus*, MacConkey sorbitol agar for *E. coli*, and Oxford agar for *L. monocytogenes*, were obtained from Biolife Italiana (Milano, Italy).

All solvents used for extraction and analyses through this study were of analytical grade. The reagent 2, 2-diphenyl-1-picrylhydrazyl (DPPH), Folin-Ciocalteu, and gallic acid were obtained from Sigma-Aldrich (*Sigma*-Aldrich, Darmstad, Germany).

2.2. Seed Flours Preparation

Fenugreek and *Moringa oleifera* seeds were purchased from a local market at Giza, Egypt. They were soaked in water for 15 min to remove impurities. The seeds were dehydrated in an air drier at 55 °C and ground to obtain their flours which passed through a 60-mesh sieve to obtain a uniform material. The flour was packed in polyethylene bags, and frozen (−18 °C) and used in yoghurt processing during 30 days.

2.3. Yoghurt Processing

Buffalo's milk contains 6.1% fat, 3.9% protein, and 14.9% total solids (TS) were used for yoghurt processing according to [16]. Fenugreek and *Moringa oleifera* seed flour were incorporated at 0.1–0.5% to buffalo milk before heat treatment for 15 min as rehydration time with stirring. Then, it heat treated until reaching 90 °C for 5 min, cooled to 42 °C, inoculated with DVS starter culture (2%), and incubated at 42 °C ± 1 °C. After achieving a pH of 4.6 (~2.5–3 h), yoghurt samples were stored at 5 °C for 14 days. Sampling points were carried out at 1, 7 and 14 days of cold storage and subjected to chemical, microbiological and organoleptic analyses at a regular interval of 7 days as well as the determination of total phenolic content (TPC), antioxidant activity (AOA %). The antibacterial activity assay was evaluated using fresh yoghurt supernatant (Yoghurt samples were centrifuged at 4 °C for 30 min at 4000 rpm (centrifuge model C-28 AC BOECO, Hamburg, Germany) and the supernatant was filtered through a 0.45-μm Millipore membrane filter. The supernatant filtrates were kept at −20 °C until analysis [17] while mineral content was estimated at 1 and 14 days of cold storage. All analyses were carried out in triplicate.

2.4. Preliminary Study

After the manufacturing of the different formulations of yoghurt, a preliminary sensory evaluation study was conducted to select the best formulations that will be used throughout the experiment. A preliminary sensory evaluation study conducted on these yoghurt formulations showed that only yoghurt formulations containing 0.1% and 0.2% of fenugreek and *Moringa oleifera* seed flours were acceptable and observed that no precipitation found in the bottom of the cup after the period of coagulation which means that the added amounts have been dissolved. Thus, only these formulations were subjected to different analyses in the current research.

2.5. Analytical Methods of Seed Flours

2.5.1. Proximate Analysis

The proximate analysis of seed flours was determined according to the methods described in AOAC [18]. Moisture, ash, and crude fiber contents were determined by the gravimetric method (AOAC 934.01), dry incineration in a muffle furnace (AOAC 942.05), and soxhlet method (AOAC 954.02), respectively. Protein content was determined by the Kjeldahl method (AOAC 976.05) and the obtained results expressed the total nitrogen content that was multiplied with factor 6.25 to obtain the total protein content. Total carbohydrate (TC) was calculated by difference (TC%) = 100 − (moisture + protein + fat + ash).

2.5.2. Individual Polyphenols

The individual polyphenols of fenugreek and moringa seed flour were identified using the HPLC technique according to [19]. These compounds have been identified using HPLC by comparison with the retention times of the mix standards of Gallic acid, Catechin, Gentisic acid, Protocatechuic acid, Vanillic acid, Caffeic acid, Syringic acid, Chlorogenic, and p-Coumaric acid. We extracted 5 g of fenugreek and moringa seed flour with methanol (50 mL) and centrifuged it for 10 min at 1000 rpm (centrifuge model C-28 AC BOECO, Hamburg, Germany). The supernatant was filtered through a 0.2 µm Millipore membrane filter. 20 µL of the filtrate was injected into HPLC on Gemini-Nx 5u, C18, 250 × 4.6 mm column operated at 30 °C. Analyses were performed on the liquid chromatography HPLC Knauer, Germany, UV detector at 284 nm. The separation is achieved using a ternary linear elution gradient with (A) HPLC grade water 0.2% H3PO4 (*v/v*) (96%), (B) methanol (2%) and (C) acetonitrile (2%). The injected volume was 20 µL.

2.5.3. TPC and AOA

For the determination of TPC and AOA, 100 mL of an aqueous methanolic solution (75%) was poured into a beaker containing 10 g of seeds flour. The beaker was covered using aluminum foil with allow stirring for 30 min. Then, it was filtered using a Whatman No. 1 filter paper. The filtrate was used in the determination of TPC and AOA. TPC was determined by using Folin-Ciocalteau reagent according to the method described by [20] and the results expressed as mg GAE/g of the sample using Gallic aid as a reference standard. While AOA% was evaluated by using DPPH according to [21].

2.6. Analytical Methods of Milk and Yoghurt Samples

2.6.1. Physicochemical Analysis

All chemical analyses of buffalo milk and yoghurt samples were carried out in triplicate. TS, fat and protein contents of milk used for the yoghurt production were determined according to [22]. The pH values of yoghurt samples were measured by pH meter (Hanna, digital pH meter, Barcelona, Spain) while titratable acidity (as lactic acid %) was determined according to [19]. Syneresis was estimated according to [23].

2.6.2. Susceptibility to Syneresis (STS)

STS of yoghurt samples was determined according to the method described by [23]. The following formula was used to calculate STS:

$$STS = (V1/V2) \times 100$$

where: V1 = Volume of whey collected after drainage; V2 = Volume of yoghurt sample.

2.6.3. Phenolic Extraction and TPC and AOA Estimation

Control yoghurt and fortified yoghurt samples were extracted for obtaining the phenolic compounds according to the method described by [24]. In brief, ten grams of yoghurt samples were extracted with 100 mL of aqueous methanolic solution (75%) in a beaker wrapped with aluminum foil with allow stirring for 15 min with vortex-mixer (VELP Scientific, Usmate Velate, Italy). Then, the mixture was centrifuged at 4 °C for 10 min at 7200 rpm. The obtained supernatants were filtered by Whatman No.1 and the collected methanolic extracts were used in TPC and AOA assay as follows:

TPC of Yoghurt

TPC of yoghurt samples was determined using the micro-scale Folin–Ciocalteau method described by [25]. TPC values were expressed as mg GAE/g based on a gallic acid standard curve.

AOA of Yoghurt

Antioxidant activity was carried out according to [26] with slight modification. In brief, an aliquot (1.5 mL) of the obtained methanolic extract (of the yoghurt sample) was mixed with 1.5 mL of DPPH solution and was kept in the dark for 30 min at ambient temperature. The absorbance of the solution was then measured by a spectrophotometer at 517 nm. The percentage decrease in absorbance of the sample relative to the control was calculated as the relative scavenging activity [27].

2.6.4. Yoghurt's Minerals Content

P, Ca, K, Fe, Zn, and Mg were determined by inductively coupled plasma–atomic emission spectrometry (ICP-OES) using iCAP 6000 Series (Thermo Scientific, New York, NY, USA) according to [28] as follow: Samples (0.5 g) were digested with 7 mL of HNO3 (65%) and 1 mL of H_2O_2 (30%) (Sigma-Aldrich, St Louis, MO, USA) in ETHOS1 advanced Microwave Digestion system (Milestone, USA) for 31 min and diluted to 100 mL with deionized water [29]. Blank digestion was carried out in the same way (digestion conditions for microwave system were: 2 min for 250 W, 2 min for 0 W, 6 min for 250W, 5 min for 400 W, 8 min for 550 W, vent: 8 min, respectively). Analysis of trace elements in yoghurt samples was performed by inductively coupled plasma-atomic emission spectrometry using iCAP 6000 Series (Thermo Scientific, New York, NY, USA). All samples were analyzed in triplicates by ICP-OES. The operational parameters were as follow: RF applied power 1150 (W), Argon external flow rate 12 (L min^{-1}), Argon intermediate flow rate 0.50 (L min^{-1}), Argon nebulizer flow rate 0.70 (L min^{-1}), Integration time 5 (s), sample uptake delay 30 (s), stabilization time 5 (s) and sample uptake rate 1.0 (mL min^{-1}).

2.6.5. Microbiological Analysis

Viable counts of *S. thermophilus* and *L. delbrueckii subsp. bulgaricus* in yoghurt samples at different sampling points were determined using the standard plate count method according to [30]. M17 and MRS agar culture media were used for the enumeration of *S. thermophilus* and *L. bulgaricus*, respectively. The plates were incubated in anaerobic conditions at 42 °C for 48 h or 37 °C for 72 h for the enumeration of *S. thermophilus* and *L. bulgaricus*, respectively. The results were expressed as log number of colony-forming units per g (cfu/g).

2.6.6. Antibacterial Activity of Yoghurt Supernatant

For studying the antibacterial activity of yoghurt supernatant, yoghurt samples were subjected to centrifugation at 4000 rpm for 30 min at 4 °C (centrifuge model C-28 AC BOECO, Hamburg, Germany). The obtained supernatants were filtered using Millipore membrane filter of 0.45-μm diameter. The supernatant filtrates were used in determining the antibacterial activity against some pathogens (*E. coli*, *S. aureus*, *L. monocytogenes* and *S. Typhimurium*). The media of agar plates used in this study were Salmonella shigella agar for *S. typhimurium*, mannitol salt agar for *S. aureus*, MacConkey sorbitol agar for *E. coli*, and Oxford agar for *L. monocytogenes* [31]. A 100-μL diluted yogurt sample was spread on agar plates. After incubation at 37 °C for 2 days, the cell colonies were counted. Antimicrobial activity was evaluated by measuring the zones of inhibition against the tested bacteria (mm). Each assay was carried out in triplicate.

2.6.7. Sensory Evaluation

Twelve trained panelists belonging to the staff member of Dairy Science Dept., Faculty of Agriculture, Cairo University were recruited for assessing the sensory descriptors of plain and fortified-yoghurts on days 1, 14 of cold storage [32]. All these trained panelists were food technology experts and were chosen based on their desire to participate and their knowledge about dairy products. They were frequent yoghurt consumers and did not have any allergies to it. Moreover, the panelists were subjected to two training sessions to study and discuss the various tested sensory descriptors including the changes in color, texture, and flavor of yoghurt samples. The panelists were instructed to wash their mouths with low sodium spring water (Safi, Egypt) during the sensory evaluation session and they were encouraged to write down any criticisms on the tested products.

Plain- and fortified-yoghurt samples were presented in plastic coded (three-digit random codes) cups. Each cup contains 50 mL of yoghurt samples that freshly removed from the refrigerator. The panelists were asked to evaluate the samples using a five-point scale where 1 = I do not like it at all, 5 = I like it extremely. The sensory evaluation of the different descriptors relied on the pre-selected descriptors: color and appearance (wheying-off, white color, reddish color), mouthfeel (ropy, uniform coagulum), body and texture (absence of curd homogeneity, lumps, bubbles), taste and flavor (sweetness, acidity, bitterness), and overall acceptability (the sum of all the character's results). The sensory evaluation was conducted using a comparative test and fresh yoghurt as a reference sample. The data were collected in specifically designed ballots.

2.7. Statistical Analysis

The obtained data were statistically analyzed with two-way ANOVA to identify the significant differences between the means of samples and storage period. All data were expressed as a mean ± standard deviation of three replicates. The means of results were compared by the Tukey test with a confidence interval set at 95%.

3. Results and Discussion

3.1. Composition, TPC and AOA of Seed Flours

The data are shown in Table 1 display the nutritional composition, total phenolic content (TPC, mg GAE/g) and antioxidant activity (AOA %) of fenugreek and *Moringa oleifera* seed flours. These data indicated that fenugreek seed flour had a higher content of moisture (5.30%) and crude fiber (5.88%) while *Moringa oleifera* seed flour had a higher protein content (33.37%), oil (42.56%), ash (4.33), TPC (140.12 mg GAE/g) and antioxidant activity (31.30%). Similar data previously reported by [33] demonstrated that *Moringa oleifera* seed has higher protein content (ranging between 27–33%) and is a good source of phytochemicals. Seeds of *Moringa oleifera* could be employed in dairy products since the seed incorporation resulted in limited color changes in the fortified products [34], especially when a lower amount of *Moringa oleifera* seed flour was incorporated. Our findings are in general agreement with those found by [35] who reported slightly higher amounts of protein, fat, ash and fiber of moringa seed powder. The later authors reported lower TPC and slightly higher antioxidant activity of moringa seed powder. The current results showed that total carbohydrates were 56.2% and 17.29% for fenugreek and Moringa seed flours, respectively. These data are in general accordance with [36,37]. The variations in the present study outcomes could be attributed to polyphenolic compound extraction methods, solvent degree polarity, plant geographic locations, and plant species. Unlike Moringa, Fenugreek is less studied especially in terms of its addition to dairy products. Generally, the nutritional composition (moisture, protein, fat, ash, and fiber) of fenugreek seed flour is in general agreement with that reported by [38] which indicated slightly higher fat content in Fenugreek seed. [39] reported a higher TPC content of fenugreek seeds (85.88 mg GAE/g) as compared to our obtained data (47.40 mg GAE/g). Thus, it has a higher antioxidant activity due to its higher content of polyphenolic compounds [40].

Table 1. Nutritional composition, total phenolic content (TPC, mg GAE/g) and antioxidant activity (AOA %) of fenugreek and *Moringa oleifera* seed flour.

Parameter	Fenugreek Seed Flour	*M. oleifera* Seed Flour
Moisture (%)	5.3 ± 0.34	2.45 ± 0.04
Protein (%)	30.7 ± 0.13	33.37 ± 0.05
Oil (%)	4.4 ± 0.26	42.56 ± 0.11
Ash (%)	3.4 ± 0.15	4.33 ± 0.04
Crude fiber (%)	7.70 ± 0.22	4.50 ± 0.2
Total carbohydrates (%)	56.2 ± 0.75	17.29 ± 0.5
TPC (mg GAE/g)	47.4 ± 0.22	140.12 ± 0.1
Antioxidant activity (%)	19.1 ± 0.66	31.3 ± 0.22

Values are mean ± SD of three independent replicates. GAE: gallic acid equivalent. Total carbohydrates (TC) = 100 − (moisture + protein + fat + ash).

3.2. Individual Phenolic Compounds of Seed Flours

Phenolic compounds of seed flours obtained from fenugreek and *Moringa oleifera* were quantified by HPLC and are shown in Table 2. It was observed that different plants have various phenolic compounds and these compounds are associated with their antioxidant activity. The current results revealed that gallic acid, catechin, and protocatechuic acid are the dominant compounds detected in both seed flour. In moringa seed flour, gallic acid showed the highest concentration (17.34 mg/100 g) followed by epicatechin, caffeic acid, p-coumaric acid, catechin, and protocatechuic acid. However, gentisic acid, vanillic acid, syringic acid and chlorogenic were not detected. On the other hand, in fenugreek seed flour, vanillic acid showed the highest concentration (57.33 mg/100 g) followed by gentisic acid, protocatechuic acid, gallic acid, chlorogenic, catechin and syringic acid. However, caffeic acid, p-coumaric acid and epicatechin were not detected.

Table 2. Individual phenolic compounds (mg/100 g) of fenugreek and *Moringa oleifera* seed flours.

Compound	Fenugreek Seed	Moringa Seed Flour
Gallic acid	1.83 ± 0.02	17.34 ± 0.17
Catechin	0.54 ± 0.01	0.343 ± 0.001
Gentisic acid	36.3 ± 0.36	BDL [1]
Protocatechuic acid	4.32 ± 0.04	0.29 ± 0.001
Vanillic acid	57.33 ± 0.57	BDL [1]
Caffeic acid	BDL [1]	3.21 ± 0.03
Syringic acid	0.41 ± 0.001	BDL [1]
Chlorogenic	0.63 ± 0.01	BDL [1]
p-Coumaric acid	BDL [1]	0.45 ± 0.001
Epicatechin	BDL [1]	7.74 ± 0.08

[1] BDL: below the detection limit.

3.3. Changes in Yoghurt's pH and Titratable Acidity

As shown in Table 3, the data revealed that the values of acidity gradually increased significantly while the values of pH gradually decreased significantly as cold storage progressed. These changes were considerable at seven days of cold storage and correlated with the progress in cold storage time as well as the added amount of fenugreek and moringa seed flours. Changes in acidity and pH values are well-known to be associated with the growth of yoghurt starter culture and other lactic acid bacteria and their ability to break down carbohydrate substances and organic acids formation [41]. Ref. [42] indicated that yoghurt culture is active even at low temperatures and can ferment lactose into lactic acid, resulting in pH reduction and acidity formation. Furthermore, the findings revealed that the incorporation of fenugreek seed flour resulted in higher pH values compared to plain yoghurt and moringa-containing yoghurt over 14 days during cold storage. This could be linked to the enhanced buffering capacity that occurred by the high protein content of fenugreek seed flour [43]. The changes reported for acidity and pH are generally associated with lactic acid bacteria contained in fermented dairy products. It is therefore of great importance to retain the viability of lactic acid bacteria and to keep their viable numbers at a higher rate to produce their health-promoting activities that reflect on consumer health.

Table 3. pH, titratable acidity (TA%) and syneresis (%) of yoghurt fortified with fenugreek and *Moringa oleifera* seed flours during cold storage.

Treatment	pH			Titratable Acidity (TA, %)			Syneresis (%)		
	d 1	d 7	d 14	d 1	d 7	d 14	d 1	d 7	d 14
C	4.59 ± 0.1 [d]	4.40 ± 0.14 [g]	4.10 ± 0.13 [m]	0.86 ± 0.05 [h]	0.95 ± 0.06 [f]	1.06 ± 0.07 [c]	9.82 ± 0.62 [f]	11.30 ± 0.71 [b]	12.39 ± 0.78 [a]
F1	4.64 ± 0.1 [c]	4.42 ± 0.1 [f]	4.13 ± 0.13 [l]	0.82 ± 0.05 [jk]	0.84 ± 0.05 [i]	0.85 ± 0.0 [j]	8.74 ± 0.5 [j]	9.60 ± 0.60 [g]	10.85 ± 0.68 [b]
F2	4.68 ± 0.15 [b]	4.44 ± 0.14 [f]	4.16 ± 0.14 [k]	0.80 ± 0.05 [k]	0.82 ± 0.05 [jk]	0.83 ± 0.05 [ij]	8.35 ± 0.52 [k]	8.64 ± 0.54 [j]	9.99 ± 0.63 [e]
M1	4.58 ± 0.15 [d]	4.33 ± 0.14 [h]	4.04 ± 0.13 [n]	0.92 ± 0.06 [g]	1.01 ± 0.06 [d]	1.09 ± 0.07 [b]	9.32 ± 0.58 [h]	9.89 ± 0.62 [ef]	11.33 ± 0.71 [b]
M2	4.74 ± 0.15 [a]	4.52 ± 0.15 [e]	4.25 ± 0.14 [i]	0.98 ± 0.06 [e]	1.06 ± 0.07 [c]	1.11 ± 0.07 [a]	9.03 ± 0.57 [i]	9.51 ± 0.60 [g]	10.66 ± 0.67 [d]

Values are means ± SD of three independent replicates. Means with different superscripts are significantly different ($p < 0.05$). C: plain yogurt, F1 and F2: yoghurt fortified with 0.1 and 0.2% fenugreek seed flour. M1 and M2: yoghurt fortified with 0.1 and 0.2% *Moringa oleifera* seed flour.

3.4. Yoghurt Syneresis

Syneresis is the main yoghurt problem that happens due to the weak protein network that contributes to a reduction in whey protein connection intensity, hence its separation from the body of yoghurt [44]. The data that existed in Table 3 indicated that whey separation of the tested samples increased significantly with the progress of cold storage time. The separation of whey reduced by adding seed flours. The data indicated that incorporation of fenugreek seed flour led to lower syneresis than that obtained by incorporation of moringa seed flour. This may be due in part to the soluble fiber content of fenugreek seed flour that provides the properties of texture and thickening. The addition of xanthan gum [45] or quince seed mucilage [46] has reported a similar reduction in yoghurt whey separation. Ref. [47] showed that low-solid yoghurt tends to be more synerestic than high-solid yoghurt. In this sense, yoghurt samples containing fenugreek or moringa seed flours had higher total solids that could bind the released whey and thus inhibit whey drainage [48].

3.5. TPC and AOA of Yoghurt

Because milk and fermented dairy products are not known to contain polyphenols, it is useful to enrich them by adding food of a plant origin. Thus, the nutritional and functional values of the resulting products will be improved. Incorporation of fenugreek and moringa seed flours increased both TPC and AOA as shown in Figure 1. Moringa seed flour-fortified yoghurt showed the highest TPC and AOA followed by fenugreek-fortified yoghurt as compared to plain yoghurt. The more seed flours incorporated, the higher TPC and AOA values obtained. The current findings showed that the values of TPC and AOA decreased with the progress of cold storage. Ref. [49] reported high radical scavenging activity of fenugreek seed flour. Ref. [50] found that *Moringa oleifera* is a great source of multiple bioactive compounds including polyphenolic antioxidant compounds. The current findings are consistent with those reported by [51], which showed that the addition of *Moringa oleifera* extract increased TPC in fortified yoghurt as compared to plain yoghurt and thus reflected as higher AOA. Interestingly, the interactions between LAB and phenolic compounds represent another factor in the formation and degradation of phenolic compounds. In this context, [52] reported that the fermentation of milk by yoghurt starter culture has produced interesting findings regarding the formation and degradation of phenolic compounds. In fact, moringa leaves are known to be rich in bioactive components including quercetin and kaempferol, among others, which display strong antioxidant properties [53]. The current data support previous findings demonstrating that food of plant origin such as fenugreek and *Moringa oleifera* seeds are rich sources of polyphenols compounds associated with its strong antioxidant activity and thus are suitable for producing bioactive yoghurt products.

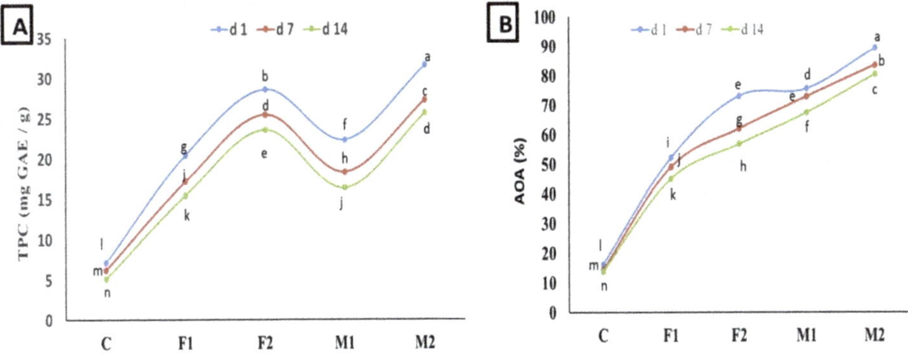

Figure 1. (**A**)Total phenolic content (TPC, mg GAE/g) and (**B**) antioxidant activity (AOA %) of yoghurt fortified with fenugreek and moringa seed flours during cold storage. Values are mean ± SD of three independent replicates. Lines chart with different letters are significantly different ($p < 0.05$). F1 and F2: yoghurt fortified with 0.1 and 0.2% fenugreek seed flour. M1 and M2: yoghurt fortified with 0.1 and 0.2% *Moringa oleifera* seed flour.

3.6. Mineral Contents of Yoghurt

The results presented in Table 4 displayed that the incorporation of fenugreek and moringa seeds flour led to increasing mineral (Ca, P, K, Mg, Zn, and Fe) content of the various products as compared to plain yoghurt. Mineral content increased with increasing the added amount of seed flours. Incorporation of 0.1% and 0.2% of moringa seed flour (M1 and M2) led to a higher content of Ca, P, K and Fe and lower content of Mg and Zn as compared to F1 and F2, respectively. This trend was observed at 1, 7 and 14 days of cold storage. Generally, mineral content increased with the progress of cold storage time. Generally, it was observed that seeds of *Moringa oleifera* and Fenugreek [54] have elevated levels of dietary minerals. Recently, it was observed that yoghurt mousses fortified with chia seeds contain higher amounts of minerals as compared with the control yoghurt mousse [55]. The supplementation of yoghurt and dairy products with the seeds flour of fenugreek and *Moringa oleifera* increased its mineral content. However, no information exists in the literature concerning the mineral content resulted from yoghurt supplementation with food of plant origin including *Moringa oleifera* or fenugreek. As far as we know, the current research is the first study which used fenugreek seed flour in formulating a novel functional yoghurt. Because of the above-mentioned, seed flours represent a rich source of several nutrients and bioactive components and can be easily incorporated into fermented dairy products in order to improve its nutritional and functional values.

Table 4. Minerals content (mg\kg) of yoghurt fortified with fenugreek and moringa seed flours during cold storage.

Sample	Storage Time (Days)	Minerals (mg/kg)					
		Ca	P	K	Mg	Zn	Fe
C	d 1	1426.66 ± 14.26 [j]	1250.00 ± 12.50 [j]	578.45 ± 5.78 [j]	140.93 ± 1.4 [j]	0.40 ± 0.004 [g]	4.90 ± 0.049 [h]
	d 14	1680.66 ± 16.80 [h]	1310.00 ± 13.10 [g]	647.46 ± 6.47 [i]	187.56 ± 1.87 [d]	0.55 ± 0.006 [d]	5.40 ± 0.054 [g]
M1	d 1	1760.00 ± 17.60 [f]	1303.00 ± 13.03 [h]	701.08 ± 7.01 [g]	142.34 ± 1.42 [i]	0.40 ± 0.004 [g]	5.94 ± 0.059 [e]
	d 14	1880.33 ± 18.80 [c]	1375.72 ± 13.75 [d]	784.72 ± 7.84 [c]	189.44 ± 1.89 [d]	0.56 ± 0.006 c	6.54 ± 0.065 [c]
M2	d 1	1840.00 ± 18.40 [d]	1333.30 ± 13.33 [e]	771.19 ± 7.71 [e]	146.57 ± 1.46 [g]	0.42 ± 0.004 [e]	6.53 ± 0.065 [c]
	d 14	2046.66 ± 20.46 [a]	1413.29 ± 14.13 [a]	863.19 ± 8.63 [a]	195.06 ± 1.95 [b]	0.57 ± 0.006 [b]	7.20 ± 0.072 [a]
F1	d 1	1640.00 ± 16.40 [i]	1300.00 ± 13.0 [i]	694.14 ± 6.94 [h]	143.75 ± 1.44 [h]	0.41 ± 0.004 [f]	5.88 ± 0.059 [f]
	d 14	1768.66 ± 17.68 [e]	1372.00 ± 13.70 [c]	776.95 ± 7.76 [d]	191.31 ± 1.91 [c]	0.56 ± 0.006 [c]	6.48 ± 0.065 [d]
F2	d 1	1753.33 ± 17.53 [g]	1330.00 ± 13.30 [f]	763.55 ± 7.63 [f]	147.98 ± 1.48 [f]	0.42 ± 0.004 [e]	6.47 ± 0.065 [d]
	d14	1933.33 ± 19.33 [b]	1409.20 ± 14.09 [b]	854.65 ± 8.54 [b]	196.938 ± 1.97 [a]	0.58 ± 0.006 [a]	7.13 ± 0.071 [b]

Values are means ± SD of three independent replicates. Means with different superscripts are significantly different ($p < 0.05$). C: plain yogurt, F1 and F2: yoghurt fortified with 0.1 and 0.2% fenugreek seed flour. M1 and M2: yoghurt fortified with 0.1 and 0.2% *Moringa oleifera* seed flour.

3.7. Viability of Yoghurt Culture

The growth and viability of probiotics and yoghurt starter culture are affected by the addition of certain compounds and this correlation has been discovered to be both species-and strain-specific. However, there has been insufficient investigation into the impact of added commercial and natural preparations on the survival and viability of bacteria [56]. In this regard, the viable counts of *S. thermophilus* and *L. bulgaricus* were determined to identify the effect of fenugreek and moringa seed flours incorporation on the viable counts of yoghurt culture. As can be seen in Figure 2, significant differences were noted in the viability of yoghurt culture in the various yoghurt samples. Unexpectedly, the obtained data displayed that the addition of moringa seed flour significantly increased the viable counts of *S. thermophilus* and *L. bulgaricus*. Moreover, fenugreek seed flour addition significantly increased the viable counts of these bacteria than control and yoghurt fortified with moringa (Figure 2). *S. thermophilus* and *L. bulgaricus* displayed higher viability at 7 days then its viability declined at final storage time (14 days) keeping higher viable counts (ranged between 5.8 to 13.3 log cfu/g for *S. thermophilus* and 4.2 to 7.9 log cfu/g for *L. bulgaricus*). Generally, [57] reported that the viable counts of *Streptococcus* in yoghurt was significantly greater than that of *Lactobacillus*. A decrease in the viable number of yoghurt bacteria during cold storage was reported by [58]. The nutrients existing in the food are among the factors that influence the viability of lactic acid bacteria. So, it is expected that using plant derivatives in yoghurt formulation led to increasing the viability of *Streptococcus* and *Lactobacillus*. This increased viability can be attributed to the polyphenols and fiber that food of a plant origin contains, among other reasons. In this sense, Fenugreek seed is well-known for its elevated fiber, gum content and other phytochemicals [6] and thus has the opportunity to improve viable counts of these bacteria as reported in the current research. This is why dietary fiber provided additional sources of carbohydrates and acts as fermentable substrate leading to lactic acid bacteria growth. Similar results previously reported by [59] demonstrated this effect of dietary fiber on the viability of lactic acid bacteria in yoghurt incorporated with pineapple dietary fiber.

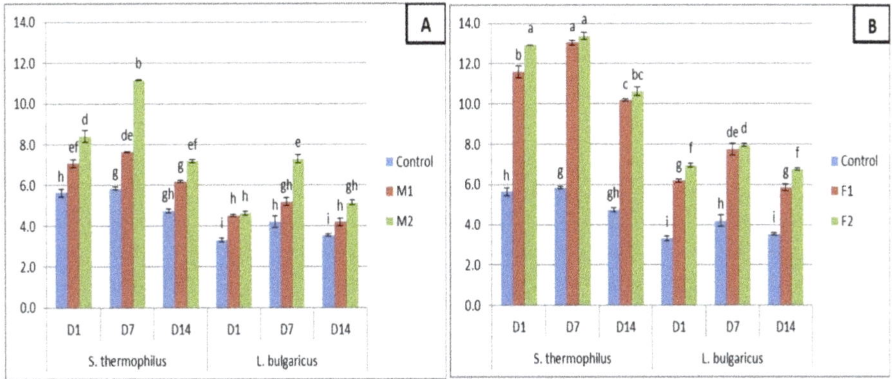

Figure 2. Viable counts (log CFU/g) of *S. thermophiles* and *L. delbrueckii subsp. bulgaricus* yogurt fortified with *Moringa oleifera* seeds flour (**A**) or Fenugreek seed flour (**B**). Values are mean ± SD of three independent replicates. Chart bars with different letters are significantly different ($p < 0.05$). F1 and F2: yoghurt fortified with 0.1 and 0.2% fenugreek seed flour. M1 and M2: yoghurt fortified with 0.1 and 0.2% *Moringa oleifera* seed flour.

Regarding the concern related with the possibility of survivability of some bacterial spores after heat treatment of yoghurt milk, it was demonstrated that the more intense heat treatment applied in yoghurt production (90–95 °C for 5 min) led to killing most vegetative microorganisms [60] while the bacterial spores survive. However, many different typologies of dairy products such as yoghurt, cheese, pasteurized milk are not suitable for the germination of the inoculated spores. This effect can be attributed to the low pH (<5) and the presence of natural microflora [61].

3.8. Antibacterial Activity

The results of the antibacterial activity were measured by the agar well diffusion method, and the results are expressed as an inhibition zone diameter (mm) (Table 5). Yoghurt fortified with moringa and fenugreek seed flour exhibited higher antibacterial activity compared to control yoghurt. Moreover, yoghurt fortified with moringa (M2) showed significantly higher antibacterial activity against all studied pathogenic microorganisms and the zones of inhibition were (12.65, 13.14, 17.23, and 14.49 mm) for *E. coli*, *S. aureus*, *L. monocytogenes* and *Salmonella* spp., respectively, compared to the yoghurt fortified with fenugreek seed flour and control yoghurt. These results are in close agreement with other findings obtained by [62]. Ref. [63] characterized a coagulant protein that showed both flocculating and antimicrobial effects of ~99% reduction of the bacterial population. Ref. [64] also identified a peptide derived from *Moringa oleifera* and has antibacterial activity against specific human pathogens.

Table 5. Antibacterial activity of yoghurts fortified with fenugreek and moringa seed flours during cold storage.

Treatment	Inhibition Zone Diameter (mm)			
	E. coli	*S. aureus*	*L. monocytogenes*	*S. typhimurium*
C	4.12 ± 0.34 [e]	5.21 ± 0.54 [e]	8.55 ± 0.53 [de]	8.39 ± 0.62 [de]
F1	5.21 ± 0.43 [e]	6.26 ± 0.42 [e]	13.45 ± 0.52 [c]	10.15 ± 0.63 [d]
F2	6.24 ± 0.46 [e]	7.77 ± 0.65 [e]	14.22 ± 0.93 [c]	12.23 ± 0.63 [cd]
M1	9.33 ± 0.52 [d]	10.33 ± 0.35 [d]	21.34 ± 0.54 [a]	18.45 ± 0.52 [ab]
M2	12.65 ± 0.54 [cd]	13.14 ± 0.54 [c]	17.23 ± 0.63 [b]	14.49 ± 0.33 [c]

Values are means ± SD of three independent replicates. Means with different superscripts are significantly different ($p < 0.05$). C: plain yogurt, F1 and F2: yoghurt fortified with 0.1 and 0.2% fenugreek seed flour. M1 and M2: yoghurt fortified with 0.1 and 0.2% *Moringa oleifera* seed flour.

3.9. Sensory Evaluation

The sensory assessment of yoghurt and other dairy products is the cornerstone of consumer acceptance. The sensory acceptance of yoghurt fortified with fenugreek and *Moringa oleifera* seeds flours was shown in Table 6. In general, the plain yoghurt sample (C) had the highest degree for the descriptors tested (color, mouthfeel, body and texture, taste and flavor, and overall acceptability) at 1 and 14 days of cold storage compared to the yoghurt samples fortified with fenugreek (F1 and F2) and *Moringa oleifera* (M1 and M2) seed flours. The various treatments gained lower degrees of sensory evaluation with the progress of cold storage time. Regarding the overall acceptability, the highest value was obtained for plain yoghurt samples and followed by M1 and F1. Despite the decline in all tested descriptors at the final stage of cold storage, the panelists retain high acceptability for the various yoghurt samples. This observed decline in the tested descriptors could be due to the increased acidity at the final storage time that prevents the formation of aromatic components [65]. It could also be clarified in part by the decline in yoghurt acetaldehyde concentration as cold storage progresses [66].

Table 6. Sensory acceptability of yoghurt fortified with fenugreek and *Moringa oleifera* seed flours during cold storage.

Treatment	Storage Time (days)	Color	Mouthfeel	Body and Texture	Taste and Flavor	Overall Acceptability
C	d1	4.8 ± 0.10 [a]	4.62 ± 0.10 [a]	4.72 ± 0.10 [a]	4.72 ± 0.10 [a]	4.7 ± 0.1 [a]
	d14	4.7 ± 0.10 [b]	4.62 ± 0.10 [a]	4.62 ± 0.10 [b]	4.52 ± 0.09 [a]	4.6 ± 0.1 [b]
F1	d1	4.3 ± 0.09 [c]	4.43 ± 0.09 [b]	4.52 ± 0.09 [b]	4.23 ± 0.09 [c]	4.4 ± 0.1 [c]
	d14	4.0 ± 0.08 [d]	4.23 ± 0.09 [c]	4.43 ± 0.09 [b]	4.13 ± 0.09 [c]	4.2 ± 0.09 [d]
F2	d1	4.1 ± 0.09 [d]	4.13 ± 0.09 [d]	4.62 ± 0.10 [b]	4.13 ± 0.09 [d]	4.3 ± 0.09 [d]
	d14	3.9±0.08 [d]	4.03±0.08 [e]	4.52±0.09 [c]	4.33±0.09 [d]	4.2±0.09 [e]
M1	d1	4.5 ± 0.09 [f]	4.33 ± 0.09 [e]	4.33 ± 0.09 [c]	4.23 ± 0.09 [e]	4.4 ± 0.1 [f]
	d14	4.1 ± 0.09 [f]	4.03 ± 0.08 [f]	4.62 ± 0.10 [d]	4.03 ± 0.08 [e]	4.2 ± 0.1 [f]
M2	d1	4.2 ± 0.09 [f]	4.13 ± 0.09 [f]	4.62 ± 0.10 [d]	4.33 ± 0.09 [e]	4.3 ± 0.1 [f]
	d14	4.1 ± 0.09 [f]	4.03 ± 0.08 [f]	4.43 ± 0.09 [e]	4.13 ± 0.09 [f]	4.2 ± 0.09 [f]

Values are means ± SD of three independent replicates. Means with different superscripts are significantly different ($p < 0.05$). C: plain yoghurt, F1 and F2: yoghurt fortified with 0.1 and 0.2% fenugreek seed flour, M1 and M2: yoghurt fortified with 0.1 and 0.2% *Moringa oleifera* seed flour.

In the light of the accumulated information, fenugreek and *Moringa oleifera* seed flours are generally recognized as safe (GRAS) by the FDA. However, it should be taken into account that fenugreek parts are not suitable for medicinal use due to the scarcity of clinical studies. Fenugreek reacts with a wide range of drugs, thus requiring medical advice regarding consumption on an individual basis. Even though no fenugreek adverse effects on human has been reported to date, testing of fenugreek toxicity effect on liver histology in animal models is the first step to open the window for future clinical trials to investigate the safety of fenugreek for applied medical uses.

4. Conclusions

In view of the findings obtained by the current research, it could be concluded that seeds flour of fenugreek and *Moringa oleifera* can be incorporated for formulating novel functional yoghurt. The fiber and phytochemical nutrients present in these seed flours can improve the viability of yoghurt culture since it acts as fermentable substrates for LAB. Its incorporation led to increased TPC and AOA without decreasing its sensory acceptability. In addition, yoghurt supplemented with moringa and fenugreek showed significantly higher antibacterial activity against all studied pathogenic microorganisms compared with control yoghurt. Moreover, mineral content increased through the incorporation of these seed flours. Finally, novel fermented milk products with high nutritional and functional values can be obtained through using seed flour of fenugreek and *Moringa oleifera*.

Author Contributions: Conceptualization, A.M.H. and E.A.; methodology, A.M.H.; formal analysis, H.S.E.-B., A.M.H. and E.A.; investigation, A.M.H. and E.A.; resources, H.S.E.-B., A.M.H. and E.A.; data curation, H.S.E.-B., A.M.H. and E.A.; writing—original draft preparation, A.M.H. and E.A.; writing—review and editing, F.D., H.S.E.-B., A.M.H. and E.A.; project administration, F.D., H.S.E.-B., A.M.H. and E.A.; funding acquisition, F.D., H.S.E.-B. All authors have read and agreed to the published version of the manuscript.

Funding: This research was supported by the Deanship of Scientific Research, King Faisal University under Nasher track (Grant No 186308).

Acknowledgments: The authors are thankful to the Chemistry Department of Cairo University Research Park (CURP) and Elham Mustafa and Eng. Salma Salman for their technical assistance in HPLC analysis at Food Safety and Quality Control Laboratory (FSQC Lab.), Faculty of Agriculture, Cairo University, and would like to thank and appreciate the Deanship of Scientific Research at King Faisal University, Saudi Arabia, for the financial support under Nasher track (Grant No 186308).

Conflicts of Interest: The authors declare no conflict of interest.

References

1. Weerathilake, W.A.; Rasika, D.M.; Ruwanmali, J.K.; Munasinghe, M.A. The evolution, processing, varieties and health benefits of yoghurt. *Int. J. Sci. Res. Pub.* **2014**, *4*, 1–10.
2. Achi, O.K.; Asamudo, N.U. Cereal-based fermented foods of Africa as functional foods. In *Bioactive Molecules in Food*; Mérillon, J.M., Ramawat, K., Eds.; Springer: Cham, Switzerland, 2019; pp. 1527–1558.
3. Jamshidi-Kia, F.; Lorigooini, Z.; Amini-Khoei, H. Medicinal plants: Past history and future perspective. *J. Herb. Pharmacol.* **2018**, *7*, 1–7. [CrossRef]
4. Iriondo-DeHond, M.; Miguel, E.; del Castillo, M. Food byproducts as sustainable ingredients for innovative and healthy dairy foods. *Nutrients* **2018**, *10*, 1358. [CrossRef]
5. Wani, S.A.; Kumar, P. Fenugreek: A review on its nutraceutical properties and utilization in various food products. *J. Saudi. Soc. Agri. Sci.* **2018**, *17*, 97–106. [CrossRef]
6. He, Y.; Ding, C.; Wang, X.; Wang, H.; Suo, Y. Using response surface methodology to optimize countercurrent chromatographic separation of polyphenol compounds from fenugreek (*Trigonella foenum-graecum* L.) seeds. *J. Liq. Chromatogr. Relat. Technol.* **2015**, *38*, 29–35. [CrossRef]
7. Ouzir, M.; El Bairi, K.; Amzazi, S. Toxicological properties of fenugreek (*Trigonella foenum graecum*). *Food Chem. Toxicol.* **2016**, *96*, 145–154. [CrossRef]
8. Abdel-Barry, J.A.; Al-Hakiem, M.H. Acute intraperitoneal and oral toxicity of the leaf glycosidic extract of *Trigonella foenum-graecum* in mice. *J. Ethnopharmacol.* **2000**, *70*, 65–68. [CrossRef]
9. Dhull, S.B.; Punia, S.; Sandhu, K.S.; Chawla, P.; Kaur, R.; Singh, A. Effect of debittered fenugreek (*Trigonella foenum-graecum* L.) flour addition on physical, nutritional, antioxidant, and sensory properties of wheat flour rusk. *Legume Sci.* **2020**, *2*, e21. [CrossRef]
10. Khorshidian, N.; Yousefi Asli, M.; Arab, M.; Adeli Mirzaie, A.; Mortazavian, A.M. Fenugreek: Potential applications as a functional food and nutraceutical. *Nutr. Food Sci. Res.* **2016**, *3*, 5–16. [CrossRef]
11. Dawit, S.; Regassa, T.; Mezgebu, S.; Mekonnen, D. Evaluation of two Moringa species for adaptability and growth performance under Bako conditions. *J. Nat. Sci. Res.* **2016**, *6*, 76–82.
12. Fejér, J.; Kron, I.; Pellizzeri, V.; Pľuchtová, M.; Eliašová, A.; Campone, L.; Konečná, M. First report on evaluation of basic nutritional and antioxidant properties of Moringa oleifera Lam. from Caribbean Island of Saint Lucia. *Plants* **2019**, *8*, 537. [CrossRef] [PubMed]
13. Anhwange, B.A.; Ajibola, V.O.; Oniye, S.J. Chemical studies of the seeds of Moringa oleifera (Lam.) and Detarium microcarpum (Guill and Sperr). *J. Biol. Sci.* **2004**, *4*, 711–715.
14. Charoensin, S. Antioxidant and anticancer activities of Moringa oleifera leaves. *J. Med. Plants Res.* **2014**, *8*, 318–325.
15. Van Tienen, A.; Hullegie, Y.; Hummelen, R.; Hemsworth, J.; Changalucha, J.; Reid, G. Development of a locally sustainable functional food for people living with HIV in Sub-Saharan Africa: Laboratory testing and sensory evaluation. *Benef. Microbes* **2011**, *2*, 193–198. [CrossRef]
16. El-Ziney, M.G.; Shokery, E.S.; Youssef, A.H.; Mashaly, R.E. Protective effects of green tea and moringa leave extracts and their bio-yogurts against oxidative effects of lead acetate in albino rats. *J. Nutr. Health Food Sci.* **2017**, *5*, 1–11. [CrossRef]

17. Tamime, A.Y.; Robinson, R.K. *Yoghurt: Science and Technology*; Woodhead Publishing: Cambridge, UK, 1999.
18. Abdel-Hamid, M.; Romeih, E.; Huang, Z.; Enomoto, T.; Huang, L.; Li, L. Bioactive properties of probiotic set-yogurt supplemented with Siraitia grosvenorii fruit extract. *Food Chem.* **2020**, *303*, 125400. [CrossRef] [PubMed]
19. AOAC. *Official Methods of Analysis of AOAC International*, 17th ed.; Association of Analytical Communities: Gaithersburg, MD, USA, 2000.
20. Proestos, C.; Chorianopoulos, N.; Nychas, G.-J.E.; Komaitis, M. RP-HPLC Analysis of the Phenolic Compounds of Plant Extracts. Investigation of Their Antioxidant Capacity and Antimicrobial Activity. *J. Agric. Food Chem.* **2005**, *53*, 1190–1195. [CrossRef] [PubMed]
21. Singleton, V.L.; Rossi, J.A. Colorimetry of total phenolics with phosphomolybdic-phosphotungstic acid reagents. *Am. J. Enol. Vitic.* **1965**, *16*, 144–158.
22. Adesegun, S.A.; Elechi, N.A.; Coker, H.A.B. Antioxidant activities of methanolic extract of Sapium elliticum. *Pak. J. Biol. Sci.* **2008**, *11*, 453–457. [CrossRef] [PubMed]
23. Tomovska, J.; Gjorgievski, N.; Makarijoski, B. Examination of pH, titratable acidity and antioxidant activity in fermented milk. *J. Mater. Sci. Eng. A* **2016**, *6*, 326–333.
24. Isanga, J.; Zhang, G. Screening of stabilizers for peanut milk based set yoghurt by assessment of whey separation, gel firmness and sensory quality of the yoghurt. *Am. J. Food Technol.* **2008**, *3*, 127–133. [CrossRef]
25. Öztürk, H.I.; Aydına, S.; Sözerib, D.; Demircia, T.; Sertc, D.; Akın, N. Fortification of set-type yoghurts with *Elaeagnus angustifolia* L. flours: Effects on physicochemical, textural, and microstructural characteristics. *LWT-Food Sci. Technol.* **2018**, *90*, 620–626. [CrossRef]
26. Waterhouse, A.L. Determination of total phenolics. *Curr. Protoc. Food Anal. Chem.* **2002**, *6*, I1.1.1–I1.1.8.
27. Aloğlu, H.Ş.; Öner, Z. Determination of antioxidant activity of bioactive peptide fractions obtained from yoghurt. *J. Dairy Sci.* **2011**, *94*, 5305–5314. [CrossRef] [PubMed]
28. Elsayed, E.M.; Hamed, A.M.; Badran, S.M.; Mostafa, A.A. A survey of selected essential and toxic metals in milk in different regions of Egypt using ICP-AES. *Int. J. Dairy Sci.* **2011**, *6*, 158–164. [CrossRef]
29. Birghila, S.; Dobrinas, S.; Stanciu, G.; Soceanu, A. Determination of major and minor elements in milk through ICP-AES. *Environ. Eng. Manag. J.* **2008**, *7*, 805–808. [CrossRef]
30. Ogunsina, B.; Radha, C.; Indrani, D. Quality characteristics of bread and cookies enriched with debittered Moringa oleifera seed flour. *Int. J. Food Sci. Nutr.* **2010**, *62*, 185–194. [CrossRef]
31. Ilyas, M.; Arshad, M.U.; Saeed, F.; Iqbal, M. Antioxidant potential and nutritional comparison of Moringa leaf and seed powders and their tea infusions. *J. Anim. Plant Sci.* **2015**, *25*, 226–233.
32. Farzamirad, V.; Aluko, R.E. Angiotensin-converting enzyme inhibition and free-radical scavenging properties of cationic peptides derived from soybean protein hydrolysates. *Int. J. Food Sci. Nutr.* **2008**, *59*, 428–437. [CrossRef]
33. Ertem, H.; Akmakçı, S. Shelf life and quality of probiotic yoghurt produced with *Lactobacillus acidophilus* and Gobdin. *Int. J. Food Sci. Technol.* **2018**, *53*, 776–783. [CrossRef]
34. Sah, B.N.P.; Vasiljevic, T.; McKechnie, S.; Donkor, O.N. Antibacterial and antiproliferative peptides in synbiotic yogurt-Release and stability during refrigerated storage. *J. Dairy Sci.* **2016**, *99*, 4233–4242. [CrossRef]
35. Mbah, B.; Eme, P.; Ogbusu, O. Effect of cooking methods (boiling and roasting) on nutrients and anti-nutrients content of Moringa oleifera seeds. *Pak. J. Nutr.* **2012**, *11*, 211–215.
36. Hooda, S.; Jood, S. Effect of soaking and germination on nutrient and antinutrient contents of fenugreek (Trigonella foenum graecum L.). *J. Food Biochem.* **2003**, *27*, 165–176. [CrossRef]
37. Abdulkarim, S.M.; Long, K.; Lai, O.M.; Muhammad, S.K.S.; Ghazali, H.M. Some physico-chemical properties of Moringa oleifera seed oil extracted using solvent and aqueous enzymatic methods. *Food Chem* **2005**, *93*, 253–263. [CrossRef]
38. Hemavathy, J.; Prabhakar, J.V. Liquid composition of Fenugreek (*Trigonella foenum-graecum L.*) seeds. *Food Chem.* **1989**, *31*, 1–7. [CrossRef]
39. Naidu, M.M.; Shyamala, B.N.; Naik, J.P.; Sulochanamma, G.; Srinivas, P. Chemical composition and antioxidant activity of the husk and endosperm of fenugreek seeds. *LWT-Food Sci. Technol.* **2011**, *44*, 451–456. [CrossRef]
40. Tung, Y.T.; Wu, J.H.; Kuo, Y.H.; Chang, S.T. Antioxidant activities of natural phenolic compounds from Acacia confusa bark. *Bioresour. Technol.* **2007**, *98*, 1120–1123. [CrossRef]

41. Yerlikaya, O. Starter cultures used in probiotic dairy product preparation and popular probiotic dairy drinks. *Food Sci. Technol.* **2014**, *34*, 221–229. [CrossRef]
42. Shah, N.P.; Lankaputhra, W.E.; Britz, M.L.; Kyle, W.S. Survival of *Lactobacillus acidophilus* and *Bifidobacterium bifidum* in commercial yoghurt during refrigerated storage. *Int. Dairy J.* **1995**, *5*, 515–521. [CrossRef]
43. Lee, W.J.; Lucey, J.A. Formation and physical properties of yoghurt. Asian-Aust. *J. Anim. Sci.* **2010**, *23*, 1127–1136.
44. Lucey, J.A. Formation and physical properties of milk protein gels. *J. Dairy Sci.* **2002**, *85*, 281–294. [CrossRef]
45. El-Sayed, E.; Abd El-Gawad, I.A.; Murad, H.; Salah, S. Utilization of laboratory-produced xanthan gum in the manufacture of yoghurt and soy yoghurt. *Eur. Food Res. Technol.* **2002**, *215*, 298–304.
46. Nikoofar, E.; Hojjatoleslami, M.; Shariaty, M.A. Surveying the effect of quince seed mucilage as a fat replacer on texture and physicochemical properties of semi-fat set yoghurt. *Int. J. Farm. Alli. Sci.* **2013**, *2*, 861–865.
47. Rani, R.; Dharaiya, C.N.; Unnikrishnan, V.; Singh, B. Factors affecting syneresis from yoghurt for preparation of chakka. *Ind. J. Dairy Sci.* **2012**, *65*, 135–140.
48. Pandey, H.; Awasthi, P. Effect of processing techniques on nutritional composition and antioxidant activity of fenugreek (*Trigonella foenum-graecum*) seed flour. *J. Food Sci. Technol.* **2015**, *52*, 1054–1060. [CrossRef]
49. Bhanger, M.I.; Bukhari, S.B.; Memon, S. Antioxidative activity of extracts from a Fenugreek seeds (*Trigonella foenum-graecum*). *Pak. J. Anal. Environ. Chem.* **2008**, *9*, 6.
50. Vergara-Jimenez, M.; Almatrafi, M.; Fernandez, M. Bioactive components in Moringa Oleifera leaves protect against chronic disease. *Antioxidants* **2017**, *6*, 91. [CrossRef]
51. Shokery, E.S.; El-Ziney, M.G.; Yossef, A.H.; Mashaly, R.I. Effect of green tea and Moringa leave extracts fortification on the physicochemical, rheological, sensory and antioxidant properties of set-type yoghurt. *J. Adv. Dairy Res.* **2017**, *5*, 2. [CrossRef]
52. Jaster, H.; Arend, G.D.; Rezzadori, K.; Chaves, V.C.; Reginatto, F.H.; Petrus, J.C.C. Enhancement of antioxidant activity and physicochemical properties of yoghurt enriched with concentrated strawberry pulp obtained by block freeze concentration. *Food Res. Int.* **2018**, *104*, 119–125. [CrossRef]
53. Wang, Y.; Gao, Y.; Ding, H.; Liu, S.; Han, X.; Gui, J.; Liu, D. Subcritical ethanol extraction of flavonoids from Moringa oleifera leaf and evaluation of antioxidant activity. *Food Chem.* **2017**, *218*, 152–158. [CrossRef]
54. El Nasri, N.A.; El Tinay, A. Functional properties of fenugreek (*Trigonella foenum-graecum*) protein concentrate. *Food Chem.* **2007**, *103*, 582–589. [CrossRef]
55. Attalla, N.R.; El-Hussieny, E.A. Characteristics of nutraceutical yoghurt mousse fortified with chia seeds. Inter. *Int. J. Environ. Agric. Biotechnol.* **2017**, *2*, 2033–2046. [CrossRef]
56. Fernandez, M.A.; Marette, A. Potential health benefits of combining yoghurt and fruits based on their probiotic and prebiotic properties. *Adv. Nutr.* **2017**, *8*, 155S–164S. [CrossRef] [PubMed]
57. Zacarchenco, P.B.; Massaguer-Roig, S. Properties of *Streptococcus thermophilus* fermented milk containing variable concentrations of *Bifidobacterium longum* and *Lactobacillus acidophilus*. *Braz. J. Microbiol.* **2006**, *37*, 338–344. [CrossRef]
58. Cais-Sokolińska, T.A.; Pikul, J. Proportion of the microflora of *Lactobacillus* and *Streptococcus* genera in yoghurts of different degrees of condensation. *Bull. Vet. Inst. Pulawy* **2004**, *48*, 443–447.
59. Sah, B.P.; Vasiljevic, T.; McKechnie, S.; Donkor, O.N. Effect of pineapple waste powder on probiotic growth, antioxidant and antimutagenic activities of yoghurt. *J. Food Sci. Technol.* **2016**, *53*, 1698–1708. [CrossRef]
60. Sfakianakis, P.; Tzia, C. Conventional and innovative processing of milk for yogurt manufacture; development of texture and flavor: A review. *Foods* **2014**, *3*, 176–193. [CrossRef]
61. Tirloni, E.; Ghelardi, E.; Celandroni, F.; Bernardi, C.; Stella, S. Effect of dairy product environment on the growth of Bacillus cereus. *J. Dairy Sci.* **2017**, *100*, 7026–7034. [CrossRef]
62. Anwar, F.; Rashid, U. Physico-chemical characteristics of Moringa Oleifera seeds and seed oil from a wild provenance of Pakistan. *Pak. J. Bot* **2007**, *39*, 1443–1453.
63. Ghebremichael, K.A.; Gunaraatna, K.R.; Henriksson, H.; Brumer, H.; Dalhammar, G. A simple purification and activity assay of the coagulant protein from Moringa oleifera seed. *Water Res.* **2005**, *39*, 2338–2344. [CrossRef]
64. Mougli, S.; Marisa, H.; Stephane, C.; Flourian, F.; Pierre, C.; Catherine, S.; Oliver, M.; Ryth, F.; Philippie, M.; Nicolas, M. Structure, function, characterization and optimization of a plant-derived antibacterial peptide. Antimicrob. *Agents Chemother.* **2005**, *49*, 3847–3857.

65. Basiri, S.; Haidary, N.; Shekarforoush, S.S.; Niakousari, M. Flaxseed mucilage: A natural stabilizer in stirred yoghurt. *Carbohydr. Polym.* **2018**, *187*, 59–65. [CrossRef] [PubMed]
66. Granato, D.; Branco, G.F.; Cruz, A.G.; Faria, J.D.A.F.; Shah, N.P. Probiotic dairy products as functional foods. *Compr. Rev. Food Sci. Food Saf.* **2010**, *9*, 455–470. [CrossRef]

Article

Development of a Multifunction Set Yogurt Using *Rubus suavissimus* S. Lee (Chinese Sweet Tea) Extract

Mahmoud Abdel-Hamid [1,2], Zizhen Huang [1], Takuya Suzuki [3], Toshiki Enomoto [4], Ahmed M. Hamed [2], Ling Li [1,*] and Ehab Romeih [2,*]

[1] Guangxi Buffalo Research Institute, Chinese Academy of Agricultural Sciences, Nanning 530001, China; mahmoud.mohamed@agr.cu.edu.eg (M.A.-H.); hzz90302@163.com (Z.H.)

[2] Dairy Science Department, Faculty of Agriculture, Cairo University, Giza 12613, Egypt; ahmed.hamed@agr.cu.edu.eg

[3] Department of Biofunctional Science and Technology, Graduate School of Biosphere Science, Hiroshima University, 1-4-4 Kagamiyama, Higashi-Hiroshima 739-8528, Japan; takuya@hiroshima-u.ac.jp

[4] Department of Food Science, Ishikawa Prefectural University, Nonoichi, Ishikawa 921-8836, Japan; enomoto@ishikawa-pu.ac.jp

* Correspondence: lling2010@163.com (L.L.); ehab.romeih@staff.cu.edu.eg (E.R.)

Received: 22 July 2020; Accepted: 21 August 2020; Published: 24 August 2020

Abstract: *Rubus suavissimus* S. Lee leaves, also known as Chinese sweet tea or Tiancha, are used in folk medicine in southern China. This study evaluated the impact of the addition of Chinese sweet tea extract (0.25%, 0.5%, and 1%) on the chemical composition, organoleptic properties, yogurt culture viability, and biological activities (i.e., antioxidant, anticancer, and antihypertensive activities) of yogurt. Seven phenolic compounds were reported in Chinese sweet tea for the first time. The numbers of the yogurt culture were similar across all yogurt treatments. The yogurt supernatant with 0.25%, 0.5%, and 1% Chinese sweet tea extract had a total phenolic content that was 3.6-, 6.1-, and 11.2-fold higher, respectively, than that of the control yogurt. The biological activities were significantly increased by the addition of Chinese sweet tea extract: Yogurt with the addition of 1% Chinese sweet tea extract had the highest biological activities in terms of the antioxidant activity (92.43%), antihypertensive activity (82.03%), and inhibition of the Caco-2 cell line (67.46%). Yogurt with the addition of 0.5% Chinese sweet tea extract received the highest aroma and overall acceptability scores. Overall, Chinese sweet tea extract is a promising food ingredient for producing functional yogurt products that may substantially contribute to reducing the risk of developing chronic diseases such as cancer and cardiovascular disease.

Keywords: *Rubus suavissimus* S. Lee (Chinese sweet tea); yogurt; antioxidant; anticancer; antihypertensive

---

1. Introduction

Phytochemicals are produced by plants as secondary metabolites to protect themselves from microbial attack and to control environmental stress. Among these phytochemicals, phenolic compounds exhibit various biological activities, including antioxidant, anticancer, antiviral, anti-allergic, antihypertensive, anti-inflammatory, and antidiabetic activities. These compounds are commonly present in vegetables, cereals, herbs, fruits, and green and black teas; due to their biological activities, they are used as natural additives in the food and pharmacology industries [1].

Rubus suavissimus S. Lee originates from the south of China and is widely planted in the Guangxi and Guizhou provinces. Due to its sweet taste, the leaves of *Rubus suavissimus* S. Lee are commonly used in Chinese sweet tea, which is locally referred to as Tiancha [2]. Its sweet taste is mainly attributed to rubusoside, a diterpene glucoside, which is 115 times sweeter than sucrose [3]. Chinese sweet tea is generally used in folk medicine for a wide range of diseases—i.e., coughs, diabetes, and atherosclerosis (high blood pressure)—as well as to promote kidney function. Moreover, different biological activities of Chinese sweet tea—i.e., antioxidant, anti-allergic, anti-inflammatory, anticancer activity, antiangiogenic, antidiabetic, and anti-obesity activities—have been reported [3–6]. These biological activities are linked to the presence of various bioactive components in Chinese sweet tea, including gallic acid, ellagic acid, rutin, kaempferol, caffeic acid, quercetin, rubusoside, and steviol monoside [3]. Recently, 14 new phenolic compounds in *Rubus suavissimus* S. Lee, including protocatechuic acid, myketin, epicatechin, vanillic acid, apigenin, catechin, ferulic acid, luteolin, 3, 3′-di-O-methylellagic acid, chlorogenic acid, 3, 3′-di-O-methylellagic acid-4′-O-β-D-glucoside, cinnamic acid, and syringate, have been identified [5].

Novel dairy products with multifunctional properties and health-promoting benefits are in high demand due to the increase in consumer health awareness. Yogurt is the most popular fermented milk, and although it has a high nutritional value and health benefits, it is a poor source of phenolic compounds. Therefore, yogurt fortified with plant materials as a natural source of phytochemicals is a new trend to increase its nutritional and functional properties. In this context, various studies have been carried out to produce yogurt supplemented with plant extracts that are rich in phytochemicals. Most of these studies have focused on the antioxidant activity of the supplemented yogurt as a direct effect of the addition of plant extracts. For instance, grape and grape callus extracts [7], moringa extract [8], cinnamon powder [9], and green and black teas [10–12] have been used to produce yogurt with a high antioxidant activity. Furthermore, grape seed and *Siraitia grosvenorii* fruit extracts, as a source of phytochemicals, have been used to prepare functional yogurts with high biological activities [13,14]. Nevertheless, only a few studies have reported the indirect effects of the addition of plant extracts on the angiotensin I-converting enzyme-inhibitory activity—for example, yogurt with *Mentha piperita, Anethum graveolens,* and *Ocimum basilicum* extracts [15], and yogurt with Raftiline HP®, a high-performance inulin material [16]. Additionally, the anticancer activity of yogurt has been shown to increase as an indirect effect of the addition of pineapple peel powder [17].

It is worth noting that studies on the anticancer and antihypertensive activities of yogurt supplemented with plant extracts are scarce and, to the best of our knowledge, the addition of Chinese sweet tea to milk products has not yet been reported. Accordingly, the aim of the present work was to evaluate the functional properties, including the antioxidant, anticancer, and antihypertensive activities, as well as the culture viability and sensory evaluation of yogurt fortified with Chinese sweet tea extract powder.

2. Materials and Methods

2.1. Chemicals

2,2′-azino-bis (3-ethylbenzothiazoline-6-sulfonic acid) (ABTS), gallic acid, and 1,1-diphenyl-2-picrylhydrazyl (DPPH) were purchased from Merck (Sigma Aldrich, Beijing, China). De Man Rogosa Sharpe (MRS) and M17 media were obtained from Thermo Fisher Biochemicals (Shanghai, China).

2.2. Preparation of the Chinese Sweet Tea Extract

Air-dried *Rubus suavissimus* S. Lee (Chinese sweet tea) leaves (local market, Nanning city, China) were extracted with boiling water (1:15 *w/v*) for 60 min, as previously described [3]. The boiled mixture was centrifuged at 5000× *g* for 15 min at room temperature and then filtered through Whatman filter paper no. 1. The final filtrate was then freeze-dried. The Chinese sweet tea extract powder was analyzed for the protein, fat, ash, and total solid contents, as previously described [18], while the carbohydrate content was calculated by the differences.

2.3. High-Performance Liquid Chromatography (HPLC) Analysis of Phenolic Compounds

The identification and quantification of the individual phenolic compounds were carried out using an Agilent-1260 HPLC (USA) equipped with a C18 column (Kinetex® 5 μm EVO 100 × 4.6 mm,

Phenomenex, Torrance, CA, USA) [19]. In brief, 5 g of Chinese sweet tea extract was dissolved in 50 mL of methanol and centrifuged at 6000× g for 10 min. After filtration (0.22 μm syringe filter, Minisart®, Sartorius Stedim Boitech, Beijing, China), 20 μL of the filtrate was injected and separated by isocratic elution of water with 0.2% phosphoric acid (*v/v*) (A; 96%), methanol (B; 2%), and acetonitrile (C; 4%) at a 0.7 mL/min flow rate. The phenolic compounds were monitored at 280 nm using a UV detector. The quantification of the phenolic compounds was calculated with external calibration through the data analysis system of the Agilent software. All the phenolic standards were provided by Merck (Sigma Aldrich, Beijing, China).

2.4. Yogurt Manufacture

Yogurt was produced according to the method of Abdel-Hamid et al. [14]. In brief, skimmed buffalo milk (Guangxi Buffalo Research Institute Farm, Nanning, China) was heat-treated at 90 °C for 10 min and then rapidly cooled to 43 °C. The heat-treated milk was inoculated with yogurt culture (YO-MIX 300, Danisco, China), a mix of *Lactobacillus delbrueckii ssp. bulgaricus*, and *Streptococcus thermophilus*, according to the manual provided by the manufacturer. The cultured milk was divided into four equal parts: one part was used as a control treatment and the other three parts were incorporated with the Chinese sweet tea extract powder at concentrations of 0.25%, 0.5%, and 1% (*w/w*). Each milk portion was poured into plastic containers (120 mL) and incubated at 42 °C until the pH was reduced to 4.6. The yogurt treatments were kept at 4 ± 1 °C.

2.5. Chemical Composition

The carbohydrate, protein, and total solid contents of the cultured milk combined with the Chinese sweet tea extract powder were measured by a MilkoScan analyzer (F120, FOSS, Hillerød, Denmark). The pH values of the yogurt samples were monitored using a digital pH meter (Methrohm AG, Herisau, Switzerland).

2.6. Bacterial Counts

Bacterial counts of the yogurt samples were performed after 24 h of storage at 4 °C using the pour plate method [14]. *S. thermophilus* was grown on M17 agar, and the plates were aerobically incubated at 37 °C for 24 h. *L. bulgaricus* was enumerated on MRS agar (pH 5.4), and the plates were anaerobically incubated at 37 °C for 24 h. The results are presented as log colony-forming units per gram yogurt (CFU/g).

2.7. Yogurt Supernatant

The yogurt supernatant was separated by centrifugation at 22,000× g for 30 min at 4 °C, followed by filtration using a 0.45 μm syringe filter (Minisart®, Sartorius Stedim Boitech, Beijing, China). The filtrate of the yogurt samples was used to evaluate the total phenolic content and the biological activities—i.e., the antioxidant, anticancer, and antihypertensive activities.

2.8. Total Phenolic Content (TPC)

The TPC of the Chinese sweet tea extract powder and the yogurt samples was measured using the Folin–Ciocalteu assay [8]. Thirty microliters of the yogurt supernatant or the Chinese sweet tea extract powder (1 mg/mL in water) were pipetted into a 96-well plate, and then distilled water (120 μL) and Folin–Ciocalteu's phenol reagent (30 μL) were added, respectively, to each well, followed by 30 μL of sodium carbonate (1N). The plates were incubated in the dark for 30 min at room temperature, and then the absorbance was read at 750 nm using a microplate reader (EPOCH, BioTek, Winooski, VT, USA). A standard curve was constructed using different concentrations of gallic acid. The results are expressed as the μg gallic acid equivalents (GAE) per milliliter.

2.9. Antioxidant Activity

ABTS and DPPH radical scavenging assays were used to evaluate the antioxidant activity of the yogurt samples according to Abdel-Hamid et al. [14]. In short, 50 μL of the yogurt supernatant was pipetted into a 96-well plate followed by ABTS$^+$ solution (200 μL). The plate was then incubated in the dark for 30 min, and the absorbance was monitored at 405 nm using a microplate reader.

For the DPPH assay, DPPH reagent (0.2 mM) was freshly prepared and added to the yogurt supernatant (1:1 *v/v*). After a 30 min reaction in the dark at 37 °C, the absorbance was measured at 517 nm.

2.10. Anticancer Activity

The anticancer activity of the yogurt samples was assessed against the Caco-2 carcinoma cell line (HTB-37; American Type Culture Collection, Manassas, VA, USA), and the cells were propagated according to Abdel-Hamid et al. [20]. Caco-2 cells were plated into 96-well plates (3000 cells/well) and incubated overnight at 37 °C under 5% CO_2. The cells were then treated with 25 μL of the yogurt supernatant and grown again for 24 h. The viability of the treated cells was evaluated by the WST assay. The antiproliferative activity was calculated using the following equation:

$$\text{Antiproliferative activity (\%)} = [1 - (A - B)/(C - B)] \times 100, \tag{1}$$

where A is the absorbance of the cells in the presence of the yogurt supernatant, B is the background absorbance (non-cell control), and C is the absorbance of the control (cells with sterilized water instead of the yogurt supernatant).

2.11. Antihypertensive Activity

The antihypertensive activity of the Chinese sweet tea extract and the yogurt samples was investigated by measuring the angiotensin-converting enzyme (ACE) inhibitory activity. The spectrophotometric method was employed to evaluate the ACE inhibition by the Chinese sweet tea extract and the yogurt samples using the ACE Kit-WST (Dojindo laboratories, Shanghai, China) [20].

2.12. Sensory Evaluation

Samples of the yogurt with and without Chinese sweet tea extract were characterized organoleptically after 24 h of cold storage at 4 ± 1 °C by seven trained panelists (researchers at Guangxi University, Nanning, China) with an interest and experience in the sensory evaluation of fermented milk. A 9-point scale was used to evaluate the yogurt samples in terms of their appearance, aroma, texture, and overall acceptability, as described by Romeih et al. [21].

2.13. Statistical Analysis

Three independent experiments were performed in this study, and the results are presented as the mean ± standard deviation (SD). A one-way analysis of variance (ANOVA) was carried out with Statistix 8.1 (Analytical Software, Tallahassee, FL, USA) using Tukey's test for pairwise comparison. The correlation between variables was evaluated by the Pearson correlation test.

3. Results and Discussion

3.1. Impact of the Chinese Sweet Tea Extract on the Chemical Composition of the Yogurt Samples

The chemical characterization of the Chinese sweet tea extract powder is presented in Table 1. The Chinese sweet tea extract powder contained 94.22 ± 1% total solids, 6.4 ± 0.25% protein, and 80.18 ± 1.2% total carbohydrates. The chemical composition of the heat-treated milk with the Chinese sweet tea extract before fermentation was measured using the MilkoScan, and the results are presented in Table 2. The addition of the Chinese sweet tea extract had no significant effect

Foods **2020**, *9*, 1163

($p > 0.05$) on the protein content of the yogurt samples, whereas the total solid and total carbohydrate contents were significantly increased ($p < 0.05$) compared with the control yogurt sample. These results might be attributed to the high contents of carbohydrates and total solids in the Chinese sweet tea extract powder. It has been reported that the crude water extract of Chinese sweet tea contains 11% polysaccharides [2]. A similar trend was observed for the total carbohydrate and total solid contents in yogurt supplemented with *Siraitia grosvenorii* fruit, moringa, and *Gnaphalium affine* extracts [8,14,22].

Table 1. Characterization of the Chinese sweet tea extract.

Characteristic	Chinese Sweet Tea Extract
Total solid (%)	94.22 ± 1.0
Protein (%)	6.40 ± 0.25
Fat (%)	0.42 ± 0.1
Ash (%)	7.21 ± 0.1
Carbohydrates (%)	80.18 ± 1.2
Total phenolic content *	21.54 ± 0.5
ACE-I activity (%)	86.85 ± 2.1
Anticancer activity (%)	89.06 ± 3.7

Values are the mean of three replicates ± standard deviation. * Total phenolic content presented as mg of gallic acid equivalents (GAE) per gram of Chinese sweet tea extract powder. ACE-I, angiotensin-converting enzyme-inhibition.

Table 2. Chemical composition of heat-treated milk fortified with Chinese sweet tea extract.

Chinese Sweet Tea Extract Concentration	Protein (%)	Carbohydrates (%)	Total Solids (%)	Total Phenolic Content *
0% (Control)	5.10 ± 0.08 [A]	6.17 ± 0.01 [D]	12.31 ± 0.02 [D]	11.64 ± 1.2 [D]
0.25%	5.15 ± 0.02 [A]	6.26 ± 0.02 [C]	12.56 ± 0.02 [C]	41.60 ± 2.0 [C]
0.5%	5.16 ± 0.02 [A]	6.35 ± 0.01 [B]	12.84 ± 0.04 [B]	70.93 ± 2.2 [B]
1%	5.18 ± 0.02 [A]	6.59 ± 0.01 [A]	13.48 ± 0.02 [A]	130.58 ± 2.2 [A]

Results are the mean of three experiments ± standard deviation. Values in the same column with different superscript letters are significantly different ($p < 0.05$). * Total phenolic content presented as µg gallic acid equivalents (GAE) per milliliter of yogurt supernatant.

Regarding the TPC, the Chinese sweet tea extract powder contained 21.54 ± 0.55 mg GAE/g (Table 1). The yogurt with the Chinese sweet tea extract had a significantly higher TPC content ($p < 0.05$) compared to the control yogurt sample, which gradually increased with an increase in the amount of Chinese sweet tea extract (Table 2). The highest TPC content (130.58 µg GAE/mL) was measured in the yogurt sample containing 1% Chinese sweet tea extract. These results are in agreement with those of Amirdivani and Baba [12], Gao et al. [22], Karaaslan et al. [7], and Zhang et al. [8], who found significantly increasing TPC contents in yogurt fortified with green tea, moringa, *Gnaphalium affine*, grape, and callus extracts. It should be noted that the TPC contents reported in these studies were lower than those detected in our study (130.58 µg GAE/mL).

3.2. Identification and Quantification of Phenolic Compounds in the Chinese Sweet Tea Extract

The phenolic compounds were determined using the HPLC assay, and the results are presented in Table 3. A total of 19 phenolic compounds were identified and quantified in the Chinese sweet tea extract powder, with concentrations ranging from 0.08 to 2.77 mg/g. Benzoic acid, quercetin, rutin, syringic acid, ellagic acid, and gallic acid were the most abundant phenolic compounds detected in the Chinese sweet tea extract. Indeed, Koh et al. [3] reported ellagic acid, rutin, and gallic acid as the major phenolic compounds of Chinese sweet tea, and they used the rubusoside content to evaluate the quality of the Chinese sweet tea. It has been reported that the concentrations of ellagic acid, rutin, and gallic acid in 14 Chinese sweet tea samples collected in different seasons ranged from 0.46% to 92%, 0.08% to 0.15%, and 0.1% to 0.16%, respectively [2]. More recently, Liu et al. [5] reported 14 new phenolic compounds in Chinese sweet tea leaves. In our study, seven new phenolic compounds were identified for the first time in Chinese sweet tea namely, *p*-coumaric acid, benzoic acid, *o*-coumaric acid, resveratrol, neringein, rosmarinic, and myricetin. These findings suggest that the phenolic compound content varies according to the region and season of growth [2].

Table 3. Identified phenolic compounds in the Chinese sweet tea extract.

Phenolics	Concentration (mg/g)
Gallic acid	1.26 ± 0.006
Catchin	0.08 ± 0.001
Chlorgenic	0.66 ± 0.003
Vanillic acid	0.96 ± 0.001
Caffeic acid	0.47 ± 0.005
Syringic acid	1.53 ± 0.006
p-Coumaric acid	1.08 ± 0.006
Benzoic acid	2.77 ± 0.007
Ferulic acid	0.11 ± 0.001
Rutin	1.63 ± 0.004
Ellagic	1.44 ± 0.007
o-Coumaric acid	0.74 ± 0.004
Resveratrol	0.52 ± 0.003
Cinnamic acid	0.11 ± 0.001
Quercetin	1.98 ± 0.005
Neringein	0.59 ± 0.004
Rosemarinic	0.55 ± 0.005
Myricetin	0.46 ± 0.005
Kampherol	0.38 ± 0.004

Values are the mean of three replicates ± standard deviation.

3.3. pH and Bacterial Count

The addition of the Chinese sweet tea extract increased the fermentation time from 5 h in the control yogurt sample to 6 h in the yogurt sample containing the 1% Chinese sweet extract. This was probably due to the antibacterial activity of the Chinese sweet tea components toward lactic acid bacteria, along with its acknowledged inhibition of pathogenic bacteria [13,23]. Additionally, Chinese sweet tea contains a high amount of benzoic acid, which is known to be an antimicrobial agent (Table 3). Likewise, the fermentation time was not affected by the addition of green and black teas [10]. In contrast, it has been reported that the addition of moringa, *Mentha piperita*, *Anethum graveolens*, or *Ocimum basilicum* extracts to yogurt shortens the fermentation time due to the enhancement of the growth of yogurt cultures [8,15].

The pH values of the yogurt samples after 24 h of cold storage are presented in Table 4. The addition of the Chinese sweet tea extract had no significant effect ($p > 0.05$) on the pH values. In agreement with this finding, the use of different types of spices (i.e., cardamom, cinnamon, or nutmeg), moringa, and cinnamon and licorice herbals in the preparation of yogurt had no significant effect on the pH values compared with the relevant control yogurt [8,23,24].

Table 4. Viability of the yogurt cultures (log CFU/g) and pH values of the yogurt fortified with Chinese sweet tea extract.

Chinese Sweet Tea Extract Concentration	pH	*L. bulgaricus* (log CFU/g)	*S. thermophilus* (log CFU/g)
0% (Control)	4.52 ± 0.02 [A]	8.65 ± 0.1 [A]	9.20 ± 0.04 [A]
0.25%	4.53 ± 0.02 [A]	8.83 ± 0.03 [A]	9.22 ± 0.03 [A]
0.5%	4.54 ± 0.03 [A]	8.90 ± 0.04 [A]	9.25 ± 0.1 [A]
1%	4.54 ± 0.02 [A]	8.86 ± 0.5 [A]	9.18 ± 0.02 [A]

Results are the mean of three experiments ± standard deviation. Values in the same column with different superscript letters are significantly different ($p < 0.05$).

The viability of the yogurt cultures of *S. thermophilus* and *L. bulgaricus* is presented in Table 4 as the log CFU/g yogurt. The viable cell counts of *L. bulgaricus* and *S. thermophilus* in the yogurt samples were 8.6–8.9 and 9.1–9.3 log CFU/g, respectively. The viable counts of *L. bulgaricus* and *S. thermophilus* in the yogurt samples are higher than the recommended dose to promote health benefits (>6 log CFU/g) [25]. Yogurt samples fortified with Chinese sweet tea had no significant influence ($p > 0.05$) on the viability of

L. bulgaricus and *S. thermophilus*. A similar trend was reported for the addition of strawberries and green or black teas before yogurt fermentation [10,26]. Behrad et al. [23] reported that yogurt mixed with cinnamon and licorice herbals had lower counts of *L. bulgaricus* and *S. thermophilus* compared to the control plain yogurt. In contrast, the addition of Japanese and Malaysian green teas or moringa extract significantly improved the viability of *L. bulgaricus* and *S. thermophilus* in yogurt [8,12]. These findings demonstrate that the effects of the addition of extracts on the fermentation time, pH values, and culture viability of yogurt depend on the plant type and the phytochemical concentrations.

3.4. Antioxidant Activity

The antioxidant activities of the yogurt samples were measured by ABTS and DPPH assays, and the results are presented in Table 5. The antioxidant activity of the yogurt samples ranged between 14.2% and 74.83% for the DPPH assay and between 32.01% and 92.43% for the ABTS assay. The addition of Chinese sweet tea extract significantly increased ($p < 0.05$) the antioxidant activity as the amount of extract added increased. The yogurt sample containing 1% Chinese sweet tea extract showed the highest ABTS (92.43%) and DPPH (74.83%) values. A positive correlation was observed between the TPC and the antioxidant activity for the DPPH ($r = 0.998$) and ABTS ($r = 0.993$) assays. Shori et al. [27] reported a similar trend between the TPC and antioxidant activity of phytomix-3+ mangosteen (a mixture of *Lycium barbarum*, *Momordica grosvenori*, and *Psidium guajava* leaves) yogurt. The increase in the antioxidant activities of yogurt fortified with Chinese sweet tea extract could be due to the higher concentrations of phytochemicals (i.e., phenols, flavonoids, and rubusoside) in Chinese sweet tea [2]. Our results are in agreement with those of Amirdivani and Baba [12] and Najgebauer-Lejko et al. [11], who found significantly higher antioxidant activities in yogurt supplemented with Japanese green tea, Malaysian green tea, green tea, or Pu-erh tea compared to the control yogurt. Furthermore, the addition of moringa, *Mentha piperita*, *Anethum graveolens*, *Ocimum basilicum*, spice oleoresins (i.e., cardamom, cinnamon, and nutmeg), grape seed, and cinnamon and licorice herbal extracts significantly increases the antioxidant activity of yogurt [8,9,13,15,23,24]. These authors concluded that the increase in antioxidant activities is a result of the concentrations of phenolic compounds in the plant extracts.

Table 5. Antioxidant activity of the yogurt fortified with Chinese sweet tea extract.

Chinese Sweet Tea Extract Concentration	Antioxidant Activity (%)	
	ABTS	**DPPH**
0% (Control)	32.01 ± 1.0 [D]	14.20 ± 0.4 [D]
0.25%	49.40 ± 0.9 [C]	29.85 ± 1.4 [C]
0.5%	68.76 ± 2.2 [B]	47.29 ± 2.2 [B]
1%	92.43 ± 0.3 [A]	74.83 ± 0.9 [A]

Results are the mean of three experiments ± standard deviation. Values in the same column with different superscript letters are significantly different ($p < 0.05$).

It is worth noting that the antioxidant activities of the yogurt samples containing Chinese sweet tea extract measured by the DPPH (29.8–74.8%) and ABTS (49.4–92.4%) assays were higher than the antioxidant activities of the yogurt produced by the above-mentioned studies. This may be attributed to the differences in the phytochemical types and their concentrations.

3.5. Anticancer Activity

The anticancer activity of the yogurt samples was examined by measuring the ability of the yogurt supernatant to inhibit the proliferation of the Caco-2 cell line, and the results are presented in Table 6. The Chinese sweet tea extract exhibited a 89.06% ± 3.7% anticancer activity at a concentration of 0.4 mg/mL (Table 1). The control yogurt sample also showed anticancer activity, which is most probably attributed to the presence of bioactive peptides. A similar finding was reported by Sah et al. [17].

Although the addition of 0.25% Chinese sweet tea extract had no significant impact ($p > 0.05$) on the anticancer activity levels compared to the control yogurt sample, yogurt samples containing 0.5% and 1% Chinese sweet tea extract exhibited significantly higher ($p < 0.05$) anticancer activity than that of the control yogurt sample (Table 5). Moreover, the supernatant of the yogurt sample containing 1% Chinese sweet tea extract showed the highest anticancer activity and inhibited the growth of the Caco-2 cell by 67.46%. It is worth noting that the anticancer activity was positively correlated ($r = 0.961$) with the TPC contents of the yogurts. This finding is probably attributed to the phytochemical content of the Chinese sweet tea extract, including phenolic compounds and rubusoside [2,3]. George Thompson et al. [28] reported that rubusoside, the main component in Chinese sweet tea (5% in dry leaves), inhibits the glucose (GLUT1) and fructose (GLUT5) transporters associated with cancer and diabetes. In addition, quercetin, rutin, and ellagic acid, which are among the major phenolic compounds in Chinese sweet tea, show anticancer activity against colon carcinoma cells, as described by Hashemzaei et al. [29] and Papoutsi et al. [30]. In accordance with our results, Sah et al. [17] reported the potential anticancer activity of probiotic yogurt supplemented with pineapple peel powder against HT29 colon cancer cells compared to the control. These authors attributed this finding to the enhanced extent of proteolysis and, consequently, the resultant bioactive peptides released by the addition of pineapple peel powder to the yogurt.

Table 6. Antiproliferative and ACE-I activities of the yogurt fortified with Chinese sweet tea extract.

Sweet Tea Extract Concentration	ACE-I (%)	Antiproliferative Activity (%)
0% (Control)	44.20 ± 0.7 [D]	18.58 ± 1.4 [C]
0.25%	50.88 ± 1.1 [C]	26.27 ± 4.5 [C]
0.5%	64.72 ± 1.6 [B]	52.41 ± 6.1 [B]
1%	82.03 ± 1.9 [A]	67.46 ± 1.9 [A]

Results are the mean of three experiments ± standard deviation. Values in the same column with different superscript letters are significantly different ($p < 0.05$).

3.6. Antihypertensive Activity

The antihypertensive activity was evaluated using the ACE inhibition method, as ACE is a key factor in the conversion of angiotensin I to angiotensin II, which raises blood pressure [31]. Chinese sweet tea extract exhibited a 86.85% ± 2.1% ACE inhibitory activity at 0.4 mg/mL (Table 1). The ACE inhibitory activity of the yogurt samples is shown in Table 6. The yogurt samples without the addition of Chinese sweet tea also showed an ACE inhibitory activity, which may due to the presence of bioactive peptides with ACE inhibitory activities already known to be present in yogurt products. In this context, Abdel-Hamid et al. [14] and Amirdivani and Baba [15] also reported ACE inhibitory activity for their control yogurts. As can be seen in Table 6, the addition of Chinese sweet tea extract significantly enhanced ($p < 0.05$) the ACE inhibitory activity of the yogurt samples. The ACE inhibitory activity was increased by increasing the amount of Chinese sweet tea extract added, where the yogurt sample containing 1% Chinese sweet tea extract showed the highest value (82.03%). This finding could be due to the bioactive compounds in the Chinese sweet tea extract. It should be noted that quercetin, rutin, and gallic acid, the major phenolic compounds in the Chinese sweet tea extract, exhibit antihypertensive activity both in vitro and in vivo, as reported by Balasuriya and Rupasinghe [32], Kang et al. [33], and Shaw et al. [34]. In addition, the ACE inhibitory activity of the yogurt samples was positively correlated with the TPC ($r = 0.994$). It has been reported that Chinese sweet tea exhibits antihypertensive activity, which can be attributed to its phytochemical content [35]. These results are in accordance with those reported by Liu and Finley [36], who concluded that phytochemicals reduce blood pressure. Furthermore, *Chrysophyllum cainito* fruit extract, rich in polyphenols, shows ACE inhibitory activity by chelating Zn^{2+} ions (enzyme cofactor) [31]. Amirdivani and Baba [15] reported that yogurt fortified with herbal water extracts (i.e., *Mentha piperita*, *Anethum graveolens*, and *Ocimum basilicum*) exhibited a higher ACE inhibitory activity compared to the control yogurt. Moreover, the addition of Raftiline

HP[®] has been shown to significantly improve the ACE inhibitory activity compared to plain yogurt [16]. These authors attributed the increase in ACE inhibitory activity after the addition Raftiline HP[®] or herbal water extracts to the higher degree of proteolysis of the fortified yogurts, resulting in bioactive peptides with ACE inhibitory activity. However, Amirdivani and Baba [15] reported that the water extract of *Mentha piperita*, *Anethum graveolens*, or *Ocimum basilicum* itself had no ACE inhibitory activity. Interestingly, the obtained ACE inhibitory activity values of yogurts prepared with Chinese sweet tea extracts were higher than those reported by Amirdivani and Baba [15] and Ramchandran and Shah [16].

It should be noted that the Chinese sweet tea extract had no effect on either the viability of the yogurt culture (Table 4) or the degree of hydrolysis (data not shown). However, the Chinese sweet tea extract exhibited potential anticancer and antihypertensive activities (Table 1). Accordingly, the biological activities (i.e., antioxidant, antihypertensive, and anticancer activities) of the yogurt samples containing Chinese sweet tea extract are most probably attributed to the phytochemicals present in Chinese sweet tea.

3.7. Sensory Evaluation

Sensory characteristics are very important for assessing the consumer acceptability of yogurt products and to confirm that the additives have no negative impacts on the organoleptic parameters of said yogurt products. The sensory evaluation of the yogurt samples fortified with Chinese sweet tea extract is shown in Table 7. The yogurt samples containing Chinese sweet tea extract received almost similar appearance scores as the control yogurt sample ($p > 0.05$). The addition of the Chinese sweet tea extract significantly increased ($p < 0.05$) the texture score. Furthermore, the addition of Chinese sweet tea significantly enhanced ($p < 0.05$) the aroma of the yogurt samples compared to the control yogurt sample, which could be due to the sweet taste of rubusoside. In particular, the yogurt sample containing 1% Chinese sweet tea extract received the lowest perceived aroma score ($p < 0.05$) compared to the samples containing 0.25% and 0.5% Chinese sweet tea extract and the control yogurt sample. This finding may be attributed to the bitter aftertaste of rubusoside and some phenolic compounds [2,3]. Similarly, the overall acceptability score of the yogurt samples containing 0.25% and 0.5% Chinese sweet extract was significantly higher ($p < 0.05$) than the control yogurt, while the yogurt sample containing 1% Chinese sweet tea extract received the lowest overall acceptability score. In this context, yogurt prepared with moringa extract at different levels (i.e., 0.05%, 0.1%, and 0.2%) had a significantly lower sensory evaluation compared to the control yogurt [8]. The authors attributed these results to the better taste of moringa. The above results demonstrate that the yogurt samples with 0.5% Chinese sweet tea extract had the best sensory evaluation.

Table 7. Sensory evaluation of yogurt fortified with Chinese sweet tea extract.

Chinese Sweet Tea Extract Concentration	Appearance	Aroma	Texture	Overall Acceptability
0% (Control)	7.7 ± 0.5 [A]	6.9 ± 0.4 [C]	6.0 ± 0.8 [C]	6.7 ± 0.8 [C]
0.25%	8.0 ± 0.7 [A]	7.7 ± 0.5 [B]	6.6 ± 0.5 [C]	7.6 ± 0.5 [B]
0.5%	8.1 ± 0.8 [A]	8.6 ± 0.5 [A]	7.6 ± 0.5 [B]	8.7 ± 0.5 [A]
1%	8.3 ± 0.8 [A]	6.0 ± 0.6 [D]	8.7 ± 0.5 [A]	5.7 ± 0.5 [D]

Results are the mean of three experiments ± standard deviation. Values in the same column with different superscript letters are significantly different ($p < 0.05$).

4. Conclusions

This study aimed to produce a multifunctional yogurt rich in phytochemicals using Chinese sweet tea extract. The addition of the Chinese sweet tea extract did not influence the viability of the yogurt culture but significantly enhanced the biological activities of the yogurt samples, including their antioxidant, anticancer, and antihypertensive activities. The addition of Chinese sweet tea at concentrations of 0.25% and 0.5% significantly improved the aroma of the yogurt samples compared to the control yogurt sample. Although the yogurt sample containing 1% Chinese sweet tea extract exhibited the highest antioxidant, anticancer, and antihypertensive activities, it received the lowest aroma score due to the bitter aftertaste of rubusoside. Overall, the yogurt sample containing 0.5% Chinese sweet tea extract had significantly higher biological activities and sensory evaluation compared to the control yogurt sample. Therefore, 0.5% Chinese sweet tea extract appears to be an efficient option and a promising ingredient for producing a multifunctional yogurt rich in phytochemicals with health-promoting properties.

Author Contributions: Investigation, formal analysis, and writing—original draft, M.A.-H.; methodology, Z.H., T.S., T.E., and A.M.H.; funding acquisition and project administration, L.L.; investigation and writing—review and editing, E.R. All authors have read and agreed to the published version of the manuscript.

Funding: The project was financed by the Guangxi Engineering Research Center (Project of Quality, Safety and Technology for Buffalo Dairy).

Acknowledgments: The authors gratefully acknowledge Wenxi Xu and Nana Takeda from Hiroshima University, Japan, for their assistance in determining the anticancer activity.

Conflicts of Interest: The authors declare that there are no conflicts of interest.

References

1. Khoshnoudi-Nia, S.; Sharif, N.; Jafari, S.M. Loading of phenolic compounds into electrospun nanofibers and electrosprayed nanoparticles. *Trends Food Sci. Technol.* **2020**, *95*, 59–74. [CrossRef]
2. Chou, G.; Xu, S.-J.; Liu, D.; Koh, G.Y.; Zhang, J.; Liu, Z. Quantitative and fingerprint analyses of Chinese sweet tea plant (*Rubus suavissimus* S. Lee). *J. Agric. Food Chem.* **2009**, *57*, 1076–1083. [CrossRef]
3. Koh, G.Y.; Chou, G.; Liu, Z. Purification of a water extract of Chinese sweet tea plant (Rubus suavissimus S. lee) by alcohol precipitation. *J. Agric. Food Chem.* **2009**, *57*, 5000–5006. [CrossRef] [PubMed]
4. Ezure, T.; Amano, S. Rubus suavissimus S. Lee extract increases early adipogenesis in 3T3-L1 preadipocytes. *J. Nat. Med.* **2011**, *65*, 247–253. [CrossRef] [PubMed]
5. Liu, M.; Li, X.; Liu, Q.; Xie, S.; Chen, M.; Wang, L.; Feng, Y.; Chen, X. Comprehensive profiling of α-glucosidase inhibitors from the leaves of Rubus suavissimus using an off-line hyphenation of HSCCC, ultrafiltration HPLC-UV-MS and prep-HPLC. *J. Food Compos. Anal.* **2020**, *85*, 103336. [CrossRef]
6. Zhang, H.; Qi, R.; Zeng, Y.; Tsao, R.; Mine, Y. Chinese sweet leaf tea (Rubus suavissimus) mitigates LPS-induced low-grade chronic inflammation and reduces the risk of metabolic disorders in a C57BL/6J mouse model. *J. Agric. Food Chem.* **2020**, *68*, 138–146. [CrossRef] [PubMed]
7. Karaaslan, M.; Ozden, M.; Vardin, H.; Turkoglu, H. Phenolic fortification of yogurt using grape and callus extracts. *LWT—Food Sci. Technol.* **2011**, *44*, 1065–1072. [CrossRef]
8. Zhang, T.; Jeong, C.H.; Cheng, W.N.; Bae, H.; Seo, H.G.; Petriello, M.C.; Han, S.G. Moringa extract enhances the fermentative, textural, and bioactive properties of yogurt. *LWT—Food Sci. Technol.* **2019**, *101*, 276–284. [CrossRef]
9. Helal, A.; Tagliazucchi, D. Impact of in-vitro gastro-pancreatic digestion on polyphenols and cinnamaldehyde bioaccessibility and antioxidant activity in stirred cinnamon-fortified yogurt. *LWT—Food Sci. Technol.* **2018**, *89*, 164–170. [CrossRef]
10. Jaziri, I.; Ben Slama, M.; Mhadhbi, H.; Urdaci, M.C.; Hamdi, M. Effect of green and black teas (*Camellia sinensis* L.) on the characteristic microflora of yogurt during fermentation and refrigerated storage. *Food Chem.* **2009**, *112*, 614–620. [CrossRef]
11. Najgebauer-Lejko, D.; Sady, M.; Grega, T.; Walczycka, M. The impact of tea supplementation on microflora, pH and antioxidant capacity of yoghurt. *Int. Dairy J.* **2011**, *21*, 568–574. [CrossRef]
12. Amirdivani, S.; Baba, A.S.H. Green tea yogurt: Major phenolic compounds and microbial growth. *J. Food Sci. Technol.* **2015**, *52*, 4652–4660. [CrossRef] [PubMed]
13. Chouchouli, V.; Kalogeropoulos, N.; Konteles, S.J.; Karvela, E.; Makris, D.P.; Karathanos, V.T. Fortification of yoghurts with grape (*Vitis vinifera*) seed extracts. *LWT—Food Sci. Technol.* **2013**, *53*, 522–529. [CrossRef]
14. Abdel-Hamid, M.; Romeih, E.; Huang, Z.; Enomoto, T.; Huang, L.; Li, L. Bioactive properties of probiotic set-yogurt supplemented with Siraitia grosvenorii fruit extract. *Food Chem.* **2020**, *303*, 125400. [CrossRef]

15. Amirdivani, S.; Baba, A.S. Changes in yogurt fermentation characteristics, and antioxidant potential and in vitro inhibition of angiotensin-1 converting enzyme upon the inclusion of peppermint, dill and basil. *LWT—Food Sci. Technol.* **2011**, *44*, 1458–1464. [CrossRef]

16. Ramchandran, L.; Shah, N.P. Growth, proteolytic, and ACE-I activities of Lactobacillus delbrueckii ssp. bulgaricus and Streptococcus thermophilus and rheological properties of low-fat yogurt as influenced by the addition of Raftiline HP®. *J. Food Sci.* **2008**, *73*, 368–374. [CrossRef]

17. Sah, B.N.P.; Vasiljevic, T.; McKechnie, S.; Donkor, O.N. Antibacterial and antiproliferative peptides in synbiotic yogurt-Release and stability during refrigerated storage. *J. Dairy Sci.* **2016**, *99*, 4233–4242. [CrossRef] [PubMed]

18. *AOAC Official Methods of Analysis of AOAC International*, 17th ed.; AOAC International: Washington, DC, USA, 2000.

19. Türköz Acar, E.; Celep, M.E.; Charehsaz, M.; Akyüz, G.S.; Yeşilada, E. Development and validation of a high-performance liquid chromatography–diode-array detection method for the determination of eight phenolic constituents in extracts of different wine species. *Turk. J. Pharm. Sci.* **2018**, *15*, 22–28. [CrossRef] [PubMed]

20. Abdel-Hamid, M.; Romeih, E.; Gamba, R.R.; Nagai, E.; Suzuki, T.; Koyanagi, T.; Enomoto, T. The biological activity of fermented milk produced by Lactobacillus casei ATCC 393 during cold storage. *Int. Dairy J.* **2019**, *91*, 1–8. [CrossRef]

21. Romeih, E.A.; Abdel-Hamid, M.; Awad, A.A. The addition of buttermilk powder and transglutaminase improves textural and organoleptic properties of fat-free buffalo yogurt. *Dairy Sci. Technol.* **2014**, *94*, 297–309. [CrossRef]

22. Gao, H.-X.; Yu, Z.-L.; He, Q.; Tang, S.-H.; Zeng, W.-C. A potentially functional yogurt co-fermentation with Gnaphalium affine. *LWT—Food Sci. Technol.* **2018**, *91*, 423–430. [CrossRef]

23. Behrad, S.; Yusof, M.Y.; Goh, K.L.; Baba, A.S. Manipulation of probiotics fermentation of yogurt by cinnamon and licorice: Effects on yogurt formation and inhibition of Helicobacter pylori growth in vitro. *World Acad. Sci. Eng. Technol.* **2009**, *36*, 590–594. [CrossRef]

24. Illupapalayam, V.V.; Smith, S.C.; Gamlath, S. Consumer acceptability and antioxidant potential of probiotic-yogurt with spices. *LWT—Food Sci. Technol.* **2014**, *55*, 255–262. [CrossRef]

25. Tripathi, M.K.; Giri, S.K. Probiotic functional foods: Survival of probiotics during processing and storage. *J. Funct. Foods* **2014**, *9*, 225–241. [CrossRef]

26. Oliveira, A.; Alexandre, E.M.C.; Coelho, M.; Lopes, C.; Almeida, D.P.F.; Pintado, M. Incorporation of strawberries preparation in yoghurt: Impact on phytochemicals and milk proteins. *Food Chem.* **2015**, *171*, 370–378. [CrossRef] [PubMed]

27. Shori, A.B.; Rashid, F.; Baba, A.S. Effect of the addition of phytomix-3+ mangosteen on antioxidant activity, viability of lactic acid bacteria, type 2 diabetes key-enzymes, and sensory evaluation of yogurt. *LWT—Food Sci. Technol.* **2018**, *94*, 33–39. [CrossRef]

28. George Thompson, A.M.; Iancu, C.V.; Nguyen, T.T.H.; Kim, D.; Choe, J. Inhibition of human GLUT1 and GLUT5 by plant carbohydrate products; insights into transport specificity. *Sci. Rep.* **2015**, *5*, 12804. [CrossRef]

29. Hashemzaei, M.; Far, A.D.; Yari, A.; Heravi, R.E.; Tabrizian, K.; Taghdisi, S.M.; Sadegh, S.E.; Tsarouhas, K.; Kouretas, D.; Tzanakakis, G.; et al. Anticancer and apoptosis-inducing effects of quercetin in vitro and in vivo. *Oncol. Rep.* **2017**, *38*, 819–828. [CrossRef]

30. Papoutsi, Z.; Kassi, E.; Chinou, I.; Halabalaki, M.; Skaltsounis, L.A.; Moutsatsou, P. Walnut extract (*Juglans regia* L.) and its component ellagic acid exhibit anti-inflammatory activity in human aorta endothelial cells and osteoblastic activity in the cell line KS483. *Br. J. Nutr.* **2008**, 715–722. [CrossRef]

31. Mao, L.M.; Qi, X.W.; Hao, J.H.; Liu, H.F.; Xu, Q.H.; Bu, P.L. In vitro, ex vivo and in vivo anti-hypertensive activity of *Chrysophyllum cainito* L. Extract. *Int. J. Clin. Exp. Med.* **2015**, *8*, 17912–17921.

32. Balasuriya, N.; Rupasinghe, H.P.V. Antihypertensive properties of flavonoid-rich apple peel extract. *Food Chem.* **2012**, *135*, 2320–2325. [CrossRef] [PubMed]

33. Kang, N.; Lee, J.; Lee, W.; Ko, J.; Kim, E.; Kim, J.; Heu, M.; Hoon, G.; Jeon, Y. Gallic acid isolated from Spirogyra sp. improves cardiovascular disease through a vasorelaxant and antihypertensive effect. *Environ. Toxicol. Pharmacol.* **2015**, *39*, 764–772. [CrossRef] [PubMed]

34. Shaw, H.; Wu, J.; Wang, M. Antihypertensive effects of Ocimum gratissimum extract: Angiotensin-converting enzyme inhibitor in vitro and in vivo investigation. *J. Funct. Foods* **2017**, *35*, 68–73. [CrossRef]

35. Zhu, X.H.; Xia, M.; Yang, C.X. Optimizing the Extraction Process of Rubusoside from the Rubus Suavissimus: A Cellulase Pretreatment Approach. *Appl. Mech. Mater.* **2014**, *707*, 144–148. [CrossRef]

36. Liu, R.; Finley, J. Potential cell culture models for antioxidant research. *J. Agric. Food Chem.* **2005**, *53*, 4311–4314. [CrossRef]

Article

Preparation and Characterization of Soy Isoflavones Nanoparticles Using Polymerized Goat Milk Whey Protein as Wall Material

Mu Tian [1], Cuina Wang [1], Jianjun Cheng [1], Hao Wang [1], Shilong Jiang [2] and Mingruo Guo [1,3,*]

[1]   Key Laboratory of Dairy Science, Northeast Agricultural University, Harbin 150030, China; tm18646175069@163.com (M.T.); wangcuina@jlu.edu.cn (C.W.); cheng577@163.com (J.C.); jlu_wh@126.com (H.W.)
[2]   HeiLongJiang FeiHe Dairy Co., Ltd., Beijing 100015, China; jiangshilong@feihe.com
[3]   Department of Nutrition and Food Sciences, College of Agriculture and Life Sciences, University of Vermont, Burlington, VT 05405, USA
*   Correspondence: mguo@uvm.edu

Received: 3 July 2020; Accepted: 23 August 2020; Published: 31 August 2020

Abstract: Soy isoflavones (SIF) are a group of polyphenolic compounds with health benefits. However, application of SIF in functional foods is limited due to its poor aqueous solubility. SIF nanoparticles with different concentrations were prepared using polymerized goat milk whey protein (PGWP) as wall material. The goat milk whey protein was prepared from raw milk by membrane processing technology. The encapsulation efficiencies of all the nanoparticles were found to be greater than 70%. The nanoparticles showed larger particle size and lower zeta potential compared with the PGWP. Fourier Transform Infrared Spectroscopy indicated that the secondary structure of goat milk whey protein was changed after interacting with SIF, with transformation of α-helix and β-sheet to disordered structures. Fluorescence data indicated that interactions between SIF and PGWP decreased the fluorescence intensity. All nanoparticles had spherical microstructure revealed by Transmission Electron Microscope. Data indicated that PGWP may be a good carrier material for the delivery of SIF to improve its applications in functional foods.

Keywords: polymerized goat milk whey protein; soy isoflavones; nanoparticle; physicochemical property

---

1. Introduction

Soy isoflavones (SIF), as bioactive substances, are the main secondary metabolites in soybeans. SIF have been found to possess several potential health benefits such as anticancer [1], reducing menopausal syndrome [2], and preventing osteoporosis [3]. Because of their health benefits, SIF have been recommended as a functional ingredient for formulation of healthy foods and pharmaceutical products. However, due to its low solubility in water, poor bioavailability, and high susceptibility to degradation under oxygen, light, and heating conditions [4], the application of SIF in the food industry is limited. Several approaches have been developed to solve these problems. Functional SIF nanoparticles were prepared by antisolvent precipitation method to improve the water dissolution rate [4]. Wang et al. [5] reported that SIF were microencapsulated in gel beads of soybean hull polysaccharides.

Goat milk whey protein has attracted increasing interest in recent years. The major components in goat milk whey protein are β-lactoglobulin (β-LG) and α-lactalbumin (α-LA), which have excellent emulsifying and foaming properties [6,7]. In recent years, membrane processing technology has been widely used in the dairy industry in an industrial scale due to its low energy consumption, low temperatures, and reduction of environmental contaminants. The goat milk whey protein was prepared from clarified cheese whey by microfiltration (MF) and ultrafiltration (UF) technology, which has good gelation and emulsifying properties [8]. The emulsifying property of goat milk whey protein can be improved by heat treatments [9]. During heating, the polypeptide chains of protein unfolded, and the sulfhydryl groups were exposed to form polymerized whey proteins (PWP) which can be used as an encapsulating material with improved functional properties [10,11]. Because of the protein surface-active property, PWP has a strong affinity towards different ligands [12]. Information about the application of polymerized goat milk whey protein (PGWP) as a bioactive compound carrier is very limited.

Therefore, the aim of this work was to prepare SIF nanoparticles using PGWP as wall material. The PGWP was prepared directly from goat milk by membrane processing technology and the physicochemical properties of nanoparticles were characterized.

2. Materials and Methods

2.1. Materials

Raw goat milk (≥8.15% nonfat solids, 3.86% protein, and 4.02% fat, *w/v*) was purchased from a local market (Feihe Dairy Industry Co., Ltd., Harbin, China). Soy isoflavones (SIF, ≥80% purity) were provided by Beijing Solarbio Science and Technology Co., Ltd. (Beijing, China). Deionized water was prepared by a water filtration device (Millipore Corp., Bedford, MA, USA).

2.2. Nanoparticle Preparation

2.2.1. Preparation of Goat Milk Whey Protein Concentrate

Raw goat milk was heated to 55 °C and skimmed with a separator (SA 10-T, Frautech SRL, Thiene, Italy) to obtain skimmed goat milk and cream. The skimmed goat milk was filtered by microfiltration (MF, 0.1 μm, 50 °C). MF permeate was ultrafiltrated (UF) by a cut-off 10 kDa spiral-wound membrane to 10-fold, and the UF retentate was electrodialyzed (ED) to remove 85% of salt [13]. Subsequently, the concentrated goat milk whey protein was freeze dried in a freeze dryer (Alpha 1-2, Marin Christ Inc., Osterode, Germany) to obtain goat milk whey protein powder (80.99% protein, 18.67% lactose, and 0.34% ash, *w/w*).

2.2.2. Preparation of Polymerized Goat Milk Whey Protein

The goat milk whey protein powder was dissolved in deionized water to obtain a 10% (*w/v*) goat milk whey protein solution and stored at 4 °C for 12 h to complete hydration. The solution was adjusted to pH 7.7 with 1 M sodium hydroxide, heated to 75 °C for 25 min with continuous stirring, and was then quickly cooled to room temperature and marked as polymerized goat milk whey protein (PGWP).

2.2.3. Preparation of Soy Isoflavones Solution

The soy isoflavones (SIF) were dissolved in 70% ethanol (5 mg/mL) by stirring using a magnetic stirrer (IKA, Staufen, Germany), and then the solution was heated to 50 °C for 1 h to obtain a clear solution. The container was wrapped with aluminum foil.

2.2.4. PGWP-SIF Nanoparticle Preparation

The PGWP-SIF solutions were prepared by combining PGWP solution, SIF solution, and deionized water to get different concentrations of SIF (2.1, 2.4, 2.7, and 3.0 mg/mL), while the content of PGWP remained at 40 mg/mL. All samples were stirred for 2 h to obtain a stable system. The mixed solution was treated to remove ethanol by nitrogen gas, and deionized water was added to maintain the original volume, and it was stored in darkness. In this study, samples of SIF-loaded PGWP nanoparticles with different SIF concentrations were termed as PGWP-SIF-A (2.1 mg/mL), PGWP-SIF-B (2.4 mg/mL),

PGWP-SIF-C (2.7 mg/mL), and PGWP-SIF-D (3.0 mg/mL), respectively. PGWP (40 mg/mL) was set as a control.

2.3. Encapsulation Efficiency Determination

The encapsulation efficiency of SIF within PGWP was measured using a previous method with some modifications [14]. The PGWP-SIF solutions were centrifuged at 5500× *g* for 20 min (25 °C), and the concentration of SIF in the supernatant was measured at 261 nm using a UV-visible spectrophotometer (UV-2550, Shimadzu, Tokyo, Japan). The linear regression of absorption versus concentration was conducted, and the regression equation was calculated. The equation is y = 0.1079x − 0.0223, R^2 = 0.9991. The encapsulation efficiency was calculated as follows in Equation (1):

$$\text{Encapsulation efficiency (\%)} = C_0/C \times 100 \tag{1}$$

where C_0 is the concentration of SIF in the supernatant after centrifugation, and C is the concentration of SIF in the nanoparticle.

2.4. Particle Size and Zeta Potential Analysis

Particle size and zeta potential were carried out using a Malvern Zetasizer Nano ZS90 (Malvern Instruments Ltd., Worcestershire, UK). The PGWP and PGWP-SIF solutions were diluted to a protein concentration of 0.1% with deionized water [15]. The refractive indexes for protein and water were 1.450 and 1.333, respectively. All measurements were performed in triplicate.

2.5. Rheological Properties Measurement

Rheological properties of the PGWP and PGWP-SIF solutions were analyzed by a rheometer (Thermo Rheometer, San Jose, CA, USA) equipped with a diameter of 35 mm plate at 25 °C, according to Khan et al. [11]. Flow ramp measurement was carried out using shear rate from 0.1 to 300 s^{-1}. For peak hold analysis, apparent viscosity data was recorded by keeping shear rate at 200 s^{-1} for 60 s.

2.6. Fourier Transform Infrared (FT-IR) Spectroscopy

Fourier Transform Infrared (FT-IR) spectra of samples were obtained using a FTIR spectrometer (Thermo Electron Scientific Instruments Corporation, San Jose, CA, USA) with a pressurized tablet method. The SIF, PGWP, and PGWP-SIF solutions were pre-frozen at −80 °C for 4 h and freeze dried at 4 °C for 12 h. Solid samples were mixed with potassium bromide (KBr) and ground into fine powder. The wavenumber ranged from 4000 to 400 cm^{-1} with a resolution of 4 cm^{-1} and 32 scans [16]. The spectral region ranges of 1600–1700 cm^{-1} were applied to calculate the secondary structure of protein using Peak FIT software. Bands between 1610–1637 cm^{-1} and 1680–1692 cm^{-1} belong to β-sheet; bands between 1638–1648 cm^{-1} belong to random coil; bands between 1649–1660 cm^{-1} belong to α-helix; and bands between 1660–1680 cm^{-1} belong to β-turn. The band area was calculated using the Gaussian function [17].

2.7. Fluorescence Spectroscopy

Fluorescence measurement was performed using an F-7000 fluorescence spectrophotometer (Hitachi Ltd., Tokyo, Japan). The excitation wavelength was set at 280 nm, and the emission was collected from 300 to 500 nm with both slit width at 2.5 nm. The emission spectra were collected at a photomultiplier tube voltage of 500 V with scan rate at 240 nm/min. The PGWP and PGWP-SIF solutions were incubated in water baths at 298, 303, and 308 K for 30 min to achieve equilibrium before measuring [18]. Synchronous fluorescence spectra were recorded at 260–320 nm (Δλ = 15 nm) and 240–320 nm (Δλ = 60 nm) [19].

2.8. Differential Scanning Calorimetry (DSC)

Thermal properties of the samples were analyzed using Differential Scanning Calorimetry (Mettler Toledo, DSC 3, Zurich, Switzerland) according to method by Khan et al. [11] with some modifications. All solid samples (about 5 mg) were sealed in aluminum pans and heated from 30 to 300 °C at 10 °C/min under a nitrogen flow rate of 50 mL/min. An empty sealed aluminum pan was used as a control.

2.9. Transmission Electron Microscopy (TEM)

Images of samples were obtained using transmission electron microscopy (H-7650, Hitachi High-Technologies, Tokyo, Japan) as described by Ghorbani Gorji et al. [20]. The PGWP and PGWP-SIF solutions were diluted to the appropriate concentration, and 10 µL of the sample was placed on a carbon copper and dyed with a negative staining method. The sample was air dried before imaging.

2.10. Statistical Analysis

Data of triplicate experiments were statistically analyzed and expressed as mean ± standard deviation. Analysis of variance ($P < 0.05$) and Tukey's test were carried out using SPSS 20 software (SPSS Inc., Chicago, IL, USA).

3. Results

3.1. Encapsulation Efficiency

Encapsulation efficiency of the nanoparticles is shown in Figure 1. The encapsulation efficiencies of all samples were greater than 70%, and the SIF concentration at 2.4 mg/mL had the highest encapsulation efficiency (89%). The encapsulation efficiency decreased as SIF content increased, suggesting that a portion of SIF were not embedded into the polymerized goat milk whey protein matrix at higher SIF concentration. Similar findings were reported by Patel et al. [21] who found that as curcumin concentration increased, encapsulation efficiency of zein-curcumin decreased.

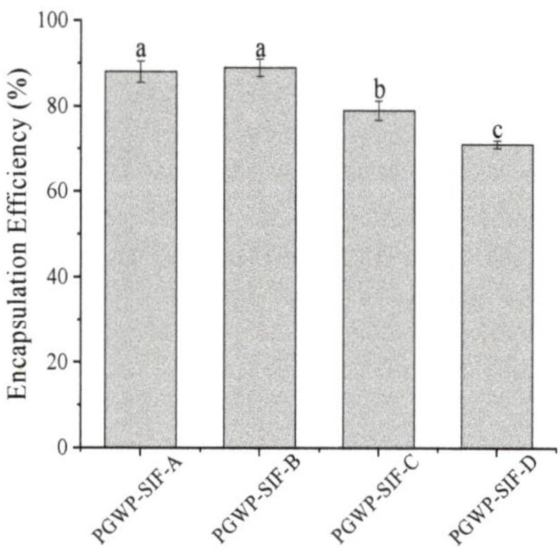

Figure 1. Effect of the soy isoflavone (SIF) contents on encapsulation efficiency of polymerized goat milk whey protein (PGWP)-SIF. Different subscript letters indicate a significant difference ($P < 0.05$). Error bars represent standard deviation of the means.

3.2. Particle Size and Zeta Potential

Particle size of PGWP and PGWP-SIF are shown in Figure 2A. All PGWP-SIF samples showed larger particle sizes than that of PGWP. This can be attributed to the fact that the PGWP had more hydrophobic groups, and SIF were entrapped in or absorbed on protein to form compact nanoparticles [22].

The particle size of nanoparticles was dependent on the concentrations of SIF in PGWP-SIF. At low SIF concentrations (at 2.1 and 2.4 mg/mL), there was no significant difference ($P > 0.05$) in particle size, which may be because the majority of SIF were entered into the hydrophobic core of protein. However, at higher SIF concentrations (the SIF concentration at 2.7 and 3.0 mg/mL), a significant ($P < 0.05$) increase in particle size was found, which suggested that more SIF were absorbed at the surface of PGWP until the surface was saturated. The phenomenon was similar to previous results reported by Rodríguez et al. [23], where at low green tea polyphenols contents, the particle size was maintained that of the β-lactoglobulin, and at large green tea polyphenols contents, the particle size was increased with the increasing of green tea polyphenol concentration.

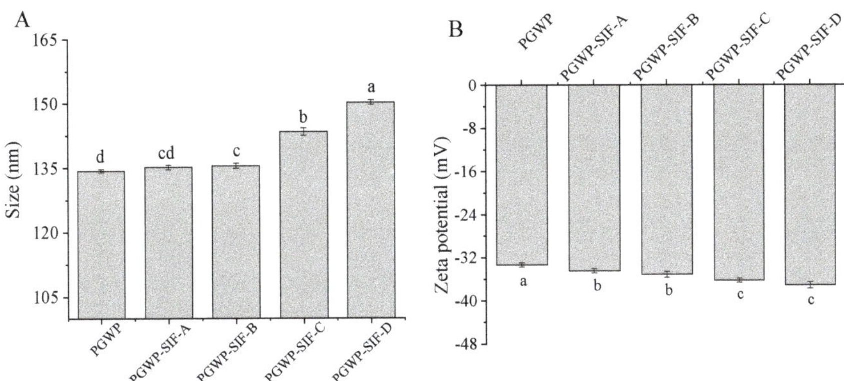

Figure 2. Effects of the SIF contents on particle size (**A**) and zeta potential (**B**) of PGWP-SIF. Different subscript letters indicate a significant difference ($P < 0.05$). Error bars represent standard deviation of the means.

Zeta potential is related to the stability of the systems. Values of zeta potential below –30 mV indicates a stable solution, which may be due to strong electrostatic repulsion [24]. Values of zeta potential in PGWP and PGWP-SIF are shown in Figure 2B. The PGWP-SIF showed lower value of zeta potential than PGWP ($P < 0.05$), suggesting that SIF adhered to PGWP and reduced the negative charge of PGWP. A similar tendency was reported by Von Staszewski et al. [25], who observed that the addition of green tea polyphenols resulted in a decrease in zeta potential values of the β-lactoglobulin-green tea polyphenols complexes. Increasing SIF content from 2.1 mg/mL to 3.0 mg/mL showed an increase of negative charge among PGWP-SIF samples, which indicated that the large number of the embedded SIF entrapped in the core had not affected the surface charge of protein [11].

3.3. Rheological Properties

The rheological properties of PGWP and PGWP-SIF are shown in Figure 3. All samples showed shear-thinning behavior (Figure 3A). This may be due to that the interaction between particles were decreased at high shear rates, causing a decrease in the size of dispersions, and thus leading to a decrease in the viscosity [26]. All PGWP-SIF showed significantly higher ($P < 0.05$) viscosity at 200 s^{-1} compared with the PGWP (Figure 3B). The increase in viscosity of the nanoparticles can be attributed to the entrapment of SIF into the PGWP networks. Values of viscosity at 200 s^{-1} for PGWP-SIF were increased ($P < 0.05$) as SIF content increased from 2.1 mg/mL to 3.0 mg/mL.

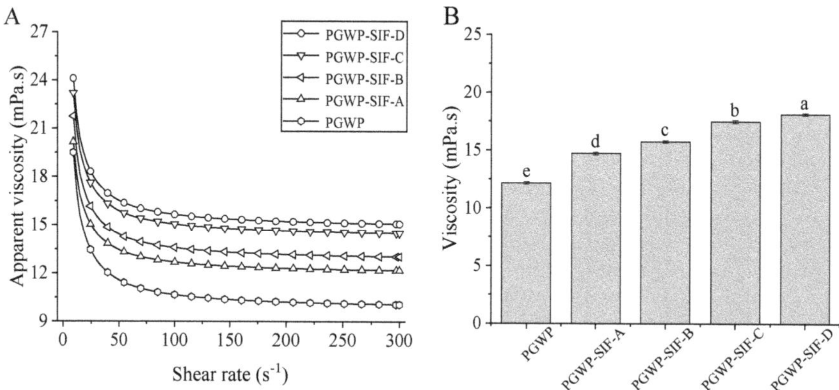

Figure 3. Effects of the SIF contents on flow behavior (**A**) and viscosity at 200 s^{-1} (**B**) of PGWP-SIF. Different subscript letters indicate a significant difference ($P < 0.05$). Error bars represent standard deviation of the means.

3.4. FT-IR Spectra

FT-IR analysis was used to determine the structural changes of PGWP when interacted with SIF. The amide I band 1600–1700 cm^{-1} (C=O stretch), amide II bands 1500–1600 cm^{-1} (C-N stretching combined with N-H bending modes), and amide A band at 3300 cm^{-1} (N-H stretching and hydrogen bonds) were used to explore the changes of the secondary structure of goat milk whey protein in PGWP-SIF [27]. FT-IR spectra of all samples were shown in Figure 4A. FT-IR spectra of the SIF showed characteristic peaks at 1623.77, 1516.31, and 3354.15 cm^{-1}. The PGWP showed bands at 1655.03, 1541.30, and 3303.87 cm^{-1}, which may belong to amide I, amide II, and amide A. After interacting with SIF, a hypsochromic shift occurred for the absorption band of PGWP-SIF. The peak in PGWP at 1655.03 cm^{-1} was shifted to 1655.15 cm^{-1} for PGWP-SIF, indicating that the addition of SIF affected the formation of the PGWP-SIF through the interaction related to C=O between PGWP and SIF. The peak at 1541.30 cm^{-1} (PGWP) shifted to 1541.95 cm^{-1} (PGWP-SIF) and may be due to the interactions between PGWP and SIF through C-N and N-H groups. The peak at 3303.87 cm^{-1} (PGWP) shifted to 3308.22 cm^{-1} (PGWP-SIF) and indicated a formation of hydrogen bonds between PGWP and SIF, suggesting that phenolic hydroxyl groups were involved in the non-covalent interaction between PGWP and SIF.

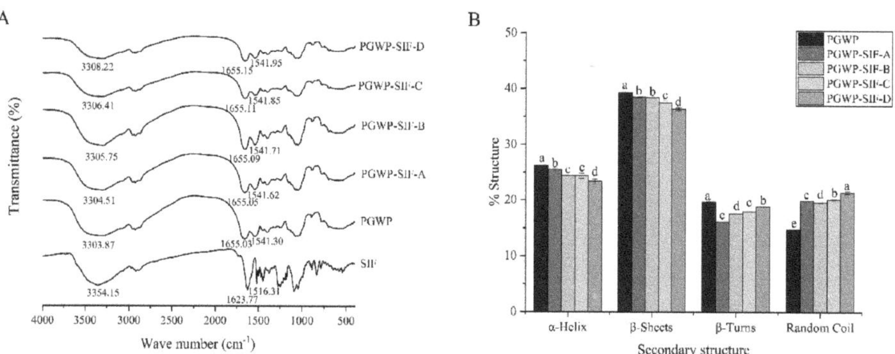

Figure 4. FT-IR spectra of SIF, PGWP, and PGWP-SIF (**A**) and the contents of secondary structures of PGWP and PGWP-SIF (**B**). Different subscript letters indicate a significant difference ($P < 0.05$). Error bars represent standard deviation of the means.

Contents of the secondary structure of proteins were calculated by using Amide I band. The results are shown in Figure 4B. When compared with PGWP, content of α-helix and β-sheet in PGWP-SIF decreased from $26.20 \pm 0.07\%$ to $23.39 \pm 0.42\%$, and $39.32 \pm 0.06\%$ to $36.36 \pm 0.25\%$, respectively. Results indicated that secondary structures of goat milk whey protein were transformed from ordered to disordered when combined with SIF.

3.5. Fluorescence Spectra

3.5.1. Inherent Fluorescence

Fluorescence intensity of PGWP and PGWP-SIF are shown in Figure 5. The PGWP-SIF showed lower fluorescence intensity than PGWP, indicating that SIF quenched the intrinsic fluorescence of PGWP. The fluorescence intensity decreased as SIF contents increased. Additionally, in the presence of SIF, the maximum peak position slightly shifted to a larger wavelength from 335 nm to 339 nm, which indicated that more fluorophores were exposed to the solvent, and thus implying a change in the polarity of tryptophan residues. The results suggested that some hydrophobic residues were buried during the interaction of SIF with PGWP, leading to the changes of structure in PGWP, and eventually improving the hydrophilicity of the medium.

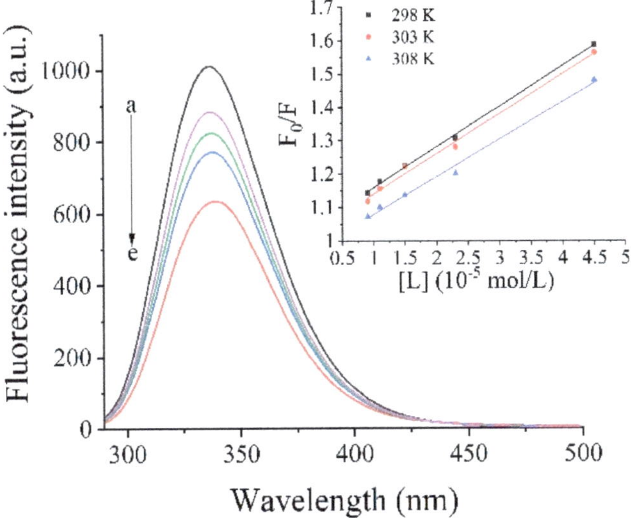

Figure 5. Fluorescence emission spectra of PGWP-SIF systems in 10 mM PB pH 7.4. a–e, PGWP, PGWP-SIF-A, PGWP-SIF-B, PGWP-SIF-C, PGWP-SIF-D. The Stern–Volmer plots for the quenching of PGWP by SIF at different temperatures.

3.5.2. Stern–Volmer Analysis of Quenching Data

To further understand the interactions between PGWP and SIF, the Stem-Volmer equation was used to analyze the data at temperatures of 298, 303, and 308 K, as follows in Equation (2):.

$$F_0/F = 1 + Ksv\,[L] = 1 + Kq\,\tau_0\,[L] \tag{2}$$

where F_0 and F are fluorescence intensities of PGWP and PGWP-SIF, respectively; [L] is the concentration of SIF; Ksv is the Stern–Volmer quenching constant; Kq is the biological macromolecules quenching constant; and τ_0 is the average lifetime of the biomolecule without quencher ($\tau_0 = 10^{-8}$ s) [28].

It was reported that Kq larger than 2.0×10^{10} L/mol/s indicated static quenching interaction, while q smaller than 2.0×10^{10} L/mol/s indicated dynamic quenching interaction. Additionally, for static quenching, higher temperature indicated lower quenching constant, while dynamic quenching had the opposite trend [29]. From Table 1, it can be seen that values of Kq were greater than 2.0×10^{10} L/mol/s, and the Ksv decreased at higher temperatures, which suggested that the interaction was static quenching.

Table 1. Relevant parameters of the Stern–Volmer quenching constant (Ksv), the biological macromolecules quenching constant (Kq), the binding constant (Ka), and the number of binding sites (n) were calculated from Stern–Volmer and double log plots at different temperatures. Thermodynamic parameters of enthalpy (ΔH), entropy (ΔS) and free energy (ΔG) changed based on the van't Hoff equation.

T (K)	Ksv (10^4 L/mol)	Kq (10^{12} L/mol)	R^2	Ka (10^5 L/mol)	n	R^2	ΔH (KJ/mol)	ΔG (KJ/mol)	ΔS (J/mol.K)
298	1.22 ± 0.1	1.22 ± 0.1	0.9988	7.80 ± 0.1	1.13 ± 0.02	0.9969	7.81 ± 0.1	-33.74 ± 0.1	139.42 ± 1.2
303	1.20 ± 0.2	1.20 ± 0.2	0.9932	8.22 ± 0.1	0.93 ± 0.01	0.9873		-34.43 ± 0.2	
308	1.13 ± 0.2	1.13 ± 0.2	0.9927	8.34 ± 0.1	0.86 ± 0.02	0.9929		-35.13 ± 0.2	

For static quenching, the binding constant (Ka) and the number of binding sites (n) conformed to the Equation (3) [30]. The slope and the intercept values of a plot of log $[(F_0 - F)/F]$ versus log [L] give n and Ka values, respectively.

$$\log ((F_0 - F)/F) = \log Ka + n \log [L] \tag{3}$$

Values of n and Ka parameters are shown in Table 1. Values of n were close to 1, indicating that PGWP had only one binding site for SIF. On the other hand, values of Ka increased as temperature increased, indicating endothermic binding reaction. The result indicated that the stability of PGWP-SIF increased as temperature increased. Our results were similar with findings of Jia et al. [19], who reported that the interaction of β-lactoglobulin with epigallocatechin-3-gallate was endothermic and the stability increased as temperature increased.

3.5.3. Thermodynamic Parameters

The four main interaction forces between polyphenols and proteins are hydrogen bonds, electrostatic forces, hydrophobic forces, and van der Waals forces [31]. The main interaction forces can be obtained by calculating the thermodynamic parameters using van't Hoff Equations (4) and (5), as follows:

$$\ln Ka = -\Delta H/RT + \Delta S/R \tag{4}$$

$$\Delta G = \Delta H - T\Delta S \tag{5}$$

where ΔH is the enthalpy change, ΔS is the entropy change, ΔG is free energy change, R is the gas constant (8.314 J/mol/K), T is the reaction temperature, and Ka is the binding constant at the temperatures of 298, 303, and 308 K. ΔH, ΔS, and ΔG could be acquired by Equations (3) and (4) [32].

Ross and Subramanian [33] reported that ΔS and $\Delta H > 0$ indicated hydrophobic forces, ΔS and $\Delta H < 0$ indicated van der Waals forces, and $\Delta H < 0$ and $\Delta S > 0$ indicated electrostatic forces. From Table 1, it can be seen that $\Delta G < 0$, indicating that the interaction between PGWP and SIF was spontaneous [34]. Both ΔH and ΔS values > 0, indicating that hydrophobic interaction was involved in the interaction between PGWP and SIF. This was similar to the findings of Xu et al. [35], who reported that the interaction between β-lactoglobulin and theaflavin/chlorogenic acid/delphinidin-3-O-glucoside was mainly maintained by hydrophobic forces.

3.5.4. Synchronous Fluorescence Spectra

Synchronous fluorescence analysis was used to study effects of SIF on structure of PGWP and obtain information about tyrosine residues or tryptophan residues at synchronous spectrum performed with $\Delta\lambda$ at 15 or 60 nm, respectively [19,35]. Some doubts about the effectiveness of the method were recently reported by Bobone et al. [36]. From Figure 6, it can be seen that as SIF content increased, the fluorescence intensity decreased in both fluorescence spectra, which indicated that the binding of SIF with PGWP exposed more chromophores into the solvent and led to a decrease in the fluorescence intensity.

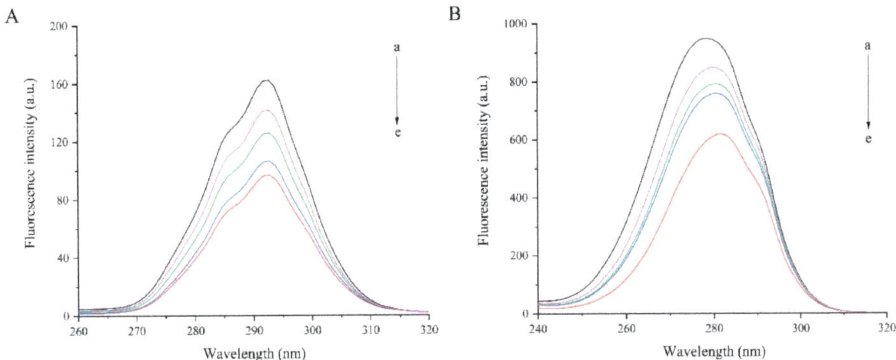

Figure 6. Synchronous fluorescence spectra of PGWP-SIF systems, a–e, PGWP, PGWP-SIF-A, PGWP-SIF-B, PGWP-SIF-C, PGWP-SIF-D. (**A**: $\Delta\lambda$ = 15 nm, **B**: $\Delta\lambda$ = 60 nm).

3.6. Differential Scanning Calorimetry (DSC)

DSC analysis was used to investigate the thermal property of nanoparticles. DSC curves of all samples were shown in Figure 7. The pure SIF showed four endothermic peaks at approximately 85.17, 136.85, 178.67, and 209.83 °C. The first three peaks may be attributed to three molecular forms in SIF, including genistein, daidzein, and glycitein [4]. The PGWP exhibited two endothermic peaks at 87.67 °C and 155.17 °C, and the PGWP-SIF showed two endothermic peaks at 105.37 °C and 157.04 °C. The result indicated that the thermal stability of the PGWP-SIF improved. Moreover, the characteristic peaks of SIF disappeared in PGWP-SIF, which may be because SIF were encapsulated into PGWP microspheres. Yang et al. [27] also reported that the disappearance of endothermic peaks of pyrogallic acid in the nanoparticle was due to the covalent interactions between pumpkin seed protein isolate and pyrogallic acid.

3.7. Microstructure

Images of PGWP and PGWP-SIF obtained by transmission electron microscope (TEM) are presented in Figure 8. The morphology of the PGWP-SIF was related to the SIF contents. At the low SIF contents (Figure 8B,C), the microspheres were homogeneously spherical in shape with smaller particle size, suggesting that the nanoparticles were homogeneously dispersed. At the large SIF contents (Figure 8D,E), irregularly larger-sized particles were formed, and this may be due to the formation of inter-surface networks between the PGWP-SIF and unencapsulated SIF. The microstructure was consistent with the mean particle size obtained from dynamic light scattering data (Figure 2A).

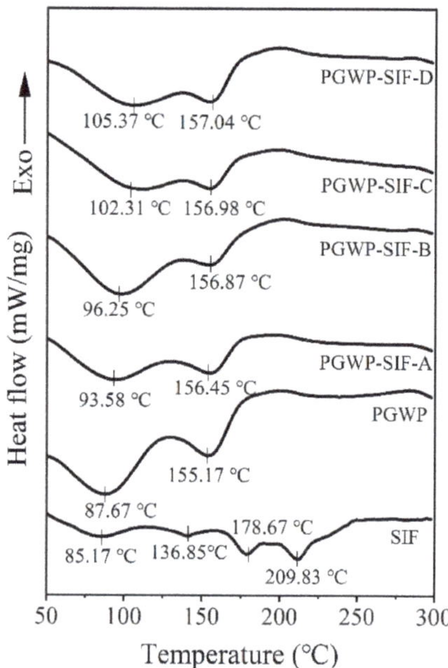

Figure 7. The differential scanning calorimeter (DSC) curves of SIF, PGWP, and PGWP-SIF.

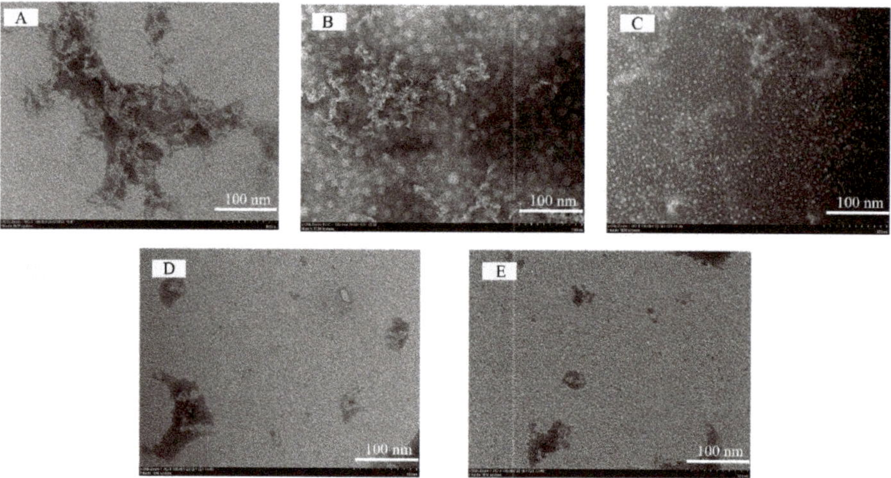

Figure 8. Transmission electron microscopy micrographs of PGWP and PGWP-SIF. PGWP (**A**); PGWP-SIF-A (**B**); PGWP-SIF-B (**C**); PGWP-SIF-C (**D**), and PGWP-SIF-D (**E**).

4. Discussion

Numerous studies have reported the beneficial properties of soy isoflavones (SIF). However, its application in the food industry and pharmaceuticals are limited owing to its low solubility in water and poor bioavailability. Goat milk whey protein has excellent emulsifying and foaming properties [6,7], which can be used to produce bioactive compounds-loaded whey protein to expand their applications in the production of functional food products as well as for designing new drug delivery systems [37].

In this study, we prepared SIF nanoparticles using polymerized goat milk whey protein (PGWP) as wall material.

The encapsulation efficiency is the index used to indicate how many SIF were loaded in the nanoparticles. The encapsulation efficiencies of SIF in the nanoparticles were higher than that obtained from gel beads of soybean hull polysaccharides by Wang et al. [5], which may be attributed to the differences in methods and conditions, and the result indicated that it was feasible to encapsulate SIF using PGWP as wall material. Particle size and zeta potential are important properties to provide valuable information on micro-encapsulated compounds regarding the formation of stable formulations. The particle size of the nanoparticles ranged from 135 nm to 155 nm due to the concentrations of SIF in PGWP-SIF. The values were similar to previous report by Khan et al. [11], who prepared whey protein isolate-DIM (3,3′-Diindolylmethane) nanoparticles to a mean particle size of 96–157 nm. In the study, concomitant observations were obtained from transmission electron microscope (TEM) images and dynamic light scattering (DLS) analyses, the microspheres were homogeneously spherical in shape. However, compared with the dynamic light scattering data, the diameter of particles appeared smaller in the TEM images, which may be due to the different principles of the two analytical methods. In addition, all the nanoparticles had values of zeta potential below −30 mV, which suggested that the samples seemed to be stabilized [24]. The nanoparticles with higher surface charge aggregated less, which suggested that they were somewhat more stable. These results suggested that PGWP could be considered as a promising encapsulation agent for the incorporation of bioactive compounds such as SIF.

Fourier transform infrared spectroscopy (FT-IR) analysis can be used to investigate the potential interactions between SIF and PGWP. FT-IR spectra of the SIF showed characteristic peaks were related to the stretching vibration of the aromatic ring and aromatic ketone [5]. The characteristic absorption peaks of SIF in nanoparticles spectrum increased from 1623.77 cm^{-1} and 1516.31 cm^{-1} to 1655.15 cm^{-1} and 1541.95 cm^{-1}, respectively, which suggested that the successful formation of PGWP-SIF nanoparticles. The results were consistent with the report described by Wang et al. [38], who investigated the formation of complexes between spiral dextrin sub-fraction and soy isoflavones. After interacting with SIF, the amide I band of PGWP changed, which may be due to hydrophobic interactions between the aromatic ring of SIF and the hydrophobic amino acids of proteins [39]. The amide I band was the most useful for infrared spectroscopic analysis of the secondary structure of proteins [40]. From Figure 4B, it could be observed that the α-helix and β-sheet contents for PGWP decreased after adding SIF, suggesting that the interactions between PGWP and SIF could lead to the alteration of PGWP secondary structure. It is presumable that SIF were non-covalently grafted onto PGWP, resulting in the partly unfolded protein, which may lead to the exposure of buried amino acids and promote hydrophobic interactions [41]. These results indicated that the secondary structure of goat milk whey protein changed after interacting with SIF, and the interaction between PGWP and SIF was probably through hydrogen bonds or hydrophobic interactions.

To confirm that the conformation of PGWP was changed after interacting with SIF, we studied the fluorescence spectrum. Fluorescence spectrum was widely used to study mechanisms of the interactions between proteins and small molecules and the microenvironment changes of proteins. The main components of whey proteins are β-lactoglobulin (β-LG) and α-lactalbumin (α-LA). Each β-LG molecule has two tryptophan residues and four tyrosine residues, while four tryptophan residues are found in α-LA molecule [42]. The fluorescence of tyrosine and tryptophan was excited by different wavelengths to analyze the structural changes of proteins. Therefore, the inherent fluorescence is a useful approach to study the structural transition and binding properties of protein. Fluorescence experiments proved that SIF quenched PGWP fluorescence strongly in static mode. The nature of the interaction forces between SIF and PGWP can be partially unveiled by studying the thermodynamic parameters of the system, and the results suggested that the main interaction force was hydrophobic interaction [33]. Synchronous fluorescence confirmed that SIF affected the conformation of PGWP by interacting with its tyrosine and tryptophan residues. In addition, tryptophan residues had a stronger fluorescence quenching effect than that of tyrosine residues, which indicated that tryptophan residues played an important role in fluorescence quenching. Combining the fluorescence spectrum and the synchronous fluorescence spectrum, the amino acids involved in the reaction were tryptophan and tyrosine, and it can be inferred that tryptophan and tyrosine residues were involved in the hydrophobic interaction. To further confirm that SIF were encapsulated inside the nanoparticles, we studied the thermal properties of SIF before and after encapsulation. The characteristic peaks of SIF disappeared in that of PGWP-SIF, which may be attributed to that SIF were entrapped in PGWP nanoparticles. These results were in agreement with the report of Wang et al. [38], who observed that the disappearance of the endotherm of soy isoflavones at 184.54 °C in the thermogram of the complexes may be due to the formation of spiral dextrin sub-fraction/soy isoflavones complexes. This work provided some comprehensive understanding about the interactions between SIF and PGWP, and these characteristics made PGWP a novel wall material for the encapsulation of SIF. However, this research has some limitations, the antioxidant activity and in vitro release behavior of the nanoparticles will be carried out in the next study, and molecular modeling study will be performed to predict the precise binding sites of soy isoflavones in goat milk whey protein. Our study indicated that the polymerized goat milk whey protein can be considered as a promising carrier for encapsulating bioactive compounds. The results may be helpful in expanding the industrial application of soy isoflavones in functional foods. The study laid the foundation for further research into the interaction between goat milk whey protein and soy isoflavones.

5. Conclusions

Soy isoflavones nanoparticles using polymerized goat milk whey protein as wall material were prepared and characterized in this study. The results suggested that polymerized goat milk whey protein prepared directly from milk was suitable to encapsulate soy isoflavones with high encapsulation efficiency, and hydrophobic interaction was considered to be the main force in the formation of the nanoparticles. The findings will be helpful for the use of polymerized goat milk whey protein as a carrier material for hydrophobic bioactive compounds.

Author Contributions: M.T.: conceptualization, formal analysis, investigation, resources and writing—original draft; C.W.: formal analysis, investigation, resources and visualization; J.C.: formal analysis and resources; H.W.: formal analysis, investigation, resources and visualization; S.J.: formal analysis and resources; M.G.: conceptualization, funding acquisition, project administration, writing—original draft and writing—review & editing. All authors approved the final version of the manuscript.

Funding: The Project was supported by the Millions of Engineering Science and Technology Major Special Program of Heilongjiang Province (Contract No. 2019ZX07B01).

Conflicts of Interest: The authors declare no conflict of interest.

References

1. Dong, X.; Xu, W.; Sikes, R.A.; Wu, C. Combination of low dose of genistein and daidzein has synergistic preventive effects on isogenic human prostate cancer cells when compared with individual soy isoflavone. *Food Chem.* **2013**, *141*, 1923–1933. [CrossRef] [PubMed]
2. Messina, M. Soy foods, isoflavones, and the health of postmenopausal women. *Am. J. Clin. Nutr.* **2014**, *100*, 423S–430S. [CrossRef] [PubMed]
3. Marini, H.R.; Bitto, A.; Altavilla, D.; Burnett, B.P.; Polito, F.; Di Stefano, V.; Minutoli, L.; Atteritano, M.; Levy, R.M.; D'Anna, R.; et al. Breast Safety and Efficacy of Genistein Aglycone for Postmenopausal Bone Loss: A Follow-Up Study. *J. Clin. Endocrinol. Metab.* **2008**, *93*, 4787–4796. [CrossRef] [PubMed]
4. Zhang, X.; Zhang, H.; Xia, X.; Pu, N.; Yu, Z.; Nabih, M.; Zhu, Y.; Zhang, S.; Jiang, L. Preparation and physicochemical characterization of soy isoflavone (SIF) nanoparticles by a liquid antisolvent precipitation method. *Adv. Powder Technol.* **2019**, *30*, 1522–1530. [CrossRef]
5. Wang, S.; Shao, G.; Yang, J.; Liu, J.; Wang, J.; Zhao, H.; Yang, L.; Liu, H.; Zhua, D.; Li, Y.; et al. The production of gel beads of soybean hull polysaccharides loaded with soy isoflavone and their pH-dependent release. *Food Chem.* **2020**, *313*, 126095. [CrossRef] [PubMed]

6. Pešić, M.B.; Barac, M.; Vrvić, M.; Ristić, N.; Macej, O.; Stanojevic, S.; Kostić, A. The distributions of major whey proteins in acid wheys obtained from caprine/bovine and ovine/bovine milk mixtures. *Int. Dairy J.* **2011**, *21*, 831–838. [CrossRef]

7. Balthazar, C.F.; Pimentel, T.; Ferrão, L.; Almada, C.; Santillo, A.; Albenzio, M.; Mollakhalili, N.; Mortazavian, A.; Nascimento, J.; Silva, M.; et al. Sheep Milk: Physicochemical Characteristics and Relevance for Functional Food Development. *Compr. Rev. Food Sci. Food Saf.* **2017**, *16*, 247–262. [CrossRef]

8. Sanmartín, B.; Díaz, O.; Rodriguez-Turienzo, L.; Cobos, Á. Properties of heat-induced gels of caprine whey protein concentrates obtained from clarified cheese whey. *Small Rumin. Res.* **2015**, *123*, 142–148. [CrossRef]

9. Mohammadi, A.; Jafari, S.M.; Assadpour, E.; Esfanjani, A.F. Nano-encapsulation of olive leaf phenolic compounds through WPC–pectin complexes and evaluating their release rate. *Int. J. Biol. Macromol.* **2016**, *82*, 816–822. [CrossRef]

10. Sun, X.; Wang, C.; Guo, M. Interactions between whey protein or polymerized whey protein and soybean lecithin in model system. *J. Dairy Sci.* **2018**, *101*, 9680–9692. [CrossRef]

11. Khan, A.; Guo, M.; Sun, X.; Killpartrick, A.; Guo, M. Preparation and Characterization of Whey Protein Isolate-DIM Nanoparticles. *Int. J. Mol. Sci.* **2019**, *20*, 3917. [CrossRef] [PubMed]

12. Guo, M.; Zhou, X.; Wang, H.; Sun, X.; Guo, M. Interactions between β-Lactoglobulin and 3,3′-Diindolylmethane in Model System. *Molecules* **2019**, *24*, 2151. [CrossRef]

13. Fang, T.; Shen, X.; Hou, J.; Guo, M. Effects of polymerized whey protein prepared directly from cheese whey as fat replacer on physiochemical, texture, microstructure and sensory properties of low-fat set yogurt. *LWT* **2019**, *115*, 108268. [CrossRef]

14. Chen, L.; Subirade, M. Alginate–whey protein granular microspheres as oral delivery vehicles for bioactive compounds. *Biomaterials* **2006**, *27*, 4646–4654. [CrossRef]

15. Salem, A.; Ramadan, A.R.; Shoeib, T. Entrapment of β-carotene and zinc in whey protein nanoparticles using the pH cycle method: Evidence of sustained release delivery in intestinal and gastric fluids. *Food Biosci.* **2018**, *26*, 161–168. [CrossRef]

16. Al-Hanish, A.; Stanić, D.; Mihailovic, J.; Prodic, I.; Minić, S.; Stojadinovic, M.; Radibratovic, M.; Milčić, M.; Veličković, T. Ćirković Noncovalent interactions of bovine α-lactalbumin with green tea polyphenol, epigalocatechin-3-gallate. *Food Hydrocoll.* **2016**, *61*, 241–250. [CrossRef]

17. Cheng, H.; Liu, H.; Bao, W.; Zou, G. Studies on the interaction between docetaxel and human hemoglobin by spectroscopic analysis and molecular docking. *J. Photochem. Photobiol. B Biol.* **2011**, *105*, 126–132. [CrossRef]

18. Ranamukhaarachchi, S.A.; Peiris, R.H.; Moresoli, C. Fluorescence spectroscopy and principal component analysis of soy protein hydrolysate fractions and the potential to assess their antioxidant capacity characteristics. *Food Chem.* **2017**, *217*, 469–475. [CrossRef]

19. Jia, J.; Gao, X.; Hao, M.; Tang, L. Comparison of binding interaction between β-lactoglobulin and three common polyphenols using multi-spectroscopy and modeling methods. *Food Chem.* **2017**, *228*, 143–151. [CrossRef]

20. Gorji, E.G.; Rocchi, E.; Schleining, G.; Bender, D.; Furtmüller, P.G.; Piazza, L.; Iturri, J.; Toca-Herrera, J.L. Characterization of resveratrol–milk protein interaction. *J. Food Eng.* **2015**, *167*, 217–225. [CrossRef]

21. Patel, A.R.; Hu, Y.; Tiwari, J.K.; Velikov, K.P. Synthesis and characterisation of zein–curcumin colloidal particles. *Soft Matter* **2010**, *6*, 6192–6199. [CrossRef]

22. Nagy, K.; Courtet-Compondu, M.-C.; Williamson, G.; Rezzi, S.; Kussmann, M.; Rytz, A. Non-covalent binding of proteins to polyphenols correlates with their amino acid sequence. *Food Chem.* **2012**, *132*, 1333–1339. [CrossRef] [PubMed]

23. Rodríguez, S.D.; Von Staszewski, M.; Pilosof, A.M. Green tea polyphenols-whey proteins nanoparticles: Bulk, interfacial and foaming behavior. *Food Hydrocoll.* **2015**, *50*, 108–115. [CrossRef]

24. Mantovani, R.A.; Fattori, J.; Michelon, M.; Cunha, R. Formation and pH-stability of whey protein fibrils in the presence of lecithin. *Food Hydrocoll.* **2016**, *60*, 288–298. [CrossRef]

25. Von Staszewski, M.; Jara, F.L.; Ruiz, A.; Jagus, R.J.; De Carvalho, J.E.; Pilosof, A.M. Nanocomplex formation between β-lactoglobulin or caseinomacropeptide and green tea polyphenols: Impact on protein gelation and polyphenols antiproliferative activity. *J. Funct. Foods* **2012**, *4*, 800–809. [CrossRef]

26. Mirhosseini, H.; Tan, C.P. Response surface methodology and multivariate analysis of equilibrium headspace concentration of orange beverage emulsion as function of emulsion composition and structure. *Food Chem.* **2009**, *115*, 324–333. [CrossRef]

27. Yang, C.; Wang, B.; Wang, J.; Xia, S.; Wu, Y. Effect of pyrogallic acid (1,2,3-benzenetriol) polyphenol-protein covalent conjugation reaction degree on structure and antioxidant properties of pumpkin (Cucurbita sp.) seed protein isolate. *LWT* **2019**, *109*, 443–449. [CrossRef]

28. Moeiniafshari, A.-A.; Zarrabi, A.; Bordbar, A.-K. Exploring the interaction of naringenin with bovine beta-casein nanoparticles using spectroscopy. *Food Hydrocoll.* **2015**, *51*, 1–6. [CrossRef]

29. Ghalandari, B.; Divsalar, A.; Saboury, A.A.; Haertlé, T.; Parivar, K.; Bazl, R.; Eslami-Moghadam, M.; Amanlou, M. Spectroscopic and theoretical investigation of oxali–palladium interactions with?-lactoglobulin. *Spectrochim. Acta Part A Mol. Biomol. Spectrosc.* **2014**, *118*, 1038–1046. [CrossRef]

30. Wang, C.; Wu, Q.H.; Wang, Z.; Zhao, J. Study of the interaction of carbamazepine with bovine whey albumin by fluorescence quenching method. *Anal. Sci.* **2006**, *22*, 435–438. [CrossRef]

31. Bordenave, N.; Hamaker, B.R.; Ferruzzi, M.G. Nature and consequences of non-covalent interactions between flavonoids and macronutrients in foods. *Food Funct.* **2014**, *5*, 18–34. [CrossRef] [PubMed]

32. Seedher, N.; Agarwal, P. Complexation of fluoroquinolone antibiotics with human serum albumin: A fluorescence quenching study. *J. Lumin.* **2010**, *130*, 1841–1848. [CrossRef]

33. Ross, P.D.; Subramanian, S. Thermodynamics of protein association reactions: Forces contributing to stability. *Biochemistry* **1981**, *20*, 3096–3102. [CrossRef] [PubMed]

34. Li, Y.; Liu, B.; Jiang, L.; Regenstein, J.M.; Jiang, N.; Poias, V.; Zhang, X.; Qi, B.; Li, A.-L.; Wang, Z.; et al. Interaction of soybean protein isolate and phosphatidylcholine in nanoemulsions: A fluorescence analysis. *Food Hydrocoll.* **2019**, *87*, 814–829. [CrossRef]

35. Xu, J.; Hao, M.; Sun, Q.; Tang, L. Comparative studies of interaction of β-lactoglobulin with three polyphenols. *Int. J. Biol. Macromol.* **2019**, *136*, 804–812. [CrossRef]

36. Bobone, S.; Van De Weert, M.; Stella, L. A reassessment of synchronous fluorescence in the separation of Trp and Tyr contributions in protein emission and in the determination of conformational changes. *J. Mol. Struct.* **2014**, *1077*, 68–76. [CrossRef]

37. Livney, Y.D. Milk proteins as vehicles for bioactives. *Curr. Opin. Colloid Interface Sci.* **2010**, *15*, 73–83. [CrossRef]

38. Wang, P.-P.; Luo, Z.-G.; Peng, X.-C. Encapsulation of Vitamin E and Soy Isoflavone Using Spiral Dextrin: Comparative Structural Characterization, Release Kinetics, and Antioxidant Capacity during Simulated Gastrointestinal Tract. *J. Agric. Food Chem.* **2018**, *66*, 10598–10607. [CrossRef]

39. Ghayour, N.; Hosseini, S.M.H.; Eskandari, M.H.; Esteghlal, S.; Nekoei, A.-R.; Gahruie, H.H.; Tatar, M.; Naghibalhossaini, F. Nanoencapsulation of quercetin and curcumin in casein-based delivery systems. *Food Hydrocoll.* **2019**, *87*, 394–403. [CrossRef]

40. Fang, Y.; Dalgleish, D.G. Conformation of β-lactoglobulin studied by FTIR: Effect of pH, temperature, and adsorption to the oil-water interface. *J. Colloid Interface Sci.* **1997**, *196*, 292–298. [CrossRef]

41. Fan, Y.; Liu, Y.; Gao, L.; Zhang, Y.; Yi, J. Oxidative stability and in vitro digestion of menhaden oil emulsions with whey protein: Effects of EGCG conjugation and interfacial cross-linking. *Food Chem.* **2018**, *265*, 200–207. [CrossRef] [PubMed]

42. Mandeville, J.S.; Tajmir-Riahi, H.A. Nanocomplexes of dendrimers with bovine whey albumin. *Biomacromolecules* **2010**, *11*, 465–472. [CrossRef] [PubMed]

Review

Comparative Proteomics of Milk Fat Globule Membrane (MFGM) Proteome across Species and Lactation Stages and the Potentials of MFGM Fractions in Infant Formula Preparation

Michele Manoni [1], Chiara Di Lorenzo [2], Matteo Ottoboni [1], Marco Tretola [3] and Luciano Pinotti [1,4,*]

1   Department of Health, Animal Science and Food Safety, VESPA, University of Milan, 20134 Milan, Italy; michele.manoni@unimi.it (M.M.); matteo.ottoboni@unimi.it (M.O.)
2   Department of Pharmacological and Biomolecular Sciences, University of Milan, 20133 Milan, Italy; chiara.dilorenzo@unimi.it
3   Agroscope, Institute for Livestock Sciences, 1725 Posieux, Switzerland; marco.tretola@agroscope.admin.ch
4   CRC I-WE (Coordinating Research Centre: Innovation for Well-Being and Environment), University of Milan, 20134 Milan, Italy
*   Correspondence: luciano.pinotti@unimi.it; Tel.: +39-02-503-15742

Received: 20 July 2020; Accepted: 4 September 2020; Published: 7 September 2020

Abstract: Milk is a lipid-in-water emulsion with a primary role in the nutrition of newborns. Milk fat globules (MFGs) are a mixture of proteins and lipids with nutraceutical properties related to the milk fat globule membrane (MFGM), which protects them, thus preventing their coalescence. Human and bovine MFGM proteomes have been extensively characterized in terms of their formation, maturation, and composition. Here, we review the most recent comparative proteomic analyses of MFGM proteome, above all from humans and bovines, but also from other species. The major MFGM proteins are found in all the MFGM proteomes of the different species, although there are variations in protein expression levels and molecular functions across species and lactation stages. Given the similarities between the human and bovine MFGM and the bioactive properties of MFGM components, several attempts have been made to supplement infant formulas (IFs), mainly with polar lipid fractions of bovine MFGM and to a lesser extent with protein fractions. The aim is thus to narrow the gap between human breast milk and cow-based IFs. Despite the few attempts made to date, supplementation with MFGM proteins seems promising as MFGM lipid supplementation. A deeper understanding of MFGM proteomes should lead to better results.

Keywords: milk fat globules; bovine milk proteins; milk fat globule membrane; comparative proteomics; infant formula preparation

---

1. Introduction

Bovine milk is an oil-in-water emulsion and is rich in nutrients and bioactive factors. Its unique composition makes it essential for the correct growth and development of newborns [1]. The main milk components are water, fat, proteins (casein micelles and serum proteins such as α-lactalbumin, β-lactoglobulin, blood serum albumin, lactoferrin, enzymes, and immunoglobulins), lactose, and minerals [2,3]. Milk fat occurs as milk fat globules (MFGs) in the water, with a size ranging from 0.1 to 15 μm. MFGs are composed of a nonpolar triglyceride (TG) core and are covered in a layer of surface-active material, which is needed to maintain their stability in the emulsion and to protect them from enzymatic degradation and coalescence. This membrane is called the milk fat globule membrane

(MFGM), of which the bovine form is the most studied and employed in the dairy industry [1,3–6]. Table 1 showcases the MFGM content in the main dairy products, such as milk, cream, and cheese.

Table 1. Comparison of milk fat globule membrane (MFGM) content in different dairy products. Data are from Dewettinck et al., 2011, and Conway et al., 2014 [7,8].

Product	MFGM (mg/100 g)
Cheese (25% fat)	150
Milk (skimmed, 0.5% fat)	15
Milk (whole, 3.5% fat)	35
Yogurt (1.5% fat)	15
Cream (38% fat)	200

Bovine MFGM is about 10–20 nm in cross-section and its mass accounts for 2–6% of the total MFG mass [9]. It is made up of many different compounds: polar lipids such as phospholipids, sphingolipids, and glycolipids, and also cholesterol, proteins, and surface glycoproteins [10]. This membrane acts as a natural emulsifier and encases the nonpolar triglyceride core of MFGs [11,12]. The complex MFGM architecture ensures stable dispersion of MFGs in milk—polar lipids and glycoproteins present in the membrane induce electrostatic and steric repulsion, preventing coalescence and aggregation of the fat globules [3,7,13]. There are several health-promoting effects of the MFGM (mainly from bovine but also from other species), such as anticarcinogenic, antimicrobial, anti-inflammatory, and anticholesterolemic activities [6,7].

The anticarcinogenic activity was assessed on HT-29 cells (a human colon cancer cell line) by three studies [14–16], which showed that MFGM could reduce the proliferation and enhance apoptosis of the cancer cells through the activation of effector caspase-3. The antimicrobial activity was observed through the inhibition of in vitro rotavirus infectivity [17] and the anti-adhesive activity exerted by a mucin 1 (MUC1)-enriched MFGM fraction against bacteria in the gut mucosa [18]. The anti-inflammatory activity was evaluated with the in vivo mitigation of lipopolysaccharide (LPS)-induced intestinal damage and inflammation in low birth weight (LBW) mice [19] and with the decrease of pro-inflammatory serum markers such as total cholesterol, low density lipoprotein (LDL)-cholesterol, along with an increased production of anti-inflammatory cytokines in obese adults challenged with a high-fat meal rich in saturated fatty acids [20]. Finally, the anticholesterolemic activity was assessed by the decrease exerted by MFGM-derived sphingomyelin of the intestinal absorption of cholesterol and fats in animal models, thus protecting the liver from fat- and cholesterol-induced steatosis and consequently preventing the inflammatory condition involved in atherosclerosis and insulin resistance [7,21]. To summarize, the MFGM could play a key nutraceutical role in many adverse health conditions, even though its effectiveness and potential claims need to be addressed properly.

The aim of this review is to provide a general overview about the formation of bovine MFGs and MFGM and to highlight the main similarities and differences across the MFGM proteomes of the most-studied species (human, cow, goat, buffalo, etc.) through the analysis of comparative proteomic studies. Moreover, the potential supplementation of MFGM fractions in infant formula (IF) is investigated in order to underline the beneficial effects exerted by MFGM bioactive components in infant feeding.

2. Bovine MFGs and the MFGM

2.1. Formation of Bovine MFGs and MFGM

The various classes of fatty acids (FAs) of milk fat derive above all from feed and rumen microbial activity. In particular, short-chain fatty acids (SCFAs) and medium-chain fatty acids (MCFAs) derive from de novo synthesis in the mammary gland, involving acetyl-coenzyme A carboxylase (ACC) and fatty acid synthase (FAS) enzymes, starting from acetate and butyrate [22,23]. These two molecules are produced in the rumen by fermentation of feed components, such as carbohydrates. Long-chain fatty acids (LCFAs) generally derive from dietary lipids or mobilization of body reserves and specifically they are released by lipoprotein lipase from TGs or from very low density lipoproteins (VLDL), or further from non-esterified fatty acids (NEFA), which are usually found in plasma bound to albumin [24–26]. Once they are synthesized, FAs pass through the basal plasma membrane of mammary gland epithelial cells via diffusion and reach the endoplasmatic reticulum (ER), where the TG droplets are synthesized starting from FA precursors. The microlipid droplets are then extruded from the ER in the cytoplasm—during the extrusion process, the droplets are encased in a surface-active inner monolayer, which surrounds the TG core and is made up of polar lipids and specific surface-associated proteins derived from the ER [3,27,28].

Once in the cytoplasm of the mammary gland epithelial cells, microlipid droplets grow in volume and then migrate through the cell cytoplasm, from the basal to the apical pole of the cell. The lipid droplets are secreted by the cells in an apocrine-like mechanism into the alveolar lumen as MFGs, surrounded by the apical plasma membrane of the cells [29,30]. The result of this process is that the MFGM is a trilayer membrane, with the inner layer composed of proteins and polar lipids from the ER, and the outer bilayer of proteins and polar lipids from the apical plasma membrane of the mammary gland epithelial cells (Figure 1) [4,5,28].

In the external layer of the MFGM there are partially embedded, loosely attached and transmembrane proteins, as well as cholesterol molecules associated with polar lipids, while glycoproteins are also present on the surface, with carbohydrate domains oriented outwards [3]. The most widely accepted model for this type of membrane is thus the fluid mosaic model [4,7]. This is because the bovine apical plasma membrane of epithelial secretory cells and the bovine MFGM membrane show a similar distribution of all their components [31].

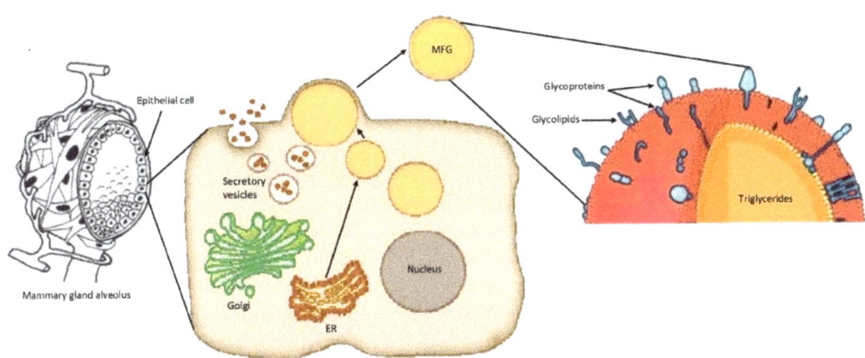

Figure 1. Schematic representation of milk fat globule (MFG) formation. The microlipid droplets (yellow circles) are extruded from the endoplasmatic reticulum (ER) in the cytoplasm of the mammary gland epithelial cell to reach the apical plasma membrane, where they are extruded in the alveolar lumen as MFGs. Adapted from Reece, 2004 [32]; Horseman et al., 2014 [33]; and Wikipedia [34].

2.2. Lipids in MFG and MFGM, and the Role of Choline

MFG core is predominantly composed of non-polar lipids, named TGs, accounting for 98% of total milk fat. Milk fat is composed of over 400 different FAs, of which 15 represent 90% of the total FA pool. Saturated FAs in bovine milk fat account for 70% of the total milk FAs, with the main forms being palmitic acid (26–32%), stearic acid (12%), and myristic acid (10%). Of the saturated FAs, MCFA (6:0–12:0) represent about 10% of total milk FAs, whereas SCFA (4:0) account for less than 3%. Mono-unsaturated FAs account for 25% of the total milk FAs, with the main form being oleic acid (20–25%). Poly-unsaturated fatty acids (PUFAs) constitute 2.5% of the total milk FAs, and the two major PUFAs are linoleic acid (1–3%) and α-linolenic acid (0.5–2%) [26]. The latter two are among the most important essential fatty acids, which exert an anti-inflammatory function and are also important to prevent cardiovascular diseases in humans [35]. Bovine MFGM is composed mainly of polar lipids that account for 0.2–1% of the total milk fat. The amount of polar lipids in milk fat is related to the amount of the MFGM and then to the size of fat globules [28,36]. The major polar lipids that build bovine MFGM are membrane glycerophospholipids, a group that include phosphatidylcholine (PC, 35–36%); phosphatidylethanolamine (PE, 27–30%); phosphatidylinositol (PI, 5–11%); phosphatidylserine (PS, 3%); and sphingolipids, in particular, sphingomyelin (SM, 25%) [3,7,29,30]. Most of these (about 60%) are choline-containing phospholipids, namely, PC, lysophospatidylcholine (lyso-PC), and sphingomyelin [37,38], showcasing why MFGM constitutes the major choline-containing component of bovine milk [37]. The amount of choline-containing phospholipids in bovine milk is about 105–210 mg/L, to which free choline should be added [38–40]. This value is calculated considering that 60% of milk phospholipids contain choline and, furthermore, that phospholipids account for 0.2–1% of total milk lipids [28,32,37,41]. Usually, choline liver reserves and also its metabolites are employed to maintain certain levels of choline secretion into milk. In 1995, Zeisel and collaborators [42] observed that rats fed with a choline-deficient diet had about 90% lower hepatic PC compared to rats fed with a choline-adequate diet. Moreover, lactating rats fed with a choline-deficient diet showed sevenfold higher levels of hepatic TG than non-mated females fed with a choline-adequate diet, whereas TG levels were fourfold higher in lactating rats fed with a choline-adequate diet [42]. According to Kinsella [43], a bovine mammary gland normally yielding 25 L of milk secretes 10 ± 3 g phospholipids per day. This quantity corresponds on average to the 5% of the total phospholipid content of the mammary tissue [43]. These findings confer to choline supplementation an important metabolic role in lipid transport to and within extra-hepatic tissues, such as the mammary gland [44–48].

This idea was confirmed more recently by Li and collaborators also [49]. In particular, the authors evaluated the effects of choline supplementation in intrauterine growth-restricted (IUGR) pigs compared to a normal-choline diet. Choline supplementation decreased hepatic free FAs and TG level, downregulated lipogenic enzyme expression, and enhanced TG export from liver, acting also on cholesterol regulation through higher high-density lipoprotein cholesterol (HDL-C) and lower total plasma cholesterol. Hence, choline supplementation improved hepatic lipid metabolism, avoiding the abnormal lipid metabolism condition of IUGR pigs [49].

Other studies confirmed the importance of choline supplementation in increasing not only milk production (yield and composition) but also choline-containing compounds in milk derived from dairy ruminants [12,50–55].

The potential of designing milk with a higher content of choline-containing compounds via choline supplementation in animal feeding is interesting from several points of view. Interestingly, there is scientific evidence about the nutraceutical benefits of phospholipids, sphingolipids, and SM-derived metabolites (ceramide and sphingosine), which have shown antiproliferative activity on cancer cells—MFGM digestion occurs along the entire length of gastrointestinal (GI) tract, with high levels of ceramide and sphingosine recovered in the small intestine and the colon, where they can directly exert their beneficial effects or where they can enter the bloodstream to reach peripheral organs. Indeed, ceramide and sphingosine are two metabolites acting as second messengers in cell signaling, exerting pro-apoptotic and antimitogenic effects [6,14,15].

2.3. Major MFGM Proteins

MFGM proteins account for 25–60% of the mass of the MFGM, 1–4% of total milk proteins, and 1% of the total globule mass. The proteins can be classified into integral proteins and peripheral proteins, whereas others are partially embedded or loosely attached to the membrane. During the secretion of the MFGs, the constituents are re-arranged within the apical plasma membrane and the MFGM [4,31]. The localization of the proteins thus varies—some are associated with the inner monolayer membrane, while others are associated with the outer bilayer membrane [7,28,29]. The main MFGM proteins are adipophilin (ADPH), butyrophilin (BTN), mucin 1 (MUC1), xanthine dehydrogenase/oxidase

(XDH/XO), CD36, periodic acid Schiff III (PAS III), PAS 6/7, lactadherin, and fatty acid-binding protein (FABP), as is shown in Figure 2 [7,56,57].

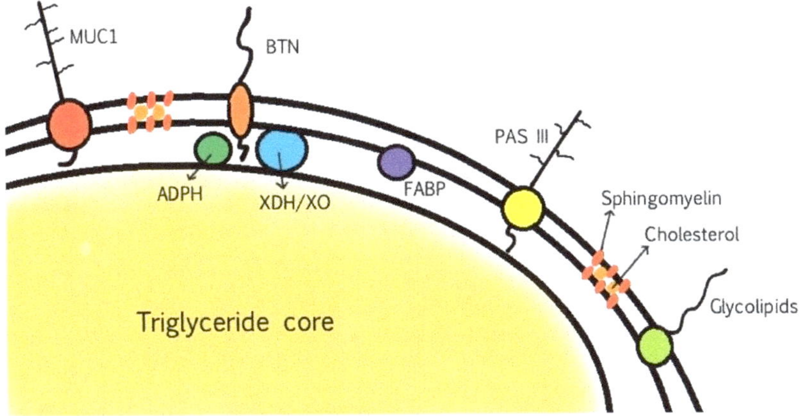

Figure 2. Structure of MFGM and localization of the main MFGM proteins.

ADPH, also known as perilipin 2, is a major constituent of the MFGM and is localized in the inner monolayer membrane. It regulates lipolysis by controlling the access of proteins to the MFG [29,56]. BTN is a transmembrane protein and is the most abundant protein in bovine MFGM. BTNs are members of the immunoglobulin (Ig) superfamily, and BTN1A1 is the form in human MFGM [28,29,56]. It has been observed that the knockout of BTN1A1 in bovine mammary epithelial cells decreased the size and the phospholipid content of lipid droplets (the precursors of MFGs), thus suggesting that BTN1A1 has a key role in regulating the synthesis of lipid droplets via a mechanism involving membrane phospholipid composition [36]. MUC1 is a glycoprotein with highly glycosylated extracellular domains localized on the outer surface of MFGs. This feature makes it resistant to digestion and potentially available to act as a decoy receptor for pathogens [18,29,56].

XDH/XO is a redox enzyme that accounts for 12% of total bovine MFGM proteins and is localized in the intermembrane space between the monolayer and the bilayer, forming a tripartite structure with BTN and ADPH (needed to interconnect the inner and outer membrane). It plays a role in antimicrobial defense of the GI tract through the production of reactive oxygen species (ROS) as well as reactive nitrogen species (RNS), which have bactericidal properties. Surface carbohydrates and XDH/XO may possibly act as decoys—pathogens can interact with receptors of the epithelial cells of the GI tract, but can also bind to similar receptors on the MFGM that themselves can act as decoys, such as MUC1 [18], to avoid bacterial interaction with their primary target (GI epithelial cells) and that can also impart an antimicrobial effect thanks to ROS/RNS production by XDH/XO [7,28,29]. Finally, FABP is a protein with a similar localization of XDH/XO and plays a key role in the synthesis of MFG lipid constituents during the intracellular transport of FAs. Indeed, it is involved in the transport of FAs through the capillary endothelium to reach the cytoplasm of mammary endothelial cells, where they cross the membrane via diffusion [24,29].

Interestingly, some authors [7,58] have demonstrated the presence of the onco-suppressors breast related cancer antigens 1/2 (BRCA1 and BRCA2) in human and bovine MFGM. These two onco-suppressors are involved in DNA repair processes [7]. The reason for their presence in the MFGM could be because MFGs are secreted by mammary gland epithelial cells and carry a fraction of their apical plasma membrane. This hypothesis could also explain the presence of these two proteins in human MFGs in secreted milk. In 2002, Vissac and collaborators evaluated BRCA1/2 expression in MFGs of women just after delivery, observing similar patterns of expression of the two proteins [58]. The nutraceutical role of the MFGM and its potential anticancer effect could be explained by the fact that after MFGM consumption, the inhibitory peptides might be released from MFGM and subsequently absorbed by the digestive tract. The absorbed peptides could enter the bloodstream and reach the organs (or tissues), where they inhibit the transforming cells [6].

In Table 2, the major components of bovine MFGM and their main functions are listed.

Table 2. Functions of the main components of bovine MFGM. Data from Lee et al., 2018 [29].

Components	Abbreviation	Functions
Polar Lipids		
Phosphatidylcholine	PC	Structural maintenance of MFGM; cholesterol regulation and lipoproteins metabolism
Phosphatidylethanolamine	PE	Structural membrane regulation
Phosphatidylinositol	PI	Cell signaling; PI3K-Akt pathway regulation
Phosphatidylserine	PS	Apoptosis regulation
Sphingomyelin	SM	Myelinization; metabolized to ceramide and sphingosine (second messengers that regulate cell growth and cell cycle)
Cholesterol	-	Structural maintenance of MFGM (lipid rafts complexes with SM)
Proteins		
Adipophilin	ADPH	Lipolysis regulation
Butyrophilin	BTN	MFG synthesis regulation
Mucin 1	MUC 1	Decoy receptor for pathogens; inhibition of in vitro rotavirus infectivity
Xanthine dehydrogenase/oxidase	XDH/XO	Structural maintenance of MFGM; antimicrobial activity (ROS/RNS production)
Fatty acid-binding protein	FABP	Fatty acid transport; MFG lipid synthesis
Breast related cancer antigens 1/2	BRCA 1/2	Onco-suppressor activity
Choline	-	Precursor of phospholipids and SM; hepatic lipid metabolism
Gangliosides	-	Cognitive development

3. Comparison of MFGM Proteome between Different Species and Lactation Stages

Bovine milk is the major substitute for human milk and the most produced animal milk in the world [59]. Bovine MFGM is thus the most studied and employed in the industry of dairy products, for example in the production of IFs [60–62]. Milk differs from species to species above all in terms of the composition and the amount of macromolecules. This review discusses the variations in MFGM across species.

Although most of the beneficial effects of MFGM are associated with its polar lipid fraction [6], studies on MFGM proteome are increasing since MFGM proteins show bioactive properties and new technologies have enabled more detailed analyses of the MFGM proteome. Comparative proteomic analyses have been performed to understand how proteomes vary between different species and between different stages of lactation [63]. Figure 3 shows the general experimental workflow.

These analyses have been performed to study and compare the proteome of various MFGMs, mainly from humans and cows [64–66], but also from other species such as the goat, yak, buffalo, horse, and donkey [61,62,67–70].

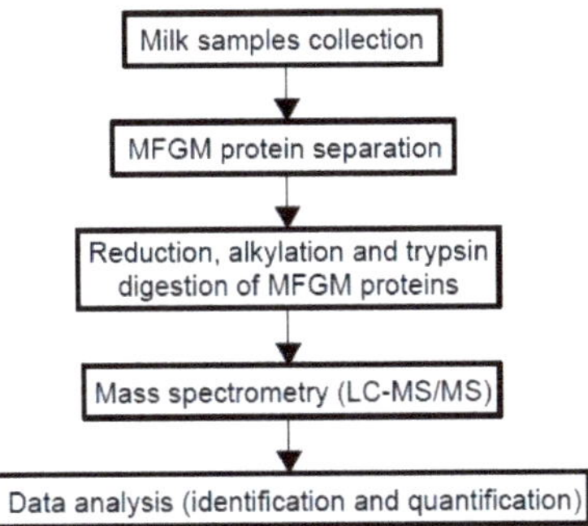

Figure 3. Experimental workflow of MFGM proteome analysis using a proteomic approach.

Human milk proteome varies across colostrum and mature milk [71], and these variations are also reflected in the MFGM proteome that varies between lactation stages [72].

This pattern of variations is also valid for other species such as cows [68,73] and goats [70]. In particular, Reinhardt and Lippolis [73] observed that the proteins associated with lipid transport synthesis and secretion (such as FABP) and MUC1 were highly upregulated in 7-day-old milk MFGM than in colostrum MFGM from cow's milk. The variation of expression of proteins such as FABP is indicative of an early developmental shift in milk fat transport, despite higher fat content in colostrum [73].

In a proteomic study focused on human colostrum MFGM, 107 proteins were identified, half of which were typical MFGM proteins, such as lactadherin and BTN [64]. A similar number of proteins (120) were detected in bovine MFGM. As with human MFGM, BTN was identified as the major MFGM protein, while membrane/protein trafficking proteins (23%) and cell signaling proteins (23%) accounted for almost half of the proteins identified [65].

However, apart from some common features, human and bovine MFGM are different in terms of the fraction of proteome involved in host defense. In 2011, Hettinga and collaborators [66] verified that the total number of host defense MFGM proteins was similar between humans (51 out of 234 proteins identified in human MFGM) and bovines (44 out of 232 proteins identified in bovine MFGM) (Figure 4). However, the human MFGM was more enriched with Igs than bovine MFGM, while bovine MFGM was more enriched with antibacterial proteins. This important information helped identify the main proteins with immunity-promoting properties for newborns [66].

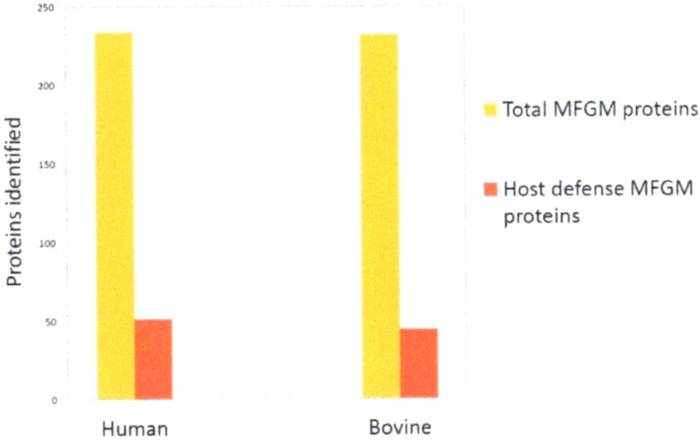

Figure 4. Comparison of total MFGM proteome and host defense MFGM proteins between humans and bovines. Data are from Hettinga et al., 2011 [66].

Human MFGM phosphoproteome has recently been studied [72]. Phosphorylation is a post-translational modification that plays a key role in regulating many signaling pathways. The authors identified 203 phosphoproteins, of which 48 were differentially phosphorylated in colostrum and mature milk. These 48 phosphoproteins were mainly associated with immune-related processes. The results showed that there were more immune system process-related phosphoproteins in human colostrum MFGM than in mature MFGM, probably because of the important role that colostrum has in building the immune system of newborns [72].

In terms of the MFGM proteome across different species, Lu and collaborators [61] identified and quantified 312, 554, 175, and 143 proteins in human, cow, goat, and yak MFGM, respectively. Fifty proteins were shared among species. Human MFGM shared the highest number of proteins with cow MFGM, whereas with goat and yak MFGM, the number was lower (Figure 5). In terms of composition, the correlation between cow and human MFGM was higher than that between goat and human MFGM, and also between yak and human MFGM. Analyses of the molecular function of proteins revealed that human MFGM was enriched in ER proteins, whereas cow MFGM was enriched in plasma membrane proteins [61]. These findings confirm that MFGM originates from the ER and the plasma membrane [31].

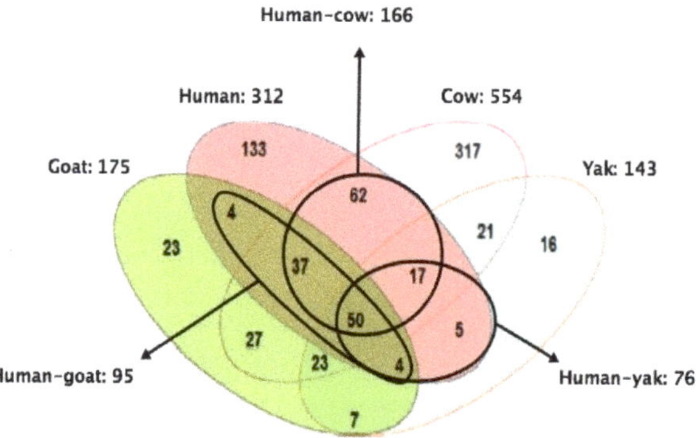

Figure 5. Shared and uniquely identified and quantified proteins in humans, cows, goats, and yaks. Adapted from Lu et al., 2016 [61].

The most shared proteins across species were involved in protein/vesicle-mediated transport, along with major MFGM proteins such as BTN, ADPH, FABP, and MUC1. The main difference regarding human MFGM proteome was a higher enrichment in enzymes involved in lipid catabolism, also reported in Liao et al. [68] and in a set of immune response proteins [61,72].

In 2015, another research team compared the proteome of cow, yak, buffalo, goat, and human MFGM [74]. The authors identified a total of 520 proteins of all species, most of which were shared among all species, although in different isoforms, as also reported by other studies [75,76], such as BTN, lactadherin, MUC1, and ADPH. These findings showed that the MFGM proteome presents a high complexity and variability among species. In terms of molecular function and Gene Ontology (GO) categories, cellular process, localization, transport, signal transduction, and response to stimulus were enriched in all the MFGM fractions [74].

Comparative proteomic analyses have been performed to compare cow and goat proteome. In a 2019 study, a total of 776 MFGM proteins were identified: 427 and 183 that are unique for goat and cow milk, respectively, and 166 proteins shared between the two species. Most of the goat MFGM proteins were related to metabolic processes (about 21%), whereas most of the cow MFGM proteins were related to disease-associated pathways (about 49%) [62]. Subsequently, the same authors evaluated the variations between goat colostrum and mature MFGM proteome. They found a higher number of proteins than in their previous study; in particular, 543 and 858 proteins in colostrum and mature milk, respectively, of which 394 are shared in colostrum and mature milk. Colostrum was found to have fewer proteins but more functions of protein processing in the ER than mature milk, whereas mature milk had more metabolism-related proteins [70].

Along with the analyses performed on proteome and phosphoproteome, the MFGM glycoproteome has also been investigated [67,68]. Cao and collaborators identified and quantified 465, 423, 334, and 176 glycoproteins in human colostrum and mature milk, and bovine colostrum and mature milk, respectively. Human colostrum and mature milk shared 362 glycoproteins, whereas bovine colostrum and mature milk shared 155 glycoproteins (Figure 6). The authors found 24.3% (156) of glycoproteins were shared between human and bovine colostrum, and 16.3% (84) of glycoproteins were shared between human and bovine mature milk. These results indicated more dramatic variations in MFGM glycosylation within species than lactation stages [68].

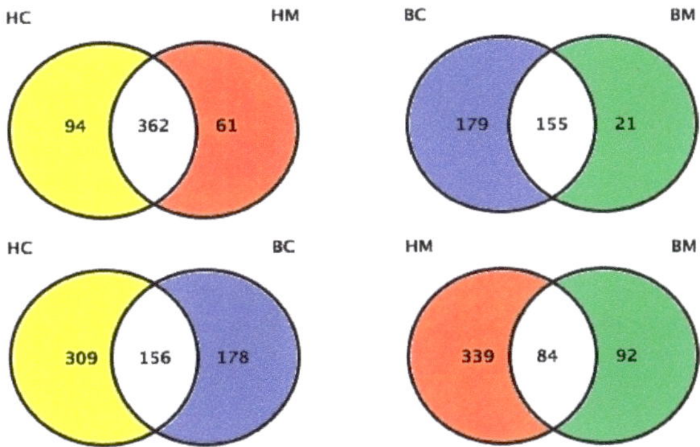

Figure 6. Quantitative comparison of MFGM proteins between human colostrum and human mature milk, bovine colostrum and bovine mature milk, human colostrum and bovine colostrum, and human mature milk and bovine mature milk (HC = human colostrum; HM = human mature milk; BC = bovine colostrum; BM = bovine mature milk). Adapted from Cao et al., 2019 [68].

In another study, Yang et al. investigated the variation among species by analyzing MFGM glycoproteome from cow, buffalo, yak, goat, and human milk. They found that the glycoproteins from the different MFGM species were mainly related to the response to stimulus, according to the GO categories, and that the fractions from ruminants (cow, buffalo, yak, goat) were more similar to each other when compared to the non-ruminant's fraction (human) [67,74].

An example of the application of these comparative studies was given recently by Ji and collaborators [16], who evaluated the antiproliferative effect of five MFGM fractions from yak, bovine, goat, camel, and buffalo milk using the HT-29 cell line. The antiproliferative effect was evaluated in terms of cell viability, cell cycle, cytomorphology, apoptosis, and mitochondrial membrane potential (MMP). The results showed that all the five MFGM fractions reduced cell growth by affecting cell cycle and inducing apoptosis, whereas MMP values were also significantly reduced by all the five MFGM fractions. Among all the tested samples, buffalo and goat MFGMs were more effective in inducing apoptosis than the other three MFGMs. These data suggest that MFGM might be a putative agent for the prevention of human colon cancer [16].

To summarize, the principal MFGM proteins have been identified in all species. However, the main molecular functions exerted by MFGM proteomes vary according to species and lactation stage due to the variations in protein expression. For example, the human MFGM proteome (in particular that contained in colostrum) has more immune response-related proteins than the MFGM proteome from other species. There are similarities between human and cow MFGM proteome and molecular functions, suggesting that bovine milk, and more specifically bovine MFGM proteins, could be used as a supplement in IFs [77,78].

The varying number of proteins identified and quantified in different studies depends on the proteomic methods performed by the authors. In any case, each study is a step forward in terms of the knowledge regarding MFGM proteome. The potential of these results could facilitate the correct management of the MFGM proteome in the design of products such as IFs supplemented with specific MFGM proteins.

4. MFGM: Potentials in Infant Formula Preparation

Breast milk is considered the gold standard for infant nutrition and is required for optimal infant growth, brain and GI tract development, as well as establishing the immune system. Some of the bioactive factors of breast milk are present in the MFGM. In order to develop products that reflect the complexity of human milk, efforts have been made to imitate the nutritional profile and composition of human breast milk. IF has been designed in order to be a suitable alternative to human breast milk [21,77–79]. Bovine milk is currently the basis for most IFs (Table 3).

Table 3. Comparison of human breast milk and cow-based infant formula composition (in terms of energy and macronutrients). Data are from the European Food Safety Authority (EFSA) Journal, 2014 [80].

Item	Human Breast Milk	Cow-Based IF
Energy (kcal/100 mL)	65	60–70
Digestible carbohydrates (g/100 kcal)	8.2–10.4	9–14
Lipids (g/100 kcal)	3.7–9.1	4.4–6
Proteins (g/100 kcal)	1.3 (0.8–2.1) [a]	1.8–2.5

[a] Mature human milk.

IFs are based on the nutrient composition of human milk in order to provide adequate levels of macronutrients (carbohydrates, lipids, and proteins) and micronutrients (vitamins and minerals) to support growth [60,81]. Despite the similarities in MFGM proteome, the composition of bovine milk differs from human milk. An example is the content of essential unsaturated fatty acids, which is higher in human milk than in bovine milk because of the high rate of biohydrogenation processes of dietary unsaturated fatty acids in the rumen—α-linolenic acid shows a lower difference (0.5–2% and 1–2% of total milk FAs for bovine and human milk fat, respectively) than linoleic acid (1–3% and 8–18% of total milk FAs for bovine and human milk fat, respectively) [25,29,82,83]. In addition to nutrients, human breast milk also contains several bioactive compounds (Igs, enzymes, hormones) and live cells (e.g., leucocytes) that cannot be easily added to IFs [79,84]. All these elements prevent IFs from having the same composition as human breast milk, although the research in this field has advanced considerably.

In fact, several studies [49,57,60,77,85] have suggested that supplementing IFs with MFGM could provide beneficial effects because of the presence of bioactive compounds such as proteins and polar lipids in the MFGM, thus narrowing the gap between human breast milk and IFs. Many of these results were obtained through the supplementation of IF with the polar lipid fraction of the MFGM (phospholipids and sphingolipids), given that these compounds are the main contributors to the nutraceutical effects of the MFGM [6].

Few studies have investigated supplementing IF with MFGM proteins [78]. One hypothesis is that the protein content of IFs is usually higher than that of human milk, and this is likely due to the lower digestibility of cow milk proteins [77,86]. Further supplementation with proteins has thus rarely been taken into account, even though MFGM proteins have health-promoting effects such as preventing pathogen adhesion and infection [18,29].

In 2014, Billeaud and collaborators [87] evaluated the safety of two IFs supplemented with a lipid-rich or a protein-rich bovine MFGM fraction in healthy infants. The authors observed no considerable differences among the two supplemented formulas and the control (standard formula) in terms of weight gain, adverse events, and morbidity rates. They concluded that MFGM enrichment, both with lipids or proteins, could improve the level of similarity between breast milk and cow-based IFs [87]. In the same year, Timby and collaborators [77] evaluated the effect of a low-energy and low-protein formula supplemented with a protein-rich bovine MFGM fraction on healthy infants. Their results showed that the cognitive score (assessed with Bayley-III tests) was 4.0 points higher in the experimental formula group than in the standard formula group (105.8 ± 9.2 vs. 101.8 ± 8.0; $p < 0.05$), but was similar to the breastfed group (105.8 ± 9.2 vs. 106.4 ± 9.5; $p > 0.05$). This suggested that the experimental formula could decrease the gap in cognitive performance between breastfed and formula-fed infants [77].

Zavaleta and collaborators [88] evaluated the efficacy of a complementary food supplemented with a MFGM-enriched protein fraction on the health status of infants. They found that the supplementation improved infants' health status by reducing episodes of diarrhea. Even though the authors did not use a supplemented IF but a supplemented complementary food, they achieved promising results, probably due to an amelioration of gut microbiota or positive changes in the developing immune system of the infants [88].

However, the relevance and the potential of MGFM is under investigation, not only as a source of several bioactive nutrients (fat-soluble vitamins including carotenoids and polar lipids) including MFGM proteins, but also for its role in fat digestion [89]. It was observed that bovine MFGM reduced the in vitro FA release from MFGs and this was probably due to the inhibitory effect of MFGM components and conformation on the pancreatic lipase activity [90]. Besides this, it is also important to mention that the composition of FAs in TG core of MFGs has a significant impact on the digestibility and the absorption of fat and other compounds, such as minerals. MCFAs are better absorbed than LCFAs, and therefore TGs made up mainly by MCFA result in being more digestible because they are better solubilized in the gut [91]. Moreover, calcium absorption is also higher in human subjects after the consumption of a MCFA-mixed meal compared to a LCFA-mixed meal [92]. In addition to these aspects, the positional distribution of FAs on the glycerol backbone is also crucial to determine FA absorption, whether as sn-2 monoglycerides or as free FAs (after the hydrolyzation by lipase of the sn-1 and sn-3 bounds). An example is palmitic acid (16:0)—in human milk it is found on the sn-2 glycerol position more than in bovine milk (>50% and 30–40%, respectively) [93]. The sn-2 position ensures higher absorption for palmitic acid than the sn-1 or sn-3 positions, also because free palmitic, being a LCFA, tends to form insoluble fatty soaps with calcium at intestinal pH conditions [94]. For this reason, the location of palmitic acid on the sn-2 position of glycerol ensures higher absorption for both the FA and the calcium, making human milk more digestible than cow-based IFs [91,95,96].

5. Conclusions

Milk has important nutritional features for newborns. Indeed, breast milk is a mixture of several bioactive compounds that modulate the GI tract and contribute to building the immune system of breast-fed infants. Moreover, breast milk is also important for brain development [81]. Bovine milk is the most used animal milk in the world [59] and it shares several features with other species' milk, such as the particular occurrence of milk fat as MFGs surrounded by the MFGM. The lipids and proteins that constitute the MFGM supply it with many of the bioactive properties of milk [6]. Along with MFGM polar lipids, MFGM proteins have important health-promoting effects such as anti-adhesive and antimicrobial functions [18,29].

The growing interest in MFGM led researchers to study the MFGM proteins from a wider approach through proteomics. Proteomic methods have been performed to better clarify the role of MFGM proteins, leading to a deeper knowledge about them. Proteomics has the potential to enable the detection, identification, and characterization of proteins, as well as to analyze a large number of proteins simultaneously [97]. Comparative proteomic studies were performed to obtain information on the variations in MFGM proteome among different species [63]. There are variations in terms of protein expression level and molecular function across species, even though the major MFGM proteins are observed among all the species considered. The properties of the MFGM proteome of each species could be exploited to design products supplemented with MFGM fractions that meet specific needs, for example, the enhancement of the immune system, the regulation of cholesterol metabolism, or the supply of beneficial polar lipids to support cognitive function.

An example of the application of the MFGM proteome is found in the dairy industry, in particular in the supplementation of IFs [77,78,89]. The promising results obtained with the supplementation of IF with MFGM proteins [50,77] and polar lipids [85] underline once again the importance of MFGM in IF preparation—since cow milk-based IF is formulated to better resemble human breast milk, the MFGM supplementation could increase the presence of bioactive compounds in IF (usually at low levels in standard formula).

Future work is likely to be addressed towards a deeper comprehension of MFGM proteome and its variations across species and lactation stages. The overall aim is to further increase the knowledge of MFGM properties and to assess the potential of the supplementation of IFs with MFGM proteins.

Author Contributions: Conceptualization, M.M. and L.P.; writing—original draft preparation, M.M., C.D.L., and L.P.; writing—review and editing, M.M., C.D.L., M.O., and L.P.; visualization, M.M. and L.P.; supervision, M.T. and L.P.; project administration, L.P.; funding acquisition, L.P. All authors have read and agreed to the published version of the manuscript.

Funding: This research received no external funding.

Conflicts of Interest: The authors declare no conflict of interest.

Abbreviations

MFG	Milk fat globule
MFGM	Milk fat globule membrane
TG	Triglyceride
LDL	Low density lipoprotein
FA	Fatty acid
SCFA	Short-chain fatty acid
MCFA	Medium-chain fatty acid
LCFA	Long-chain fatty acid
PE	Phosphatidylethanolamine
PC	Phosphatidylcholine
PI	Phosphatidylinositol
PS	Phosphatidylserine
SM	Sphingomyelin
GI	Gastrointestinal
ADPH	Long-chain fatty acid
BTN	Butyrophilin
MUC1	Mucin 1
XDH/XO	Xanthine dehydrogenase/oxidase
PAS III	Periodic acid Schiff III
FABP	Fatty acid-binding protein

References

1. Bernard, L.; Bonnet, M.; Delavaud, C.; Delosiere, M.; Ferlay, A.; Fougere, H.; Graulet, B. Milk Fat Globule in Ruminant: Major and Minor Compounds, Nutritional Regulation and Differences Among Species. *Eur. J. Lipid Sci. Tech.* **2018**, *120*, 1700039. [CrossRef]
2. Mulder, H.; Walstra, P. The Milk Fat Globule. Emulsion Science as Applied to Milk Products and Comparable Foods. In *Commonwealth Agricultural Bureau*; Farnham Royal and Centre for Agricultural Publishing and Documentation: Wageningen, The Netherlands, 1974.
3. Arranz, E.; Corredig, M. Invited review: Milk phospholipid vesicles, their colloidal properties, and potential as delivery vehicles for bioactive molecules. *J. Dairy Sci.* **2017**, *100*, 4213–4222. [CrossRef]

4.   Evers, J.M. The milkfat globule membrane—Compositional and structural changes post secretion by the mammary secretory cell. *Int. Dairy J.* **2004**, *14*, 661–674. [CrossRef]

5.   Evers, J.M. The milkfat globule membrane—Methodologies for measuring milkfat globule (membrane) damage. *Int. Dairy J.* **2004**, *14*, 747–760. [CrossRef]

6.   Spitsberg, V.L. Invited review: Bovine milk fat globule membrane as a potential nutraceutical. *J. Dairy Sci.* **2005**, *88*, 2289–2294. [CrossRef]

7.   Dewettinck, K.; Rombaut, R.; Thienpont, N.; Le, T.T.; Messens, K.; Van Camp, J. Nutritional and technological aspects of milk fat globule membrane material. *Int. Dairy J.* **2008**, *18*, 436–457. [CrossRef]

8.   Conway, V.; Gauthier, S.; Pouilot, Y. Buttermilk: Much more than a source of milk phospholipids. *Anim. Front.* **2014**, *4*, 44–51. [CrossRef]

9.   Singh, H. The milk fat globule membrane-A biophysical system for food applications. *Curr. Opin. Colloid Interface Sci.* **2008**, *11*, 154–163. [CrossRef]

10.   Danthine, S.; Blecker, C.; Paquot, M.; Innocente, N.; Deroanne, C. Évolution des connaissances sur la membrane du globule gras du lait: Synthèse bibliographique. *Lait* **2000**, *80*, 209–222. [CrossRef]

11.   Pinotti, L.; Baldi, A.; Dell'Orto, V. Comparative mammalian choline metabolism with emphasis on the high-yielding dairy cow. *Nutr. Res. Rev.* **2002**, *15*, 315–331. [CrossRef]

12.   Pinotti, L.; Baldi, A.; Politis, I.; Rebucci, R.; Sangalli, L.; Dell'Orto, V. Rumen-protected choline administration to transition cows: Effects on milk production and vitamin E status. *J. Vet. Med. A* **2003**, *50*, 18–21. [CrossRef] [PubMed]

13.   Michalski, M.C.; Briard, V.; Michel, F.; Tasson, F.; Poulain, P. Size distribution of fat globules in human colostrum, breast milk, and infant formula. *J. Dairy Sci.* **2005**, *88*, 1927–1940. [CrossRef]

14.   Zanabria, R.; Tellez, A.M.; Griffiths, M.; Corredig, M. Milk fat globule membrane isolate induces apoptosis in HT-29 human colon cancer cells. *Food Funct.* **2013**, *4*, 222–230. [CrossRef] [PubMed]

15.   Zanabria, R.; Griffiths, M.W.; Corredig, M. Does structure affect biological function? Modifications to the protein and phospholipids fraction of the milk fat globule membrane after extraction affect the antiproliferative activity of colon cancer cells. *J. Food Biochem.* **2020**, *44*, 13104. [CrossRef]

16.   Ji, X.; Xu, W.; Cui, J.; Ma, Y.; Zhou, S. Goat and buffalo milk fat globule membranes exhibit better effects at inducing apoptosis and reduction the viability of HT-29 cells. *Sci. Rep.* **2019**, *9*, 2577. [CrossRef]

17.   Fuller, K.L.; Kuhlenschmidt, T.B.; Kuhlenschmidt, M.S.; Jiménez-Flores, R.; Donovan, S.M. Milk fat globule membrane isolated from buttermilk or whey cream and their lipid components inhibit infectivity of rotavirus in vitro. *J. Dairy Sci.* **2013**, *96*, 3488–3497. [CrossRef]

18.   Struijs, K.; Van de Wiele, T.; Le, T.T.; Debyser, G.; Dewettinck, K.; Devreese, B.; Van Camp, J. Milk fat globule membrane glycoproteins prevent adhesion of the colonic microbiota and result in increased bacterial butyrate production. *Int. Dairy J.* **2013**, *32*, 99–109. [CrossRef]

19.   Huang, S.; Wu, Z.; Liu, C.; Han, D.; Feng, C.; Wang, S.; Wang, J. Milk Fat Globule Membrane Supplementation Promotes Neonatal Growth and Alleviates Inflammation in Low-Birth-Weight Mice Treated with Lipopolysaccharide. *Biomed. Res. Int.* **2019**, *2019*, 4876078. [CrossRef]

20.   Demmer, E.; Van Loan, M.D.; Rivera, N.; Rogers, T.S.; Gertz, E.R.; German, J.B.; Smilowitz, J.T.; Zivkovic, A.M. Addition of a dairy fraction rich in milk fat globule membrane to a high-saturated fat meal reduces the postprandial insulinaemic and inflammatory response in overweight and obese adults. *J. Nutr. Sci.* **2016**, *5*, e14. [CrossRef]

21.   Hernell, O.; Timby, N.; Domellöf, M.; Lönnerdal, B. Clinical benefits of milk fat globule membranes for infants and children. *J. Pediatr.* **2016**, *173*, S60–S65. [CrossRef]

22.   Hanuš, O.; Samková, E.; Křížová, L.; Hasoňová, L.; Kala, R. Role of Fatty Acids in Milk Fat and the Influence of Selected Factors on Their Variability-A Review. *Molecules* **2018**, *23*, 1636. [CrossRef] [PubMed]

23.   Markiewicz-Keszycka, M.; Grażyna-Runowska, G.; Lipińska, P.; Wójtowski, J. Fatty Acid Profile of Milk—A Review. *Bull. Vet. Inst. Pulawy* **2013**, *57*, 135–139. [CrossRef]

24.   Bernard, L.; Leroux, C.; Chilliard, Y. Nutritional regulation of mammary lipogenesis and milk fat in ruminant: Contribution to sustainable milk production. *Rev. Colomb. Cienc. Pecu.* **2013**, *26*, 292–302.

25.   Neville, M.C.; Picciano, M.F. Regulation of milk lipid secretion and composition. *Annu. Rev. Nutr.* **1997**, *17*, 159–183. [CrossRef]

26.   Månsson, H.L. Fatty Acids in Bovine Milk Fat. *Food Nutr. Res.* **2008**, *52*, 1–3.

27. Hintze, K.J.; Snow, D.; Burtenshaw, I.; Ward, R.E. Nutraceutical properties of milk fat globular membrane. In *Biotechnol. Biopolymers*; Elnashar, M., Ed.; Springer: Paris, France, 2011; pp. 321–342.

28. Lopez, C. Milk fat globules enveloped by their biological membrane: Unique colloidal assemblies with a specific composition and structure. *Curr. Opin. Colloid Interface Sci.* **2011**, *16*, 391–404. [CrossRef]

29. Lee, H.; Padhi, E.; Hasegawa, Y.; Larke, J.; Parenti, M.; Wang, A.; Hernell, O.; Lönnerdal, B.; Slupsky, C. Compositional Dynamics of the Milk Fat Globule and Its Role in Infant Development. *Front. Pediatr.* **2018**, *6*, 313. [CrossRef]

30. Argov-Argaman, N. Symposium review: Milk fat globule size—Practical implications and metabolic regulation. *J. Dairy Sci.* **2019**, *102*, 2783–2795. [CrossRef]

31. Keenan, T.W.; Morré, D.J.; Olson, D.E.; Yunghans, W.N.; Patton, S. Biochemical and morphological comparison of plasma membrane and milk fat globule membrane from bovine mammary gland. *J. Cell Biol.* **1970**, *44*, 80–93. [CrossRef]

32. Reece, W.O. *Dukes' Physiology of Domestic Animals*, 12th ed.; Melvin, J., Swenson, W., Reece, O., Eds.; Cornell University Press: Ithaca, NY, USA, 2004.

33. Horseman, N.D.; Collier, R.J. Serotonin: A local regulator in the mammary gland epithelium. *Annu. Rev. Anim. Biosci.* **2014**, *2*, 353–374. [CrossRef]

34. Wikipedia. Available online: https://en.wikipedia.org/wiki/Milk_fat_globule_membrane (accessed on 19 August 2020).

35. Sokoła-Wysoczańska, E.; Wysoczański, T.; Wagner, J.; Czyż, K.; Bodkowski, R.; Lochyński, S.; Patkowska-Sokoła, B. Polyunsaturated Fatty Acids and Their Potential Therapeutic Role in Cardiovascular System Disorders-A Review. *Nutrients* **2018**, *10*, 1561. [CrossRef] [PubMed]

36. Han, L.; Zhang, M.; Xing, Z.; Coleman, D.N.; Liang, Y.; Loor, J.J.; Yang, G. Knockout of butyrophilin subfamily 1 member A1 (BTN1A1) alters lipid droplet formation and phospholipid composition in bovine mammary epithelial cells. *J. Anim. Sci. Biotechnol.* **2020**, *11*, 72. [CrossRef] [PubMed]

37. McPherson, A.V.; Kitchen, B.J. Reviews of the progress of dairy science: The bovine milk fat globule membrane-its formation, composition, structure and behaviour in milk and dairy products. *J. Dairy Res.* **1983**, *50*, 107–133. [CrossRef]

38. Rohlfs, E.M.; Garner, S.C.; Mar, M.H.; Zeisel, S.H. Glycerophosphocholine and phosphocholine are the major choline metabolites in rat milk. *J. Nutr.* **1993**, *123*, 1762–1768. [CrossRef] [PubMed]

39. Kaufmann, W.; Hagemeister, H. Composition of milk. In *World Animal Science-C Production-System Approach, Dairy Cattle Production*; Gravert, H.O., Ed.; Elsevier Applied Science Publishers: Amsterdam, The Netherlands, 1987; pp. 107–171.

40. Noble, R.C. Digestion, absorption and transport of lipids in ruminant animals. In *Lipid Metabolism in Ruminant Animals*; Christie., W.W., Ed.; Pergamon Press: Oxford, UK, 1981; pp. 57–93.

41. Bitman, J.; Wood, D.L. Changes in milk fat phospholipids during lactation. *J. Dairy Sci.* **1990**, *73*, 1208–1216. [CrossRef]

42. Zeisel, S.H.; Mar, M.; Zhou, Z.; Da Costa, K.A. Pregnancy and lactation are associated with diminished concentrations of choline and its metabolites in rat liver. *J. Nutr.* **1995**, *125*, 3049–3054.

43. Kinsella, J.E. Preferential labelling of phosphatidylcholine during phospholipid synthesis by bovine mammary tissue. *Lipids* **1973**, *8*, 393–400. [CrossRef]

44. Piepenbrink, M.S.; Overton, T.R. Liver metabolism and production of periparturient dairy cattle fed rumen-protected choline. *J. Dairy Sci.* **2000**, *83*, 257.

45. Pinotti, L.; Campagnoli, A.; Sangalli, L.; Rebucci, R.; Dell'Orto, V.; Baldi, A.. Metabolism of periparturient dairy cows fed rumen-protected choline. *J. Anim. Feed Sci.* **2004**, *13*, 551–554. [CrossRef]

46. Baldi, A.; Pinotti, L. Choline metabolism in high-producing dairy cows: Metabolic and nutritional basis. *Can. J. Sci.* **2006**, *86*, 207–212. [CrossRef]

47. Cooke, R.F.; Silva Del Río, N.; Caraviello, D.Z.; Bertics, S.J.; Ramos, M.H.; Grummer, R.R. Supplemental choline for prevention and alleviation of fatty liver in dairy cattle. *J. Dairy Sci.* **2007**, *90*, 2413–2418. [CrossRef] [PubMed]

48. Baldi, A.; Bruckmaier, R.; D'ambrosio, F.; Campagnoli, A.; Pecorini, C.; Rebucci, R.; Pinotti, L. Rumen-protected choline supplementation in periparturient dairy goats: Effects on liver and mammary gland. *J. Agric. Sci.* **2011**, *150*, 1–7. [CrossRef]

49. Li, W.; Li, B.; Lv, J.; Dong, L.; Zhang, L.; Wang, T. Choline supplementation improves the lipid metabolism of intrauterine-growth-restricted pigs. *Asian Australas. J. Anim. Sci.* **2018**, *31*, 686–695. [CrossRef] [PubMed]

50. Pinotti, L.; Campagnoli, A.; Dell'Orto, V.; Baldi, A. Choline: Is there a need in lactating dairy cow? *Livest. Prod. Sci.* **2005**, *98*, 149–152. [CrossRef]

51. Pinotti, L.; Campagnoli, A.; D'Ambrosio, F.; Susca, F.; Innocenti, M.; Rebucci, R.; Fusi, E.; Cheli, F.; Savoini, G.; Dell'Orto, V.; et al. Rumen-Protected Choline and Vitamin E Supplementation in Periparturient Dairy Goats: Effects on Milk Production and Folate, Vitamin B12 and Vitamin E Status. *Animals* **2008**, *2*, 1019–1027. [CrossRef] [PubMed]

52. Pinotti, L.; Polidori, C.; Campagnoli, A.; Dell'Orto, V.; Baldi, A. A meta-analysis of the effects of rumen protected choline supplementation on milk production in dairy cows. In *Energy and Protein Metabolism and Nutrition*; EAAP Publication No. 127; Crovetto, G.M., Ed.; Wageningen Academic Publishers: Wageningen, The Netherlands, 2010; pp. 321–322.

53. Pinotti, L. Vitamin-Like Supplementation in Dairy Ruminants: The Case of Choline. In *Milk Production-An. Up-to-Date Overview of Animal Nutrition, Management and Health*; Chaiyabutr, N., Ed.; IntechOpen: London, UK, 2012; pp. 65–86.

54. Pinotti, L.; Manoni, M.; Fumagalli, F.; Rovere, N.; Tretola, M.; Baldi, A. The role of micronutrients in high-yielding dairy ruminants: Choline and vitamin E. *Ank. Univ. Vet. Fak. Derg.* **2020**, *67*, 209–214. [CrossRef]

55. Arshad, U.; Zenobi, M.G.; Staples, C.R.; Santos, J.E.P. Meta-analysis of the effects of supplemental rumen-protected choline during the transition period on performance and health of parous dairy cow. *J. Dairy Sci.* **2020**, *103*, 282–300. [CrossRef]

56. Mather, I.H. A review and proposed nomenclature for major proteins of the milk-fat globule membrane. *J. Dairy Sci.* **2000**, *83*, 203–247. [CrossRef]

57. Brink, L.R.; Herren, A.W.; McMillen, S.; Fraser, K.; Agnew, M.; Roy, N.; Lönnerdal, B. Omics analysis reveals variations among commercial sources of bovine milk fat globule membrane. *J. Dairy Sci.* **2020**, *103*, 3002–3016. [CrossRef]

58. Vissac, C.; Lemery, D.; Le Corre, L.; Fustier, P.; Dechelotte, P.; Maurizis, J.C.; Bignon, Y.J.; Bernard-Gallon, D.J. Presence of BRCA1 and BRCA2 proteins in human milk fat globules after delivery. *Biochim. Biophys. Acta* **2002**, *1586*, 50–56. [CrossRef]

59. OECD/FAO. *Agricultural Outlook 2019–2028*; OECD Publishing: Paris, France; Food and Agriculture Organization of the United Nations: Rome, Italy, 2019. [CrossRef]

60. Gallier, S.; Vocking, K.; Post, J.A.; Van De Heijning, B.; Acton, D.; van der Beek, E.M.; Van Baalen, T. A novel infant milk formula concept: Mimicking the human milk fat globule structure. *Colloid Surf. B* **2015**, *136*, 329–339. [CrossRef] [PubMed]

61. Lu, J.; Wang, X.; Zhang, W.; Liu, L.; Pang, X.; Zhang, S.; Lv, J. Comparative proteomics of milk fat globule membrane in different species reveals variations in lactation and nutrition. *Food Chem.* **2016**, *196*, 665–672. [CrossRef] [PubMed]

62. Sun, Y.; Wang, C.; Sun, X.; Guo, M. Comparative Proteomics of Whey and Milk Fat Globule Membrane Proteins of Guanzhong Goat and Holstein Cow Mature Milk. *J. Food Sci.* **2019**, *84*, 244–253. [CrossRef] [PubMed]

63. Cavaletto, M.; Giuffrida, M.G.; Conti, A. The proteomic approach to analysis of human milk fat globule membrane. *Clin. Chim. Acta.* **2004**, *347*, 41–48. [CrossRef] [PubMed]

64. Fortunato, D.; Giuffrida, M.G.; Cavaletto, M.; Garoffo, L.P.; Dellavalle, G.; Napolitano, L.; Giunta, C.; Fabris, C.; Bertino, E.; Coscia, A.; et al. Structural proteome of human colostral fat globule membrane proteins. *Proteomics* **2003**, *3*, 897–905. [CrossRef] [PubMed]

65. Reinhardt, T.A.; Lippolis, J.D. Bovine milk fat globule membrane proteome. *J. Dairy Res.* **2006**, *73*, 406–416. [CrossRef]

66. Hettinga, K.; van Valenberg, H.; de Vries, S.; Boeren, S.; van Hooijdonk, T.; van Arendonk, J.; Vervoort, J. The host defense proteome of human and bovine milk. *PLoS ONE* **2011**, *6*, e19433. [CrossRef]

67. Yang, Y.; Zheng, N.; Wang, W.; Zhao, X.; Zhang, Y.; Han, R.; Ma, L.; Zhao, S.; Li, S.; Guo, T.; et al. N-glycosylation proteomic characterization and cross-species comparison of milk fat globule membrane proteins from mammals. *Proteomics* **2016**, *16*, 2792–2800. [CrossRef]

68. Cao, X.; Zheng, Y.; Wu, S.; Yang, N.; Yue, X. Characterization and comparison of milk fat globule membrane N-glycoproteomes from human and bovine colostrum and mature milk. *Food Funct.* **2019**, *10*, 5046–5058. [CrossRef]

69. Li, W.; Li, M.; Cao, X.; Yang, M.; Han, H.; Kong, F.; Yue, X. Quantitative proteomic analysis of milk fat globule membrane (MFGM) proteins from donkey colostrum and mature milk. *Food Funct.* **2019**, *10*, 4256–4268. [CrossRef]

70. Sun, Y.; Wang, C.; Sun, X.; Jiang, S.; Guo, M. Characterization of the milk fat globule membrane proteome in colostrum and mature milk of Xinong Saanen goats. *J. Dairy Sci.* **2020**, *103*, 3017–3024. [CrossRef] [PubMed]

71. Froehlich, J.W.; Chu, C.S.; Tang, N.; Waddell, K.; Grimm, R.; Lebrilla, C.B. Label-free liquid chromatography-tandem mass spectrometry analysis with automated phosphopeptide enrichment reveals dynamic human milk protein phosphorylation during lactation. *Anal. Biochem.* **2011**, *408*, 136–146. [CrossRef] [PubMed]

72. Yang, M.; Deng, W.; Cao, X.; Wang, L.; Yu, N.; Zheng, Y.; Wu, J.; Wu, R.; Yue, X. Quantitative Phosphoproteomics of Milk Fat Globule Membrane in Human Colostrum and Mature Milk: New Insights into Changes in Protein Phosphorylation during Lactation. *J. Agric. Food Chem.* **2020**, *68*, 4546–4556. [CrossRef]

73. Reinhardt, T.A.; Lippolis, J.D. Developmental changes in the milk fat globule membrane proteome during the transition from colostrum to milk. *J. Dairy Sci.* **2008**, *91*, 2307–2318. [CrossRef] [PubMed]

74. Yang, Y.; Zheng, N.; Zhao, X.W.; Zhang, Y.D.; Han, R.W.; Ma, L.; Zhao, S.; Li, S.; Guo, T.; Wang, J. Proteomic characterization and comparison of mammalian milk fat globule proteomes by iTRAQ analysis. *J. Proteom.* **2015**, *116*, 34–43. [CrossRef]

75. Liao, Y.; Alvarado, R.; Phinney, B.; Lönnerdal, B. Proteomic characterization of human milk fat globule membrane proteins during a 12 month lactation period. *J. Proteome Res.* **2011**, *10*, 3530–3541. [CrossRef]

76. Spertino, S.; Cipriani, V.; De Angelis, C.; Giuffrida, M.G.; Marsano, F.; Cavaletto, M. Proteome profile and biological activity of caprine, bovine and human milk fat globules. *Mol. Biosyst.* **2012**, *8*, 967–974. [CrossRef]

77. Timby, N.; Domellöf, E.; Hernell, O.; Lönnerdal, B.; Domellöf, M. Neurodevelopment, nutrition, and growth until 12 mo of age in infants fed a low-energy, low-protein formula supplemented with bovine milk fat globule membranes: A randomized controlled trial. *Am. J. Clin. Nutr.* **2014**, *99*, 860–868. [CrossRef]

78. Demmelmair, H.; Prell, C.; Timby, N.; Lönnerdal, B. Benefits of Lactoferrin, Osteopontin and Milk Fat Globule Membranes for Infants. *Nutrients* **2017**, *9*, 817. [CrossRef]

79. Institute of Medicine (IOM) of the National Academics. *Infant Formula Evaluating the Safety of New Ingredients*; The National Academics Press: Washington, DC, USA, 2004.

80. EFSA NDA Panel (EFSA Panel on Dietetic Products, Nutrition and Allergies). Scientific Opinion on the essential composition of infant and follow-on formulae. *EFSA J.* **2014**, *12*, 3760. [CrossRef]

81. Martin, C.R.; Ling, P.R.; Blackburn, G.L. Review of Infant Feeding: Key Features of Breast Milk and Infant Formula. *Nutrients* **2016**, *8*, 279. [CrossRef] [PubMed]

82. Jensen, R.G.; Ferris, A.M.; Lammi-Keefe, C.J.; Henderson, R.A. Lipids of bovine and human milks: A comparison. *J. Dairy Sci.* **1990**, *73*, 223–240. [CrossRef]

83. Bzikowska-Jura, A.; Czerwonogrodzka-Senczyna, A.; Jasińska-Melon, E.; Mojska, H.; Olędzka, G.; Wesołowska, A.; Szostak-Węgierek, D. The Concentration of Omega-3 Fatty Acids in Human Milk Is Related to Their Habitual but Not Current Intake. *Nutrients* **2019**, *11*, 1585. [CrossRef] [PubMed]

84. Corkins, K.G.; Shurley, T. What's in the Bottle? A Review of Infant Formulas. *Nutr. Clin. Pr.* **2016**, *31*, 723–729. [CrossRef] [PubMed]

85. Gurnida, D.A.; Rowan, A.M.; Idjradinata, P.; Muchtadi, D.; Sekarwana, N. Association of complex lipids containing gangliosides with cognitive development of 6-month-old infants. *Early Hum. Dev.* **2012**, *88*, 595–601. [CrossRef]

86. Lönnerdal, B. Infant formula and infant nutrition: Bioactive proteins of human milk and implications for composition of infant formulas. *Am. J. Clin. Nutr.* **2014**, *99*, 712S–717S. [CrossRef]

87. Billeaud, C.; Puccio, G.; Saliba, E.; Guillois, B.; Vaysse, C.; Pecquet, S.; Steenhout, P. Safety and tolerance evaluation of milk fat globule membrane-enriched infant formulas: A randomized controlled multicenter non-inferiority trial in healthy term infants. *Clin. Med. Insights Pediatr.* **2014**, *8*, 51–60. [CrossRef] [PubMed]

88. Zavaleta, N.; Kvistgaard, A.S.; Graverholt, G.; Respicio, G.; Guija, H.; Valencia, N.; Lönnerdal, B. Efficacy of an MFGM-enriched complementary food in diarrhea, anemia, and micronutrient status in infants. *J. Pediatr. Gastroenterol. Nutr.* **2011**, *53*, 561–568. [CrossRef]

89. Givens, I. *Milk and Dairy Foods*, 1st ed.; Academic Press: London UK, 2020; pp. 1–417.

90. Ye, A.; Cui, J.; Singh, H. Effect of the fat globule membrane on in vitro digestion of milk fat globules with pancreatic lipase. *Int. Dairy J.* **2010**, *20*, 822–829. [CrossRef]

91. Ramírez, M.; Amate, L.; Gil, A. Absorption and distribution of dietary fatty acids from different sources. *Early Hum. Dev.* **2001**, *65*, 95–101. [CrossRef]

92. Agnew, J.E.; Holdsworth, C.D. The effect of fat on calcium absorption from a mixed meal in normal subjects, patients with malabsorptive disease, and patients with a partial gastrectomy. *Gut* **1971**, *12*, 973–977. [CrossRef] [PubMed]

93. Petit, V.; Sandoz, L.; Garcia-Rodenas, C.L. Importance of the regiospecific distribution of long-chain saturated fatty acids on gut comfort, fat and calcium absorption in infants. *Prostaglandins Leukot. Essent. Fat. Acids.* **2017**, *121*, 40–51. [CrossRef] [PubMed]

94. Ayala-Bribiesca, E.; Turgeon, S.L.; Britten, M. Effect of calcium on fatty acid bioaccessibility during in vitro digestion of Cheddar-type cheeses prepared with different milk fat fractions. *J. Dairy Sci.* **2017**, *100*, 2454–2470. [CrossRef] [PubMed]

95. Tomarelli, R.M.; Meyer, B.J.; Weaber, J.R.; Bernhart, F.W. Effect of Positional Distribution on the Absorption of the Fatty Acids of Human Milk and Infant Formulas. *Nutr. J.* **1968**, *95*, 583–590. [CrossRef] [PubMed]

96. German, J.B.; Dillard, C.J. Composition, structure and absorption of milk lipids: A source of energy, fat-soluble nutrients and bioactive molecules. *Crit. Rev. Food Sci. Nutr.* **2006**, *46*, 57–92. [CrossRef] [PubMed]

97. O'Donnell, R.; Holland, J.W.; Deeth, H.C.; Alewood, P.F. Milk proteomics. *Int. Dairy J.* **2004**, *14*, 1013–1023. [CrossRef]

 *foods*

Article

Evaluation of Biological, Textural, and Physicochemical Parameters of Panela Cheese Added with Probiotics

Karina A. Parra-Ocampo, Sandra T. Martín-del-Campo, José G. Montejano-Gaitán, Rubén Zárraga-Alcántar and Anaberta Cardador-Martínez *

Tecnologico de Monterrey, School of Engineering and Sciences, Epigmenio González 500, Fracc, San Pablo 76130, Querétaro, Mexico; karina.parraocampo@hotmail.com (K.A.P.-O.); smartinde@tec.mx (S.T.M.-d.-C.); gmonteja@tec.mx (J.G.M.-G.); rzarragaa@tec.mx (R.Z.-A.)
* Correspondence: mcardador@tec.mx; Tel.: +52-442-238-3224

Received: 25 September 2020; Accepted: 14 October 2020; Published: 21 October 2020

Abstract: Biological, physicochemical and textural parameters of a Panela cheese with and without probiotics (LSB-c and C-c) were analyzed during 15 days of storage at 4 °C. Changes in cohesiveness, hardness, springiness, and chewiness were measured by texture profile analysis. Additionally, moisture, pH, nitrogenous fractions (nitrogen soluble in pH 4.6, non-protein nitrogen, 70% ethanol-soluble nitrogen, and water-soluble extract) were evaluated. The peptide profile of nitrogenous fractions was also analyzed. Finally, biological activity was evaluated by ABTS (2,2′-Azino-bis(3-ethylbenzothiazoline-6-sulfonic acid) diammonium salt) and DPPH (2,2-diphenyl-1-picrylhydrazyl), as well as the Inhibition of Angiotensin-Converting Enzyme. Analysis of variance showed significant differences for most of the evaluated parameters. By principal component analysis (PCA), two groups were separated, one corresponding to LSB-c and the other corresponding to C-c. The separation was given mostly by hardness, chewiness, and ABTS of all nitrogenous fractions. LSB-c showed higher biological activities than C-c.

Keywords: panela cheese; angiotensin-converting enzyme inhibition; probiotic addition; antioxidant activity; DPPH; ABTS

1. Introduction

Bioactive peptides are genuine or generated components of ready-to-eat foods that may exert a regulatory activity in the human organism, regardless of their nutritive functions [1].

It is known that bovine milk is the most significant source of food-derived bioactive peptides [2]. The existence of bioactive peptides in fermented milk products and ripened cheese has been described [3]. During proteolysis, various peptides are released from the milk proteins; they are inactive while encrypted in the milk proteins. Proteolysis takes place during food processing, e.g., milk fermentation and cheese maturation, or during gastrointestinal transit. Some of the bioactive properties reported in peptides derived from milk products are antihypertensive, antioxidant, antimicrobial, immunomodulatory, and mineral binding [4–6]. The amount and type of bioactive peptides in cheese are affected by the starter culture and ripening conditions [7].

In 2019, the production of fluid milk in Mexico was approximately 12.6 million metric tons, of which almost 50% was utilized for cheese elaboration [7]. According to SAGARPA (Mexican Ministry of Agriculture and Sustainable Development) [8], in 2018, Mexican cheese production was 418,650 tons, where panela represented the third most-produced cheese and represented 11.7% of total production.

Panela cheese is a very popular handcrafted Mexican white, soft, fresh cheese manufactured from pasteurized cow's skim or partially skimmed milk [9], with little or no starter culture acidification [9,10].

According to the manufacture characteristics of this type of cheese, it is expected some degree of proteolysis and thus a release of bioactive peptides during manufacturing procedures, storage, and post-consumption.

On the other hand, hypertension is a state of a sustained increase in blood pressure (BP), related to cardiovascular diseases. Hypertension is the mortality most related factor around the world [11]. This is a chronic disease derived from many factors such as genetics, excessive sodium intake, age, smoking, sedentary lifestyle, and chronic diseases such as diabetes and obesity [12,13]. According to W.H.O. [14], more than one of every five adults suffers from hypertension.

Within the organism, the regulation of blood pressure is related to the hormone "renin-angiotensin system" (RAS). The angiotensin-converting enzyme (ACE) is key within RAS because it converts the peptide angiotensin I to the vasoconstrictor angiotensin II, which tightens the blood vessel and increase the BP. ACE-inhibitors are competitive substrates for ACE, and among them are milk-derived bioactive peptides. The C-terminal of the inhibitor is the primary feature governing the inhibition of ACE [11].

Oxidative stress is a condition of imbalance between reactive oxygen species (ROS) with unpaired electrons and the body's ability to detoxify and repair the damage of the reactive components. It is widely related to the illnesses of the human body, including hypertension and other chronic diseases. Milk protein-derived peptides are among the natural dietary sources of antioxidants. Peptides from β-casein and α_{s1}-casein are potent anion radical scavengers [11].

This work aimed to measure antioxidant, and ACE inhibition of a Panela cheese added (LSB-c)/not added (C-c) with probiotics. Physicochemical and textural parameters were also monitored during 15 days of storage (4 ± 0.6 °C) to evaluate their effect and relationship upon structural changes in both types of cheeses.

2. Materials and Methods

2.1. Cheese Manufacture

For this study, one 80-L batch of whole milk was obtained in the Tecnológico de Monterrey experimental agricultural field (CAETEC) (Querétaro, Mexico) and transported to the Tecnológico de Monterrey, Querétaro, Mexico under controlled temperature conditions. In the CAETEC, milk production is controlled to avoid composition variation throughout the year. To achieve this, the calf diet is standardized with a feed formulate by Tecnologico de Monterrey. The herd has about 100 milking cows. Different national associations and industrial clients have certified the homogeneity and quality of CAETEC milk.

Whole milk (3.11% protein, 3.19% fat, initial pH 6.69) was cooled and stored at 4 °C for 24 h before cheese making. For cheese making, the Querataro's traditional panela making procedure was followed. Milk was pasteurized in a big pot (63 °C for 30 min) previously to cheese manufacturing and was split into two portions of 40-L at the food engineering facilities of Tecnológico de Monterrey, Querétaro, Mexico. The experimental design was a unifactorial design where the factor evaluated was storage time with four levels (0, 5, 10, and 15 days). Cheeses with probiotics (LSB-c) and without probiotics (C-c) where considered as independent blocks.

For the cheese added with probiotics (LSB-c), after pasteurization, the 40 L of milk was heated gradually to 32 °C, and 10^9 CFU/L of commercial type MM101 (Lyofast®, Sacco, Via Manzoni, Italy) culture was added, which consisted of *Lactococcus lactis*, *Lactococcus cremoris*, *Lactococcus diacetylactis*, and *Streptococcus thermophilus*. Then, 10^9 CFU/mL *Bifidobacterium animalis* ssp lactis (Lyofast®, Sacco, Via Manzoni, Italy) was added to the mix. Inoculated milk was kept at 32 °C for three h until it reached a pH value of 6.4. $CaCl_2$ (Cal-Sol501, Industrias Cuamex, San Miguel Iztapalapa, CDMX, Mexico) was added by diluting 7 mL in 35 L of milk and kept in slow agitation for 1 min. Liquid calf rennet (Strength 1:7500. Qualact®, Altecsa SA, Mexico City, Mexico) was added to milk (1.95 mL + 30 mL water). Coagulation of milk was completed in 30–35 min; then, the curd was cut into cubes (1 cm^3) and allowed to rest in whey for 5 min before draining approximately 2/3 of whey. Salt was added to curd at 1% and mixed manually for 5 min. Then cheeses were molded in 100 g plastic molds. Molds were slightly pressed for 20 min (by each side) by piling them up on one another, allowing natural whey drainage at room temperature. For C-c (control cheese, without any added culture), 40 L of milk were used following the above procedure without culture addition. Both kinds of cheeses were made simultaneously to avoid changes in milk quality. For each kind of cheese, 34×100 g-pieces were obtained, giving a total of 68×100 g pieces of cheese.

Individual cheeses were packed in plastic bags. Then, they were stored in refrigeration at $93.8 \pm 1\%$ of relative humidity (RH) and 4 ± 0.6 °C temperature for 15 days.

2.2. Sampling

Two whole 100-g pieces of cheese (LSB-c and C-c, respectively) were removed after 24 h (denoted as day 0), 5, 10, and 15 days of storage for biological, physicochemical, and textural determinations. Analysis was performed by triplicate on the same cheese, and two cheesemaking trials were done.

2.3. Physicochemical Analysis

Moisture was determined by oven drying cheese samples at 100 ± 2 °C, according to NOM-116-SSA1-1994 [15], by triplicate. The pH values of cheeses were determined on the surface by triplicate, according to NMX-F-317-S-1978 [16], with an Oakton pH meter (Eutech Instruments, Vernon Hills, IL, USA).

2.4. Instrumental Texture Profile Analysis

For texture properties evaluation, six replicates were made in Panela cheese samples with a CT3 Texture Analyzer® (Brookfield, AMETEK, Middleborough, MA, USA). The texture was evaluated using a two-bite compression test. Cylindrical samples (1.8 cm of diameter and 1.5 cm height) were tested by using a 50 N load cell and two parallel plates (10 cm diameter). The compression ratio was established at 50% deformation from the original height and a rate of 200 mm/min, similar to a deformation rate between fingers during squeezing [17]. Cheeses were left at room temperature for 15 min after being removed from refrigeration before obtaining cheese cylinders and proceeding with the texture profile analysis (TPA). Parameters measured were cohesiveness, hardness, springiness, and chewiness, and were obtained from the force-time plots of Tension Zero version 1.0 [18]. Cohesiveness, defined as the strength of the internal bonds making up the body of the product, was calculated by the ratio between the area under the second-bite curve and the area under the first-bite curve. Hardness, defined as the maximum force required to compress the cheese sample 50% from its original height during the first compression. Springiness, defined as the distance regained by the sample during the time between the end of the first compression and the beginning of the second compression. The chewiness was defined as the product of hardness, cohesiveness, and springiness, as described by Bourne [19].

2.5. Nitrogenous Fractions Obtention

Nitrogenous fractions were obtained by crude fractionation and were used to evaluate biological activities, antioxidant, and inhibition of the angiotensin-converting enzyme (ACEI) and to evaluate the peptide profile among each fraction. Nitrogen soluble in pH 4.6 (ASN), nonprotein nitrogen (NPN), 70% ethanol-soluble nitrogen (EtOH-SN), and a water-soluble extract (WSE) were obtained. For ASN and NPN the method described by Leclercq-Perlat, et al. [20] was used with minor modifications. A cheese suspension was prepared with 10 g of ground cheese and 100 mL of NaCl solution (9 g/L). This was homogenized (10 min, 25 °C) using an ULTRA-TURRAX IKA T18 basic (Interscience, Wilmington, NC, USA). ASN was obtained by adjusting the pH of the suspension to 4.6 by adding 2N HCl. After pH adjustment, the samples were incubated 20 min at 25 °C. Then, they were centrifuged during 30 min at 6000 rpm. The soluble fraction was recovered after filtration through Whatman No. 42 paper. For NPN, an aliquot of 25 mL of cheese suspension was mixed with 15 mL water. Then 10 mL of 60% (*w/v*) trichloroacetic acid (TCA) was added to achieve a final TCA concentration of 12%. Samples were homogenized and incubated at 25 °C for 20 min. Then they were filtered through Whatman No. 42 paper. EtOH-SN was prepared according to the method described by Guerra Martínez, et al. [21]. WSE was prepared according to Rohm, et al. [22], 20 g of cheese was added to 40 mL of distilled water and homogenized for 2 min using an ULTRA-TURRAX IKA T18 basic. The homogenate was held at 40 °C for 1 h and centrifuged at 3000 *g* for 30 min at 4 °C. The fat was removed, then the supernatant was filtered through Whatman No. 42 paper. Nitrogenous fractions were held at −80 °C until analysis.

2.6. Evaluation of Biological Activities of Nitrogenous Fractions

2.6.1. Antioxidant Capacity by ABTS

Antioxidant capacity was measured using the methodology of Re, et al. [23], through the decolorization of the radical 2,2′-azino-bis(3-ethylbenzothiazoline-6-sulphonic acid) (ABTS) detected spectrophotometrically at 734 nm.

A solution of 7 mM radical cation ABTS in a 2.45 mM potassium persulfate solution was prepared and allowed to stand in darkness at room temperature for 16 h. ABTS solution was diluted with ethanol in a 1:20 ratio to get an absorbance of 0.70 (± 0.02) at 734 nm. 20 µL of sample/Trolox and 200 µL of ABTS solution were added to each well, and after 6 min of reaction, absorbance was recorded at 734 nm. Results are expressed as µM equivalents of Trolox.

2.6.2. Antioxidant Activity by DPPH

Anti-free radical activity using 2,2-diphenyl-1-picrylhydrazyl (DPPH) was determined by the method described by Pyrzynska and Pękal [24]. A solution of 125 µM DPPH with 80% methanol was prepared. 20 µL of sample/standard and 200 µL of DPPH were plated in each well and incubated for 90 min in darkness. Absorbance was measured at 520 nm. Results are expressed in % discoloration.

2.6.3. Angiotensin-Converting Enzyme Inhibitory Activity

Evaluation of inhibition of the angiotensin-converting enzyme (ACE) was developed according to the methodology proposed by Wang et al. [25]. ACE (0.1 U/mL mM) and hippuryl histidyl leucine (HHL, 5 mM) were dissolved in borate buffer (100 mM, pH 8.3, 300 mM NaCl).

The reaction mixture (10 µL HHL, 10 µL ACE, 40 µL sample, and 40 µL borate buffer) was incubated at 37 °C for 30 min, and then 250 µL HCl 1N was added to stop the reaction. Samples were analyzed in an HPLC (1200 Agilent, Milford, MA, USA) equipped with an Eclypse XDB-C18 column (4.6 × 150 mm, 5 µm, Agilent). The mobile phase consisted of solvent A, 0.05% TFA (trifluoroacetic acid) and 0.05% TEA (triethylamine) in water; solvent B, 100% ACN (acetonitrile); the ratio of solvent A/solvent B was 7/3 with a gradient of 5–60% of B the first 10 min, 2 min at 60% of B and 1 min of 5% of B. The flow rate was 0.5 mL/min, and the injection volume was 10 µL. The detector was set at 226 nm. The column temperature was held at 30 °C. The inhibitory rate was calculated by:

$$\%I = \frac{(A - B)}{A} \times 100 \tag{1}$$

where *A* was the peak area of HA without adding ACE inhibitors, *B* was the peak area of HA with adding ACE inhibitors.

2.7. Peptide Profile by HPLC

The peptide profile was analyzed with the methodology of Abadía-García, et al. [26]. All nitrogenous fractions were analyzed by RP-HPLC. Peptides separation was performed at 25 °C in an Agilent 1200 series system (Agilent Technologies, Palo Alto, Santa Clara, CA, USA) using a Zorbax 300 SB column (C18 5 µm, 4.6 × 150 mm). Mobile phase was solvent A, 10% ACN with 0.05% TFA; solvent B, 60% ACN with 0.05% TFA. The flow rate was 0.75 mL/min. The gradient consisted of 100%

of A for 10 min; 0–49% of B from minute 11–98; 50.80% of B from 99–108 min; 81–100% of B from 109–114 min and from 115–120 min 100% of B. The detector was set at 215 nm.

Peaks in each fraction were coded by retention time, and consecutive numbers were assigned for further statistical analysis. Quantification was done using the peak integrated area.

2.8. Statistical Analysis

The statistical analyses were carried out using Statistica v13 (TIBCO Software Inc., Palo Alto, CA, USA). One-way analysis of variance (ANOVA) was used to determine significant differences ($p < 0.05$) between the sampling days for antioxidant activity, ACEI, moisture, pH, and TPA parameters of cheeses. General linear model was used to obtain the least square average. For each significant variable, differences between means were detected using Tukey's honest significant difference (HSD) test with $\alpha = 0.05$. Correlation analyses between physicochemical and textural parameters and among biological activities and the peaks obtained from the peptide profile were done.

Finally, a principal component analysis (PCA) was applied using all the response variables. PCA is a multivariate statistical method that replaces the original variables with new ones called principal components, making it possible to obtain an overview of the data set information.

3. Results and Discussion

3.1. Physicochemical and Textural Parameters of Cheeses

Mean values of pH, moisture content, and textural properties are given in Table 1. Overall, pH in LSB-c decreased significantly ($p < 0.05$) from day 0 until day 10, and remained the same until day 15; in C-c, pH decreased significantly ($p < 0.05$) constant until day 15 (pH 4.89). C-c final pH was significantly ($p < 0.05$) lower than LSB-c pH. Hayaloglu, et al. [27], confirmed that *Lactococcus* spp., *S. thermophilus*, and *B. animalis* ssp. *lactis* have extensive activity in cheese acidification. For a cheese without added probiotics, results are similar to those previously reported by Guerra Martínez, et al. [21].

Table 1. Mean values of physicochemical and textural parameters of LSB-c and C-c both studied during 15 days of storage under refrigeration at 4 ± 0.6 °C.

		Treatments [b]						
		LSB-c [¥]				C-c [¥]		
Parameter [c]	p [a]	Day 0	Day 5	Day 10	Day 15	Day 0	Day 5	Day 10	Day 15
pH	0.00 ***	4.99 [a]	4.95 [b]	4.93 [c]	4.92 [c]	6.5 [d]	6.16 [e]	6.13 [e]	4.89 [f]
Moisture [s]	0.00 ***	72.11 [a]	61.10 [b]	58.95 [c]	58.12 [d]	60.95 [d]	60.53 [d]	59.77 [e]	59.17 [f]
Hardness [u]	0.00 ***	955.83 [a]	618.33 [b]	500.83 [b]	625.00 [b]	1720.83 [e,f]	1883.33 [d,e]	2068.33 [d]	1654.17 [f]
Springiness	0.58	0.92	0.71	0.74	0.69	0.88	0.83	0.84	0.83
Cohesiviness	0.00 ***	0.35 [b]	0.64 [a]	0.66 [a]	0.58 [a]	0.64 [d]	0.36 [e]	0.42 [e,f]	0.60 [d,e]
Chewiness [u]	0.00 ***	308.42 [a]	275.57 [a]	243.96 [a]	247.83 [a]	971.24 [d]	565.72 [e]	741.44 [d,e]	822.19 [d,e]

[¥] LSB-c: Probiotic cheese; C-c: Control cheese. [a] Significant at *** $p < 0.001$. [b] Means of each cheese with different letters within the same row are significantly different ($p < 0.05$). [c] Expressed as: [s] %. [u] (N).

Moisture decreased significantly, along with storage in both LSB-c and C-c. The acidification could have enhanced the expulsion of whey and, in the specific case of LSB-c, internal metabolism of added microorganisms [28].

In general, hardness in LSB-c presented significant differences ($p < 0.05$) from day 0 to day 5 but remained constant until the end of storage. Hardness in C-c was greater ($p < 0.05$) than LSB-c. Panela cheese presents a porous structure, and as moisture decreases, the size of porous spaces increases, causing a decrease in instrumental hardness. Additionally, the decrease in hardness in LSB-c could be explained with the casein matrix hydrolysis caused by the added culture [29]. Souza and Saad [30] reported that Minas fresh cheese supplemented with mesophilic culture presented a significant increase in hardness throughout storage and lower hardness values compared with a control cheese. Buriti, et al. [31], found that Minas fresh cheese added with probiotics presented an increase in hardness. Additionally, Dinakar and Mistry [32], reported that probiotics added to Cheddar cheese showed significant changes in texture without affecting either flavor or appearance in the sensorial analysis. Hardness in C-c remained the same until day 10, and decreased at day 15, which could be related to cheese proteolysis that could be determined by the organoleptic characteristics of the cheese.

3.2. Correlations Between Physicochemical and Textural Parameters of Cheeses

The correlations observed between texture and physicochemical parameters for LSB-c and C-c are shown in Tables 2 and 3, respectively. Hardness showed a high positive correlation with moisture, springiness, and chewiness in LSB-c, it can be attributed to early casein matrix hydrolysis by residual enzymes present in the rennet and also to the proteolytic system of added culture [33]. Springiness and hardness are affected by proteolytic enzymes that act mainly over α_{s1} casein [29,34]. A decrease in hardness also contributes to a decrease in chewiness since it is defined as the effort used to chew food to reduce it to the consistency necessary to swallow it.

Table 2. Correlation coefficients for the physicochemical and textural parameters analyzed in LSB-c and studied during storage under refrigeration at 4 °C.

Variables [a]	Moisture [s]	pH	Hardness	Springiness	Cohesivity
pH	0.23				
Hardness [u]	0.76 *	0.08			
Springiness	0.97 *	0.29	0.73 *		
Cohesivity	−0.80 *	−0.41 *	−0.79 *	−0.80 *	
Chewiness [u]	0.36	−0.30	0.65 *	0.33	−0.10

LSB-c: probiotic cheese. * Correlations are significant at $p < 0.05$. [a] Expressed as: [s] %; [u] (N).

Table 3. Correlation coefficients for the physicochemical and textural parameters analyzed in C-c and studied during storage under refrigeration at 4 °C.

Variables [a]	Moisture [s]	pH	Hardness	Springiness	Cohesivity
pH	0.83 *				
Hardness [u]	0.06	0.37			
Springiness	0.69 *	0.61 *	−0.25		
Cohesivity	0.06	−0.18	−0.33	0.44 *	
Chewiness [u]	0.15	0.02	0.00	0.45 *	0.94 *

C-c: control cheese. C-c stands for control cheese. * Correlations are significant at $p < 0.05$. [a] Expressed as: [s] %; [u] (N).

Moisture is ($p < 0.05$) negatively correlated with cohesivity (Table 2) and is explained by the fact that as the cheese ripens, it becomes a more cohesive material [35]. Springiness and moisture showed a positive correlation. This result matches with those reported by Osorio Tobón, et al. [35], in Edam cheese. Cohesivity is significantly ($p < 0.05$) and negatively correlated with springiness (Table 2). Thus, if the cheese is more cohesive, proteins within it are hydrolyzed, so it becomes more difficult for the cheese to restore its initial shape after compression.

In Table 3, a positive correlation ($p < 0.05$) between pH and moisture content in C-c was observed. This could be attributed to the influence of water on the ionic environment of the cheese, which induces ionization of the calcium phosphate complexes and the functional groups of the amino acids [36].

Springiness is significantly correlated ($p < 0.05$) with moisture and pH. The crumbling characteristics at high pH of the C-c give it a higher capacity to restore its initial shape after compression when more water is available within the cheese matrix. This agrees with the results reported by Osorio Tobón, et al. [35].

Cohesivity is significantly correlated ($p < 0.05$) with chewiness; an increase in curd particle fusion leads to a firmer and closer structure of the cheese; thus, more bites are needed to disrupt the whole structure of cheese before swallowing [21].

3.3. Biological Activities of Nitrogenous Fractions

The results of biological activities presented by nitrogenous fractions are given in Table 4. All the biological activities within all nitrogenous fractions presented significant differences ($p < 0.001$). In NPN fraction, LSB-c and C-c ABTS values were significantly different ($p < 0.05$).

3.3.1. Antioxidant Activity of Nitrogenous Fractions

Both types of cheeses presented significant differences in antioxidant activity. ABTS values oscillated significantly during storage time. This could be attributed to the rate of formation of peptides during proteolysis. Gupta, et al. [37], observed the same behavior of a cheddar cheese added with Lactobacilli. NPN showed the highest ($p < 0.05$) percentage of DPPH discoloration during storage. In LSB-c, it increased significantly ($p < 0.05$) from day 0 to day 5, and it remained the same until the end of storage (50.81%); in C-c, it increased significantly ($p < 0.05$) until a 49.49% at day 15. These results are similar to Hernández Galán, et al. [38], who reported the highest discolorations in the NPN fraction of a Cotija hard cheese.

ASN presented the best ($p < 0.05$) ABTS antioxidant activity compared to all the fractions in both LSB-c and C-c; ABTS values remained the same ($p < 0.05$) from day 0 to day 10, and it increased significantly on day 15 for both kinds of cheese (Table 4). C-c ABTS values were lower than those for LSB-c (1511.12 and 1272.08 µM Trolox equivalents, respectively). ASN did not show DPPH free radical scavenging activity. Floegel et al. [39], reported ABTS as the best method for detecting antioxidant capacity in a variety of foods.

ETOH-SN ABTS values in both kinds of cheese were 250 µM Trolox, approximately. These results were similar to those obtained by Abadía-García, et al. [26], who reported an overall antioxidant activity of 300 µM Trolox equivalents in Cottage cheese.

In WSE, ABTS activity for both LSB-c and C-c showed a similar tendency than the other nitrogenous fractions.

3.3.2. Angiotensin Converting Enzyme Activity (ACEI) of Nitrogenous Fractions

NPN fraction also showed significant ACEI activity ($p < 0.001$). In LSB-c, ACEI remained the same ($p < 0.05$) during storage until it increased considerably ($p < 0.05$) at day 15, reaching 90.21%; C-c showed the same tendency ($p < 0.05$) but with a lower final value (80.99%). The ACEI activity could also be related to the release of bioactive peptides during cheese proteolysis. Hernández Galán, et al. [38], evaluated Cotija hard cheese during ripening, and they correlated ACE inhibitory activity with cheese proteolysis.

In ASN, LSB-c presented an average ACEI activity of 45.90% throughout storage. In C-c, ACEI activity was observed after day 5 and remained at 40.06% until the end of storage.

WSE of LSB-c had the highest ACE inhibitory activity ($p < 0.05$) among all the nitrogenous fractions. It is supposed that WSE contains all the peptides produced during proteolysis. This is confirmed by Gorostiza, et al. [40], who reported that the water-soluble extract refers to all the nitrogenous fractions.

3.4. Peptide Profile

Figure 1 shows a representative HPLC chromatogram of the whole experiment, which corresponds to ASN fraction in LSB-c. Table 5 shows peaks and their corresponding peak area count (which is a measure of the concentration of the compound it represents) during storage for both LSB-c and C-c. For each nitrogenous fraction, the significant statistical correlations with biological activities are shown in the corresponding tables (Tables 6–9).

Table 4. Mean values of biological activities of nitrogenous fractions analyzed in LSB-c and C-c both studied during storage under refrigeration at 4 °C.

		Nitrogenous Fractions [b]											
		NPN			ASN			ETOH-SN			WSE		
p [a]		0.00 ***	0.00 ***	0.00 ***	0.00 ***	0.00 ***	0.00 ***	0.00 ***	0.00 ***	0.00 ***	0.00 ***	0.00 ***	0.00 ***
Tx [c]	Day	ABTS [s]	DPPH [u]	ACE [w]	ABTS [s]	DPPH [u]	ACE [w]	ABTS [s]	DPPH [u]	ACE [w]	ABTS [s]	DPPH [u]	ACE [w]
LSB-c [y]	0	708.75 [c]	32.11 [b]	60.97 [b]	1343.33 [b]	0.00 [a]	53.11 [a]	258.33 [b]	2.51 [a]	0.00 [a]	1136.25 [a,b]	0.00 [a]	87.35 [a]
	5	942.50 [b]	49.11 [a]	61.16 [b]	1473.33 [a,b]	0.00 [a]	39.04 [a]	231.25 [b]	0.00 [b]	0.00 [a]	1107.08 [a,b]	0.00 [a]	60.17 [b]
	10	604.58 [d]	48.09 [a]	53.87 [b]	1479.17 [a,b]	0.00 [a]	39.42 [a]	580.83 [a]	0.00 [b]	0.00 [a]	1062.08 [b]	0.00 [a]	94.32 [a]
	15	1152.5 [a]	50.81 [a]	90.21 [a]	1511.12 [a]	0.00 [a]	52.07 [a]	230.58 [b]	2.57 [a]	0.00 [a]	1336.25 [a]	0.00 [a]	76.39 [a,b]
C-c [z]	0	959.58 [a]	37.33 [c]	49.79 [b]	924.17 [b]	0.00 [b]	0.00 [b]	239.58 [b]	0.00 [b]	0.00 [a]	584.17 [b]	6.18 [b]	56.86 [b]
	5	890.83 [b]	44.56 [b,c]	63.96 [b]	1082.92 [a,b]	0.00 [b]	46.37 [a]	242.08 [b]	0.00 [b]	0.00 [a]	994.58 [a]	0.00 [b]	66.06 [a,b]
	10	927.92 [a,b]	48.25 [a,b]	45.49 [b]	1073.33 [a,b]	1.10 [b]	29.82 [a,b]	256.67 [b]	0.00 [b]	0.00 [a]	1060.00 [a]	36.81 [a]	75.22 [a]
	15	596.67 [c]	49.49 [a]	80.99 [a]	1272.08 [a]	4.29 [a]	57.79 [a]	354.17 [a]	3.76 [a]	0.00 [a]	1200.00 [a]	0.00 [b]	0.00 [c]

LSB-c: probiotic cheese; C-c: control cheese. [a] Significant at *** $p < 0.001$. [y/z] Means with different letters within the same column are significantly different ($p < 0.05$). [b] NPN: Non-protein nitrogen, ASN: Nitrogen soluble at pH 4.6, ETOH-SN: 70% ethanol-soluble nitrogen, WSE: Water-soluble extract. [c] Tx: Treatment. Expressed as: [s] ABTS (2,2′-Azino-bis(3-ethylbenzothiazoline-6-sulfonic acid) diammonium salt) uM Trolox equivalents; [u] DPPH (2,2-diphenil-1-picrylhydrazyl) % discoloration; [w] ACE (Angiotensin converting enzyme) % inhibition.

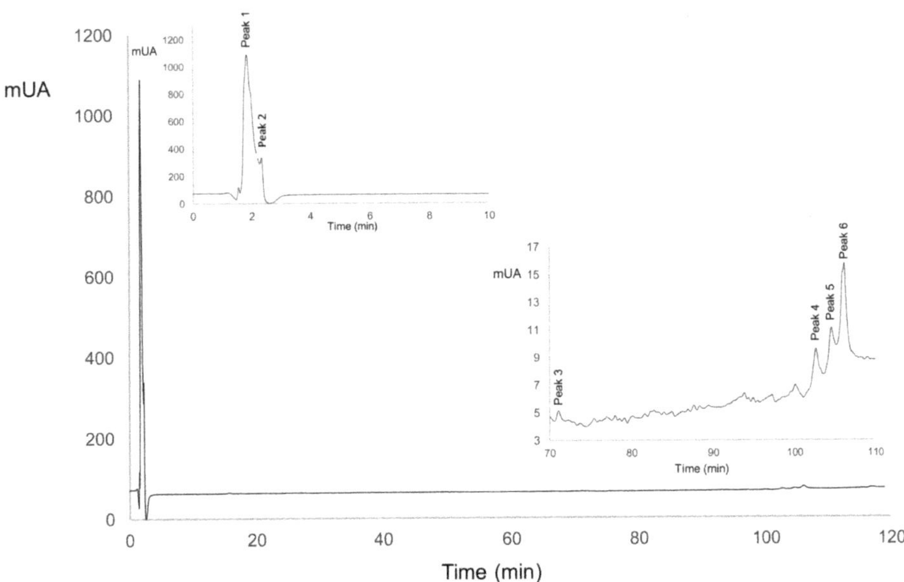

Figure 1. Peptide profile in ASN (nitrogen-soluble at pH 4.6) fraction for LSB-c (probiotic cheese).

Table 5. Peak Area within each nitrogenous fraction was analyzed in LSB-c and C-c during storage under refrigeration at 4 °C.

			Peak Area [b]							
			Day 0		Day 5		Day 10		Day 15	
NF [c]	Peak number	Retention time [a]	LSB-c [Y]	C-c [Y]	LSB-c	C-c	LSB-c	C-c	LSB-c	C-c
ASN	1	2.005	5562	0	8707	5030	10,022	6123	0	7231
	2	2.245	3772	0	2432	0	2300	763	2158	659
	3	72.428	58	0	0	0	0	0	0	0
	4	102.088	374	0	0	0	0	0	0	0
	5	102.935	11	0	339	0	333	0	128	0
	6	105.52	841	0	0	0	0	0	0	0
NPN	1	4.468	5137	5019	5076	5239	5296	5155	4588	4049
	2	116.899	0	221	220	345	384	420	610	568
ETOH-SN	1	1.536	21	20	39	25	34	35	112	74
	2	1.742	0	0	0	0	0	2101	0	6463
	3	1.842	5780	2547	5800	3504	6549	0	6960	0
	4	2.597	14	12	13	14	15	17	17	26
	5	116.653	377	220	180.6	178	203	230	451	345
WSE	1	1.12	533	509	556	570	483	475	596	576
	2	1.889	52,003	18,050	58,595	51,901	70,450	30,852	75,957	60,113
	3	2.286	3800	1759	0	0	0	5223	1688	4305
	4	117.284	690	676	754	680	621	325	345	301

[a] Expressed in minutes. [b] Expressed as units of chromatogram area. [c] NF: Nitrogenous fraction: ASN: Nitrogen soluble at pH 4.6; NPN: Non-protein nitrogen; ETOH-SN: Nitrogen soluble in Ethanol; WSE: Water-soluble extract. [Y] LSB-c: probiotic cheese; C-c: control cheese.

Table 6. Correlation coefficients for the peaks, day, treatment (LSB-c and C-c), and the biological activity parameters analyzed in ASN fraction of Panela cheeses studied during storage under refrigeration at 4 °C.

	Peak					
Parameter [a]	1 (2.005) [r]	2 (2.245)	3 (72.428)	4 (102.088)	5 (102.935)	6 (103.520)
Treatment	−0.28 *	−0.61 ***	−0.32 *	−0.32 *	−0.5 ***	−0.32 **
Day	−0.28 *	−0.14	−0.43 ***	−0.43 ***	−0.01	−0.43 ***
ABTS [s]	0.39 **	0.5 ***	0.38 **	0.37 **	0.43 ***	0.39 **
DPPH [u]	−0.08	−0.24	−0.14	−0.14	−0.23	−0.15
ACE [w]	0.52 ***	0.3 *	0.34 **	0.31 *	0.11	0.32 **

LSB-c: Probiotic cheese; C-c: Control cheese; ASN: Nitrogen soluble at pH 4.6. [a] Correlations are significant at * $p < 0.05$, ** $p < 0.01$, *** $p < 0.001$. Expressed as: [s] ABTS (2,2′-Azino-bis(3-ethylbenzothiazoline-6-sulfonic acid) diammonium salt) uM Trolox equivalents; [u] DPPH (2,2-diphenil-1-picrylhydrazyl) % discoloration; [w] ACE (Angiotensin converting enzyme) % inhibition. [r] Numbers in brackets: retention time.

Table 7. Correlation coefficients for the peaks area, day, treatment (LSB-c and C-c), and the biological activity parameters analyzed in NPN fraction of Panela cheeses studied during storage under refrigeration at 4 °C.

	Peak [r]	
Parameter [a]	1 (4.468)	2 (116.899)
Treatment	−0.05	0.19
Day	−0.51 ***	0.92 ***
ABTS [s]	−0.07	−0.05
DPPH [u]	−0.35 **	0.75 ***
ACE [w]	−0.55 ***	0.61 ***

LSB-c: probiotic cheese; C-c: control cheese; NPN: Non protein nitrogen. [a] Correlations are significant at ** $p < 0.01$, *** $p < 0.001$. Expressed as: [s] uM Trolox equivalents; [u] % discoloration; [w] ACE: Angiotensin converting enzyme % inhibition. [r] Numbers in brackets: retention time.

Table 8. Correlation coefficients for the peaks, day, treatment (LSB-c and C-c), and the biological activity parameters analyzed in ETOH-SN fraction of Panela cheeses studied during storage under refrigeration at 4 °C.

	Peak [r]				
Parameter [a]	1 (1.536)	2 (1.742)	3 (1.842)	4 (2.597)	5 (116.653)
Treatment	−0.23	0.43 ***	−0.66 ***	0.41 **	−0.19
Day	0.6 ***	0.53 ***	−0.05	0.49 ***	0.32 *
ABTS [s]	0.1	0.54 ***	−0.03	0.37 **	0.24
DPPH [u]	0.51 ***	0.64 ***	−0.28 *	0.58 ***	0.67 ***

LSB-c: Probiotic cheese; C-c: Control cheese; ETOH-SN: Nitrogen soluble in ethanol. [a] Correlations are significant at * $p < 0.05$, ** $p < 0.01$, *** $p < 0.001$. Expressed as: [s] ABTS (2,2′-Azino-bis(3-ethylbenzothiazoline-6-sulfonic acid) diammonium salt) uM Trolox equivalents; [u] DPPH (2,2-diphenil-1-picrylhydrazyl) % discoloration;. [r] Numbers in brackets: Retention time.

Table 9. Correlation coefficients for the peaks, day, treatment (LSB-c and C-c) and the biological activity parameters analyzed in WSE fraction of Panela cheeses studied during storage under refrigeration at 4 °C.

	Peak [r]			
Parameter [a]	1 (1.120)	2 (1.889)	3 (2.286)	4 (117.284)
Treatment	0.42 ***	−0.45 ***	0.38 **	−0.15
Day	−0.02	0.34 **	−0.02	−0.75 ***
ABTS [s]	−0.13	0.54 ***	−0.22	−0.6 ***
DPPH [u]	0.03	−0.57 ***	0.25	−0.05
ACE [w]	−0.39 **	0	−0.36 **	0.54 ***

LSB-c: Probiotic cheese; C-c: Control cheese; WSE: Water soluble extract. [a] Correlations are significant at ** $p < 0.01$, *** $p < 0.001$. Expressed as: [s] ABTS (2,2′-Azino-bis(3-ethylbenzothiazoline-6-sulfonic acid) diammonium salt) uM Trolox equivalents; [u] DPPH (2,2-diphenil-1-picrylhydrazyl) % discoloration; [w] ACE (Angiotensin converting enzyme) % inhibition. [r] Numbers in brackets: Retention time.

For ASN, Table 5 shows that in LSB-c, peak 1 and 5 increased their peak area and disappeared on day 15. For C-c, peak 1 appeared at day 5 and increased its peak area until the end of storage. Peak 5 was not present in C-c. Both peaks are positively correlated with ABTS, and Peak 1 is also positively correlated with ACEI (Table 6). Peak 5 could be related to LSB-c ABTS activity. Peak 1 could also be related to ACEI activity since its increase in area is related to an increase in ACEI. Peak 2 decreased over time, but it showed a positive correlation with ABTS and ACE. Peak 3, 4, and 6 were only present on day 0, and in Table 6, it could be observed their positive correlation with ABTS and ACEI. It is suggested that probiotics hydrolyzed those peptides into shorter ones during storage, and thus biological activities increased.

Correlations for the NPN fraction (Table 7) showed a negative correlation of peak 1 with DPPH and ACEI. This peak decreased during storage days (Table 5), while biological activities increased. It could be suggested that probiotics continued enhancing proteolysis, thus even though peak 1 decreased, other peptides were released, and biological activities increased. The positive correlation of day and peak 2 with DPPH and ACEI is confirmed in Table 5, where an increase can be observed during storage.

Table 8 shows the peak's correlation of ETOH-SN fraction; peak 1, 3, 4, and 5 increased over time (Table 5) and are positively correlated with DPPH; peak 4 is also positively correlated with ABTS. This suggests that their increase during storage is attributed to the increase in antioxidant activity. Peak 2 was not in LSB-c (Table 5), but its positive correlation with ABTS and DPPH suggests that when it appeared in C-c (day 10) was in big quantity (area under the curve), and its increase contributed to the same behavior of antioxidant activities.

In Table 9, it can be observed the correlations of the peaks present within WSE fraction. Peak 1 showed a negative correlation with ACEI; this peak oscillated over time and increased at the end of storage. This suggests that it became partially hydrolyzed, and it increased was not enough to enhance ACEI activity. Peak 2 increased over time (Table 5) and showed a significant positive correlation with ABTS and negative with DPPH. This could suggest that this peak had hydrophilic characteristics; thus, the DPPH was diminished. Peak 3 disappeared from day 5 and 10 and increased on day 15 with a lower value than the initial (Table 5). In Table 9, a negative correlation with ACEI is shown, which is suggested by the hydrolysis of the peptide in shorter ones with no ACEI activity and at day 15, with an increment of peak 3, it is suggested they all coexisted, peptide 3 and the shorter ones, thus even the ACEI activity increased—it was not the same as the beginning of storage. Peak 4 decreases over time (Table 5) and its positive correlation with ABTS, and a negative one with ACEI suggests that this peptide had a hydrophilic profile, which enabled it to show the antioxidant but not the inhibitory activity of ACE.

3.5. Principal Component Analysis (PCA)

The principal component analysis (PCA) plots for the first two principal components is shown in Figure 2a Factorial map and Figure 2b Eigenvectors.

As observed in Figure 2a, the factorial map formed by PC1 and PC2 explained 71.61% and 12.38% of the total variance, respectively. Samples were separated into two well-defined groups by PC1. On the positive side, cheeses with probiotics LSB-c (coded b), and in the negative side, the control cheeses C-c (coded c).

Figure 2b shows that for eigenvector 1, hardness and chewiness showed the highest capacity to separate cheeses in two well-defined groups. Those two variables showed significant differences ($p < 0.05$) within ANOVA (Table 1). ABTS-ASN, ABTS-ETOH, and ABTS-WSE also showed the capacity to separate LSB-c into the positive axis suggesting that probiotic cheese has higher antioxidant activity compared to C-c.

PCA did not make it possible to classify cheeses according to the storage time. Solieri, et al. [41], properly used PCA when studying ripened Parmigiano Reggiano cheese.

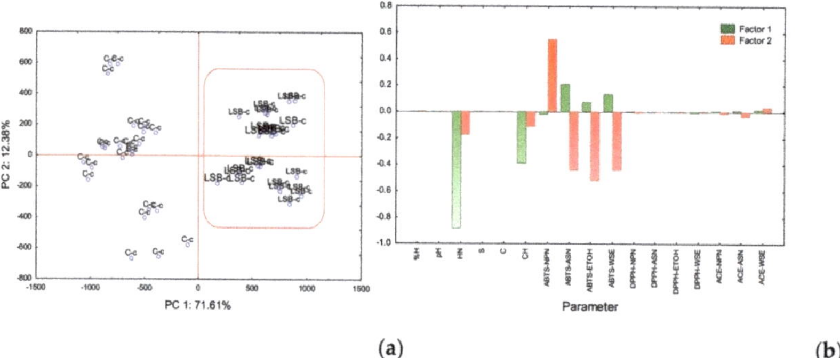

Figure 2. Principal component analysis (PCA) plots for the first two principal components. (**a**) Factorial map: and (**b**) Eigenvectors. %M: % Moisture; HN: Hardness; S: Springiness; C: Cohesivity; CH: Chewiness; ABTS-NPN: ABTS antioxidant activity of Non-Protein Nitrogen; ABTS-ASN: ABTS antioxidant activity of Nitrogen soluble in pH 4.6; ABTS-ETOH: ABTS antioxidant activity of Ethanol-Soluble Nitrogen; ABTS-WSE: ABTS antioxidant activity of Water-Soluble Extract; DPPH-NPN: DPPH radical scavenging of Non-Protein Nitrogen; DPPH-ASN: DPPH radical scavenging of Nitrogen soluble in pH 4.6; DPPH-ETOH: DPPH radical scavenging of Ethanol-Soluble Nitrogen; DPPH-WSE: DPPH radical scavenging of Water-Soluble Extract; ACE-NPN: Angiotensin-Converting Enzyme of Non-Protein Nitrogen; ACE-ASN: Angiotensin-Converting Enzyme of Nitrogen soluble in pH 4.6; ACE-WSE: Angiotensin-Converting Enzyme of Water-Soluble Extract.

4. Conclusions

ANOVA made it possible to determine that parameters were affected by the addition of probiotics to Panela cheese. PCA separated samples into two different groups corresponding to LSB-c and C-c, which could be explained by the textural, physicochemical, and biological changes during storage. The addition of probiotics made it possible to increase the biological activities that could have a benefit in the consumer's health, not only because of the probiotics but also by bioactive peptides released. Correlation between the peaks of the peptide profile and biological activities could be made. A deeper study is necessary to obtain detailed information regarding the identity of peptides, free fatty acids profile, aroma compounds, and sensory attributes. It would also be interesting to evaluate how consumers perceive changes evaluated in this study in a sensory panel evaluation.

Author Contributions: Conceptualization, S.T.M.-d.-C., and A.C.-M.; methodology, S.T.M.-d.-C., J.G.M.-G., R.Z.-A. and A.C.-M.; validation, S.T.M.-d.-C., and A.C.-M.; formal analysis, S.T.M.-d.-C. and A.C.-M.; investigation, K.A.P.-O.; resources, S.T.M.-d.-C., J.G.M.-G., R.Z.-A. and A.C.-M.; data curation, K.A.P.-O., S.T.M.-d.-C., and A.C.-M.; writing—original draft preparation, K.A.P.-O., S.T.M.-d.-C., J.G.M.-G., and A.C.-M.; writing—review and editing, S.T.M.-d.-C., J.G.M.-G., and A.C.-M.; supervision, S.T.M.-d.-C. and A.C.-M.; project administration, S.T.M.-d.-C. and A.C.-M.; funding acquisition, S.T.M.-d.-C. and A.C.-M. All the authors have read and agreed to the published version of the manuscript. All authors have read and agreed to the published version of the manuscript.

Funding: This research received no external funding.

Acknowledgments: First author acknowledges CONACyT for the scholarship CVU number 748518.

Conflicts of Interest: The authors declare no conflict of interest.

References

1. Gobbetti, M.; Minervini, F.; Rizzello, C.G. Angiotensin I-converting-enzyme-inhibitory and antimicrobial bioactive peptides. *Int. J. Dairy Technol.* **2004**, *57*, 173–188. [CrossRef]
2. Moller, N.P.; Scholz-Ahrens, N.R.; Schrezenmeir, J. Bioactive peptides and proteins from foods: Indication for health effects. *Eur. J. Nutr.* **2008**, *47*, 171–182. [CrossRef]

3. Korhonen, H.; Pihlanto, A. Bioactive peptides: Production and functionality. *Int. Dairy J.* **2006**, *16*, 945–960. [CrossRef]

4. Korhonen, H. Milk-derived bioactive peptides: From science to applications. *J. Funct. Foods* **2009**, *1*, 177–187. [CrossRef]

5. Rasmusson, K. Bioactive Peptides in long-Time Ripened Open Texture Semi-Hard Cheese. Master's Thesis, Swedish University of Agricultural Sciences, Uppsala, Sweden, 2012.

6. Park, Y.W.; Nam, M.S. Bioactive Peptides in Milk and Dairy Products: A Review. *Korean J. Food Sci. Anim. Resour.* **2015**, *35*, 831–840. [CrossRef] [PubMed]

7. O'Brien, N.M.; O'Connor, T.P. Nutritional Aspects of Cheese. In *Cheese*, 4th ed.; McSweeney, P.L.H., Fox, P.F., Cotter, P.D., Everett, D.W., Eds.; Academic Press: San Diego, CA, USA, 2017; pp. 603–611. [CrossRef]

8. SAGARPA. *Panorama de la Lechería en México [Overview of the Dairy in Mexico]*; SAGARPA: Mexico City, Mexico, 2018; p. 12.

9. Cervantes Escoto, F. *Los Quesos Mexicanos Genuinos. Patrimonio Cultural Que Debe Rescatarse*; Mundi Prensa: Mexico City, Mexico, 2008; p. 186.

10. Farkye, N.Y.; Vedamuthu, E.; Products, M. Microbiology of soft cheeses. In *Dairy Microbiology Handbook: The Microbiology of Milk and Milk Products*, 3th ed.; Robinson, R.K., Ed.; John Wiley and Sons, Inc.: New York, NY, USA, 2002; pp. 479–513.

11. Hayes, M.; Stanton, C.; Fitzgerald, G.F.; Ross, R.P. Putting microbes to work: Dairy fermentation, cell factories and bioactive peptides. Part II: Bioactive peptide functions. *Biotechnol. J.* **2007**, *2*, 435–449. [CrossRef]

12. Campos-Nonato, I.; Hernández-Barrera, L.; Pedroza-Tobías, A.; Medina, C.; Barquera, S. Hipertensión arterial en adultos mexicanos: Prevalencia, diagnóstico y tipo de tratamiento. Ensanut MC 2016. *Salud Pública de México* **2018**, *60*, 233–243. [CrossRef]

13. Baró, L.; Jiménez, J.; Martínez-Férez, A.; Bouza, J. Péptidos y proteínas de la leche con propiedades funcionales. *Ars. Pharm. (Internet)* **2017**, *42*, 135–145.

14. W.H.O. Cardiovascular Diseases (CVDs) Fact Sheet. Available online: http://www.who.int/mediacentre/factsheets/fs317/en/ (accessed on 15 January 2020).

15. NOM-116-SSA1-1994. Bienes y servicios. In *Determinación de Humedad en Alimentos por Tratamiento Térmico. Método por Arena o Gasa*; Diario Oficial de la Federación: Mexico City, Mexico, 1995; p. 5.

16. NMX-F-317-S-1978. Determinación de pH en alimentos. In *Determinatizon of pH in Fozods*; Diario Oficial de la Federación: Mexico City, Mexico, 1978; p. 3.

17. Voisey, P.W.; Crete, R. A technique for establishing instrumental conditions for measuring food firmness to simulate consumer evaluations. *J. Texture Stud.* **1973**, *4*, 371–377. [CrossRef]

18. Hernández, M.; Sariñana, A. Diseño Integral de Máquina General de Ensayos de Materiales. In *5° Congreso Internacional Sobre Investigación y Desarrollo Tecnológico (CIINDET)*; IEEE (Institut of Electrical and Electronics Engineers), Sección Morelos: Cuernavaca, Morelos, Mexico, 2007.

19. Bourne, M. *Food Texture and Viscosity: Concept and Measurement*, 2th ed.; Academic Press: San Diego, CA, USA, 2002; p. 416.

20. Leclercq-Perlat, M.N.; Oumer, A.; Bergère, J.L.; Spinnler, H.E.; Corrieu, G. Growth of Debaryomyces hansenii on a bacterial surface-ripened soft cheese. *J. Dairy Res.* **1999**, *66*, 271–281. [CrossRef]

21. Guerra Martínez, J.A.; Montejano, J.G.; Martin del Campo, S.T. Evaluation of proteolytic and physicochemical changes during storage of fresh Panela cheese from Queretaro, Mexico and its impact in texture. *CYTA J. Food* **2012**, *10*, 296–305. [CrossRef]

22. Rohm, H.; Jaros, D.; Rockenbauer, C.; Riedler-Hellrigl, M.; Uniacke-Lowe, T.; Fox, P. Comparison of ethanol and trichloracetic acid fractionation for measurement of proteolysis in Emmental cheese. *Int. Dairy J.* **1996**, *6*, 1069–1077. [CrossRef]

23. Re, R.; Pellegrini, N.; Proteggente, A.; Pannala, A.; Yang, M.; Rice-Evans, C. Antioxidant activity applying an improved ABTS radical cation decolorization assay. *Free Radic. Bio Med.* **1999**, *26*, 1231–1237. [CrossRef]

24. Pyrzynska, K.; Pękal, A. Application of free radical diphenylpicrylhydrazyl (DPPH) to estimate the antioxidant capacity of food samples. *Anal. Methods* **2013**, *5*, 4288–4295. [CrossRef]

25. Wang, W.; Wang, N.; Zhang, Y.; Cai, Z.; Chen, Q.; He, G. A Convenient RP-HPLC Method for Assay Bioactivities of Angiotensin I-Converting Enzyme Inhibitory Peptides. *ISRN Biotechnol.* **2013**, *2013*, 453910. [CrossRef]

26. Abadía-García, L.; Cardador, A.; Martín del Campo, S.T.; Arvízu, S.M.; Castaño-Tostado, E.; Regalado-González, C.; García-Almendarez, B.; Amaya-Llano, S.L. Influence of probiotic strains added to Cottage Cheese on potentially-antioxidant peptides generation, anti-Listerial activity, and survival of probiotic microorganisms in simulated gastrointestinal conditions. *Int. Dairy J.* **2013**, *33*, 191–197. [CrossRef]

27. Hayaloglu, A.A.; Guven, M.; Fox, P.F.; Hannon, J.A.; McSweeney, P.L.H. Proteolysis in Turkish White-brined cheese made with defined strains of *Lactococcus*. *Int. Dairy J.* **2004**, *14*, 599–610. [CrossRef]

28. Gomez, M.J.; Gaya, P.; Nunez, M.; Médina, M. *Streptococcus thermophilus* as adjunct culture for a semi-hard cows' milk cheese. *Lait* **1998**, *78*, 501–511. [CrossRef]

29. Lawrence, R.C.; Creamer, L.K.; Gilles, J. Texture Development During Cheese Ripening. *J. Dairy Sci.* **1987**, *70*, 1748–1760. [CrossRef]

30. Souza, C.H.B.; Saad, S.M.I. Viability of *Lactobacillus acidophilus* La-5 added solely or in co-culture with a yoghurt starter culture and implications on physico-chemical and related properties of Minas fresh cheese during storage. *LWT Food Sci. Technol.* **2009**, *42*, 633–640. [CrossRef]

31. Buriti, F.C.A.; da Rocha, J.S.; Assis, E.G.; Saad, S.M.I. Probiotic potential of Minas fresh cheese prepared with the addition of *Lactobacillus paracasei*. *LWT Food Sci. Technol.* **2005**, *38*, 173–180. [CrossRef]

32. Dinakar, P.; Mistry, V.V. Growth and Viability of *Bifidobacterium bifidum* in Cheddar Cheese1. *J. Dairy Sci.* **1994**, *77*, 2854–2864. [CrossRef]

33. Bertola, N.C.; Califano, A.N.; Bevilacqua, A.E.; Zaritzky, N.E. Textural Changes and Proteolysis of Low-Moisture Mozzarella Cheese Frozen under Various Conditions. *LWT Food Sci. Technol.* **1996**, *29*, 470–474. [CrossRef]

34. Lucey, J.A.; Johnson, M.E.; Horne, D.S. Invited Review: Perspectives on the Basis of the Rheology and Texture Properties of Cheese. *J. Dairy Sci.* **2003**, *86*, 2725–2743. [CrossRef]

35. Osorio Tobón, J.F.; Ciro Velásquez, H.J.; Mejía Restrepo, L.G. Caracterización textural y fisicoquímica del queso Edam. *Rev. Fac. Nac. Agron. Medellín* **2004**, *57*, 2269–2278.

36. Lee, S.K.; Anema, S.; Klostermeyer, H. The influence of moisture content on the rheological properties of processed cheese spreads. *Int. J. Food Sci. Technol.* **2004**, *39*, 763–771. [CrossRef]

37. Gupta, A.; Mann, B.; Kumar, R.; Sangwan, R.B. Antioxidant activity of Cheddar cheeses at different stages of ripening. *Int. J. Dairy Technol.* **2009**, *62*, 339–347. [CrossRef]

38. Hernández Galán, L.; Cardador Martínez, A.; Picque, D.; Spinnler, H.E.; López del Castillo Lozano, M.; Martín del Campo Barba, S.T. Angiotensin converting enzyme inhibitors and antioxidant peptides release during ripening of Mexican Cotija hard cheese. *J. Food Res.* **2016**, *5*, 85–91. [CrossRef]

39. Floegel, A.; Kim, D.-O.; Chung, S.-J.; Koo, S.I.; Chun, O.K. Comparison of ABTS/DPPH assays to measure antioxidant capacity in popular antioxidant-rich US foods. *J. Food Compos. Anal.* **2011**, *24*, 1043–1048. [CrossRef]

40. Gorostiza, A.; Cichoscki, A.J.; Valduga, A.T.; Valduga, E.; Bernardo, A.; Fresno, J.M. Changes in soluble nitrogenous compounds, caseins and free amino acids during ripening of artisanal prato cheese; a Brazilian semi-hard cows variety. *Food Chem.* **2004**, *85*, 407–414. [CrossRef]

41. Solieri, L.; Bianchi, A.; Mottolese, G.; Lemmetti, F.; Giudici, P. Tailoring the probiotic potential of non-starter *Lactobacillus* strains from ripened Parmigiano Reggiano cheese by in vitro screening and principal component analysis. *Food Microbiol.* **2014**, *38*, 240–249. [CrossRef] [PubMed]

Publisher's Note: MDPI stays neutral with regard to jurisdictional claims in published maps and institutional affiliations.

MDPI

St. Alban-Anlage 66

4052 Basel

Switzerland

Tel. +41 61 683 77 34

Fax +41 61 302 89 18

www.mdpi.com

Foods Editorial Office

E-mail: foods@mdpi.com

www.mdpi.com/journal/foods